4th Edition

ENERGY

IN THE 21ST CENTURY

4th Edition

ENERGY

IN THE 21ST CENTURY

John R. Fanchi
Texas Christian University, USA

and

Christopher J. Fanchi
Fanchi Enterprises, USA

World Scientific

NEW JERSEY • LONDON • SINGAPORE • BEIJING • SHANGHAI • HONG KONG • TAIPEI • CHENNAI • TOKYO

Published by

World Scientific Publishing Co. Pte. Ltd.
5 Toh Tuck Link, Singapore 596224
USA office: 27 Warren Street, Suite 401-402, Hackensack, NJ 07601
UK office: 57 Shelton Street, Covent Garden, London WC2H 9HE

British Library Cataloguing-in-Publication Data
A catalogue record for this book is available from the British Library.

ENERGY IN THE 21ST CENTURY
Fourth Edition

ISBN 978-981-3144-77-4
ISBN 978-981-3144-78-1 (pbk)

Printed in Singapore by Mainland Press Pte Ltd.

PREFACE TO THE FOURTH EDITION

Energy in the 21st Century is a valuable source of information for students, decision makers, opinion leaders, and the general public. Oil and natural gas price volatility continue to affect both the supply and demand for energy. Advances in other technologies, such as nuclear, wind, solar, and tidal technology, are altering the comparative economics of competing energy sources. New government policies are changing the landscape of the global energy marketplace. The fourth edition updates data and includes more discussion of recent advances. Some of the highlights of the fourth edition are expanded discussion of climate change and anthropogenic climate change; the 2015 COP21 Paris Agreement on Climate Change; nuclear fusion reactor prototypes (tokomak ITER and stellarator W7-X); advances in solar thermal and solar photovoltaic power plants, space based solar power, transparent photovoltaic cells, and hybrid solar wind technology; tidal and wave energy converters; oil from algae; the EU Supergrid; and the Goldilocks Policy and Grand Energy Bargain.

Energy in the 21st Century has been used as the text for an introductory energy course for the general college student population. It has also been used to provide an overview of energy topics for MBA students. Several features enhance the value of the fourth edition as a textbook. The book includes learning objectives at the beginning of each chapter, end of chapter activities, a comprehensive index, a glossary, and an Appendix to help with converting units. Points to Ponder are abbreviated as P2P in the Learning Objectives boxes and are provided throughout the text. They are designed to encourage the reader to consider the material from different perspectives.

Most of the statistics presented in the book are from the United States Energy Information Administration (www.eia.doe.gov). The U.S. EIA has been selected because it is widely used by policy makers. Other sources are used when appropriate. Published statistical data are subject to revision, even if the data are historical data that have been published by a credible source. Data revisions may change specific numbers as database corrections are made and new information is received.

We want to thank students, guest speakers, and colleagues in academia and industry for their comments and suggestions. Kathy Fanchi and the editorial staff at World Scientific were again instrumental in preparing this book for publication.

John R. Fanchi
July 1, 2016

Christopher J. Fanchi
July 1, 2016

To the pioneers in the emerging energy industry –
for the benefit of future generations.

ABOUT THE AUTHORS

John R Fanchi is a Professor in the Department of Engineering and Energy Institute at Texas Christian University, Fort Worth, Texas where he teaches courses in energy and engineering. Previously, Dr. Fanchi taught petroleum and energy engineering courses at the Colorado School of Mines and worked in the technology centers of four energy companies (Getty, Cities Service, Marathon, and Chevron). Dr. Fanchi is the author of many articles and books in the areas of energy, physics, engineering, earth science, and mathematics. His books include **Energy in the 21st Century, 3rd Edition** (World Scientific, 2013), **Integrated Reservoir Asset Management** (Elsevier, 2010), **Energy: Technology and Directions for the Future** (Elsevier-Academic Press, 2004), **Principles of Applied Reservoir Simulation, 3rd Edition** (Elsevier, 2006), **Math Refresher for Scientists and Engineers, 3rd Edition** (Wiley, 2006), **Shared Earth Modeling** (Elsevier, 2002), **Integrated Flow Modeling** (Elsevier, 2000), and **Parametrized Relativistic Quantum Theory** (Kluwer, 1993). Dr. Fanchi has a Ph.D. in physics from the University of Houston, has been President of the International Association for Relativistic Dynamics, and was Vice President of the Board of Trustees of Littleton Public Schools in Colorado. He co-edited the General Engineering volume of the **Petroleum Engineering Handbook** (SPE, 2006, Volume 1). Some of his books have been translated into Arabic, Chinese, and Russian.

Christopher J. Fanchi has a B.A. in Business Administration and Economics from Colorado State University, and is in the MBA program at Texas Christian University, Fort Worth, Texas. He co-authored **Energy in the 21st Century, 3rd Edition** (World Scientific, 2013). He has worked for Fanchi Enterprises since 2000, including contributions to **Integrated Reservoir Asset Management** (Elsevier, 2010), **Math Refresher for Scientists and Engineers, 3rd Edition** (Wiley, 2006) and **Integrated Flow Modeling** (Elsevier, 2000).

CONTENTS

CHAPTER 1

A BRIEF HISTORY OF ENERGY CONSUMPTION

After reading this chapter, you should be able to...

- Define energy and how it is used
  - *P2P*: What does an energy unit mean to me?*
- Define energy density and explain its significance
  - *P2P: Why is energy density important?*
- Describe historical energy consumption
- Connect energy consumption to quality of life
  - *P2P: Using quality of life to forecast energy consumption*
- Discuss the transition from wood to coal to oil to a broad energy mix
  - *P2P: Why should I care about the global distribution of energy?*

* Point to Ponder

We all make decisions about energy. We decide how much electricity we will use to heat or cool our homes. We decide how far we will travel every day and the mode of transportation we will use. Those of us in democratic societies choose leaders who create budgets that can support new energy initiatives or maintain a military capable of defending energy supply lines. Each of these decisions and many others impact the global consumption of energy and the demand for available natural resources. The purpose of this book is to give you the information you need to help you understand the issues and make informed decisions.

The choices we make today will affect generations to come. What kind of future do we want to prepare for them? What kind of future is possible? We can make the best decisions by being aware of our options and the consequences of our choices. In this book, we consider the location, quantity and accessibility of energy sources. We discuss ways to distribute available energy, and examine how our choices will affect the economy, society, and the environment. We begin by defining energy and reviewing our history of energy consumption.

1.1 WHAT IS ENERGY?

Energy is the ability to do work. It can be classified as stored (potential) energy, and working (kinetic) energy. Potential energy is the ability to produce motion, and kinetic energy is the energy of motion. Forms of energy include energy of motion (kinetic energy), heat (thermal energy), light (radiant energy), photosynthesis (biological energy), stored energy in a battery (chemical energy), stored energy in a capacitor (electrical energy), stored energy in a nucleus (nuclear energy), and stored energy in a gravitational field (gravitational energy).

Sources of energy with some common examples include biomass (firewood), fossil fuels (coal, oil, natural gas), flowing water (hydroelectric dams), nuclear materials (uranium), sunlight, and geothermal heat (geysers). Energy sources may be classified as renewable or nonrenewable. Nonrenewable energy is energy that is obtained from sources at a rate that exceeds the rate at which the sources are replenished. Examples of nonrenewable energy sources include fossil fuels and nuclear fission material such as uranium. Renewable energy is energy that is obtained from sources at a rate that is less than or equal to the rate at which the sources are replenished. Examples of renewable energy include solar energy and wind energy.

Renewable and nonrenewable energy sources are considered primary energy sources because they provide energy directly from fuels. A fuel is a material which contains one form of energy that can be transformed into another form of energy. Primary energy is energy that has not been obtained by anthropogenic conversion or transformation. The term "anthropogenic" refers to human activity or human influence. Primary energy is often converted to secondary energy for more convenient use in human systems. Secondary energy sources are produced from primary sources of energy and can be used to store and deliver energy in a useful form. Hydrogen and electricity are considered secondary sources of energy, or "carriers" of energy.

1.1.1 Energy Transformation

Modern civilization depends on the observation that energy can change from one form to another. If you hold a book motionless above a table and then release it, the book will fall onto the table. The book has potential energy when it is being held above the table. The potential energy is energy associated with the position of the book in a gravitational field. When you drop the book, the energy of position is transformed into energy of motion, or kinetic energy. When the book hits the table, some of the kinetic energy is transformed into sound (sonic energy), and the rest of the kinetic energy is transformed into energy of position (potential energy) when the book rests on the table top.

Energy transformation is needed to produce commercial energy. As an illustration, suppose we consider a coal-fired power plant. Coal stores energy as chemical energy. Combustion, or burning the coal, transforms chemical energy into heat energy. In steam power plants, the heat energy changes water into steam and increases the energy of motion, or kinetic energy, of the steam. Flowing steam spins a turbine in a generator. The mechanical energy of the spinning turbine is converted to electrical energy in the generator. In a real system, energy is lost so that the efficiency of electrical energy generation from the combustion of coal is less than 100%. A measure of the energy that is available for doing useful work is called exergy.

Real power systems transform energy into useful work, but some of the energy is wasted. The energy efficiency of a system is the amount of energy needed by the system to perform a specific function divided by the amount of energy that is supplied to the system. Energy efficiency has a value between 0% and 100% and is a measure of the lost as non-useful energy. For example, suppose we have two light bulbs A and B. Both light bulbs provide the same amount of light, but light bulb B produces less heat than light bulb A and uses less energy than light bulb A. Consequently, light bulb B has higher energy efficiency than light bulb A because light bulb B uses less energy to achieve its intended purpose, to provide light.

In the light bulb example, we can reduce energy consumption by adopting a more energy efficient technology. Another way to reduce energy consumption is to turn off the light when it is not needed. In this case, we are conserving energy by changing our behavior. Energy conservation can be achieved by adopting a behavior that results in the use of less energy. An improvement in energy efficiency or conservation can be viewed as increasing the supply of useful energy because the improvement lets us get more value from existing energy sources.

Point to Ponder: What does an energy unit mean to me?

To get an idea of the meaning of an energy unit such as kilocalorie or megajoule, it is helpful to compare the energy consumed by the operation of modern devices. For example, a 1200 Watt hair dryer uses approximately one megajoule of energy in 15 minutes. A megajoule is 1 million Joules, which can be written as 10^6 Joules or 10^6 J. A 100 Watt light bulb uses approximately one megajoule of energy in about three hours. [Fanchi, 2004, Exercise 1-3]

If we run the 1200 Watt hair dryer for one hour, we will use 1.2 kilowatt-hours of energy. We abbreviate 1.2 kilowatt-hours as 1.2 kWh. A kilowatt (kW) is 1000 Watts. One kWh equals 1 kW times 1 hr, or about 3.6×10^6 J of energy. A typical American household will use between 20 and 50 kWh per day. Energy consumption in a household depends on such factors as how appliances are used, heating or cooling, etc.

A typical power plant provides approximately 1000 megawatts of power, which is abbreviated as 1000 MW of power. The power plant can provide power to approximately 1.2 million households that use 20 kWh per day for each household.

A unit of energy that is commonly used for discussing energy on a national scale is the quad. One quad equals one quadrillion British Thermal Units (BTU) or 10^{15} BTU. A BTU is approximately 1000 Joules, so one quad is approximately 10^{18} Joules. A quad is comparable in magnitude to global energy values. For example, in 2010 the United States consumed about 100 quads of energy and the world consumed about 500 quads of energy.

Units and scientific notation are reviewed in Appendix A. A set of statistics that summarizes energy production and consumption in the United States and the world are tabulated in Appendices B and C, respectively. Appendix D provides a set of commonly used unit conversion factors to enable the conversion of one unit into another.

1.1.2 Energy Density

The amount of energy that can be produced from a material depends on the heat content of the material. Energy density is the amount of energy in a material divided by the amount of material. Energy density can be defined on a mass basis or a volume basis. For a volume basis, energy density is defined as the energy that can be obtained from a material per unit volume of material. Similarly, for a mass basis, energy density is defined as the energy that can be obtained from a material per unit mass of material. Table 1-1 shows energy densities for several common materials.

Table 1-1
Energy Density of Common Materials*

Material	Energy Density	
	MJ/kg	MJ/m^3
Crude oil	42	37,000
Coal	32	42,000
Dry wood	12.5	10,000
Hydrogen, gas	120	10
Hydrogen, liquid	120	8,700
Methanol	21	17,000
Ethanol	28	22,000
Methane, gas**	54	37.7
Natural gas**	55	38.4

*Source: [Sørensen, 2004, page 555]
**Source: [Engineering Toolbox FGHV, 2015]

Historically, energy density was one of the most important factors considered in selecting a fuel. Coal and oil have relatively large energy densities and were often preferentially chosen as the raw fuel that was input to power plants. For example, a kilogram (2.2 pounds mass) of crude oil provides 42 megajoules of energy, while one kilogram of sawmill scrap (dry wood) provides 12.5 megajoules of energy. Thus, a kilogram of crude oil provides over three times as much energy during the combustion process as the same mass of sawmill scrap.

Point to Ponder: Why is energy density important?
Energy density tells us how much energy we can obtain from a given amount of material and it is a quantity we can use to compare different fuels. For example, a larger energy density in fossil fuels than wood was one of the reasons wood was replaced by coal and oil as a fuel in the 19th century. A smaller volume of fossil fuel could provide more energy than wood, which meant less storage was needed for fossil fuel. This was an especially valuable advantage in transportation, since reducing the storage volume for fuel on ships and trains decreased the cost of transporting goods and passengers.

1.2 Historical Energy Consumption

The history of energy consumption shows how important energy is to the quality of life for each of us. Societies have depended on different types of energy in the past, and societies have been forced to change from one energy type to another. Global energy consumption can be put in perspective by considering the amount of energy consumed by individuals.

E. Cook [1971] provided estimates of daily human energy consumption at six different periods of societal development. The six periods from oldest to most recent are the Primitive Period, the Hunting Period, the Primitive Agricultural Period, the Advanced Agricultural Period, the Industrial Period, and the Technological Period. Cook's estimates are given in Table 1-2 for each period. The table shows that personal energy consumption was relatively constant until the Advanced Agricultural period when it increased substantially.

Table 1-2
Historical Energy Consumption [Cook, 1971]

Period	Era	Daily per capita Consumption (1000 kcal)				
		Food	H & C*	I & A**	Trans.***	Total
Primitive	1 million B.C.	2				2
Hunting	100,000 B.C	3	2			5
Primitive Agricultural	5000 B.C.	4	4	4		12
Advanced Agricultural	1400	6	12	7	1	26
Industrial	1875	7	32	24	14	77
Technological	1970	10	66	91	63	230

* H & C = Home and Commerce
** I & A = Industry and Agriculture
*** Trans. = Transportation

Energy is essential for life, and food was the first source of energy. Cook assumed the only source of energy consumed by a person living during the period labeled "Primitive" was food. Cook's energy estimate was for an East African about one million years ago. Humans require approximately 2000 kilocalories (about eight megajoules) of food per day. One food Calorie (with an upper case "C") is equal to one kilocalorie, or 1000 calories. One calorie is the amount of energy required to raise the temperature of one gram of water by one degree Centigrade (1.0°C). A change in temperature of 1.0°C is equal to a change in temperature of 1.8 degrees Fahrenheit (1.8°F). For comparison, one BTU is the amount of heat energy needed to raise the temperature of one pound of water by 1.0°F.

The ability to control fire during the Hunting period let people use wood to heat and cook. Fire provided light at night and could illuminate caves. Firewood was the first source of energy for consumption in a residential setting. Cook's estimate of the daily per capita energy consumption for Europeans about 100,000 years ago was 5,000 kilocalories (about 21 megajoules).

The Primitive Agricultural period was characterized by the domestication of animals. Humans were able to use animals to help them grow crops and cultivate their fields. The ability to grow more food than you needed became the impetus for creating an agricultural industry. Cook's estimate of the daily per capita energy consumption for people in the Fertile Crescent circa 5000 B.C. was 12,000 kilocalories (about 50 megajoules). Humans continue to use animals to perform work (Figure 1-1).

Figure 1-1. Animal Labor in Ahmedabad, India (© Fanchi 2000)

More energy was consumed during the Advanced Agricultural period when people learned to use coal, and built machines to harvest the wind and water. By the early Renaissance, people were using wind to push sailing ships, water to drive mills, and wood and coal to generate heat. Transportation became a significant component of energy consumption by humans. Cook's estimate of the daily per capita energy consumption for people in northwestern Europe circa 1400 was 26,000 kilocalories (about 109 megajoules).

The steam engine ushered in the Industrial period. It provided a means of transforming heat energy to mechanical energy. Wood was the first source of energy for generating steam in steam engines. Coal, a fossil fuel, eventually replaced wood and hay as the primary energy source in industrialized nations. Coal was easier to store and transport than wood and hay, which are bulky and awkward. Coal was useful as a fuel source for large vehicles, such as trains and ships, but of limited use for personal transportation. Oil, another fossil fuel, was a liquid and contained about the same amount of energy per unit mass as coal. Oil could flow through pipelines and be stored in tanks. People just needed a machine to convert the energy in oil to a more useful form. Cook's estimate of the daily per capita energy consumption for people in England circa 1875 was 77,000 kilocalories (about 322 megajoules).

The modern Technological period is associated with the development of internal combustion engines, and applications of electricity. Internal combustion engines can vary widely in size and were designed to burn fossil fuels. The internal combustion engine could be scaled to fit on a wagon and create a "horseless carriage." The transportation system in use today evolved as a result of the development of internal combustion engines. Electricity, by contrast, is generated from primary energy sources such as fossil fuels. Electricity generation and distribution systems made the widespread use of electric motors and electric lights possible. One advantage of electricity is that it can be transported easily, but electricity is difficult to store. Cook's estimate of the daily per capita energy consumption for people in the United States circa 1970 was 230,000 kilocalories (about 962 megajoules).

Figure 1-2 shows the consumption of energy in the United States from 1650 to 2000. The figure shows that wood was the primary energy source for most of history. The transition from wood (a renewable energy source) to fossil fuels (a nonrenewable energy source) began in the middle of the 19th century. Fossil fuels became the dominant energy source from the mid-19th century through the end of the 20th century. A similar scenario applies to other developed nations.

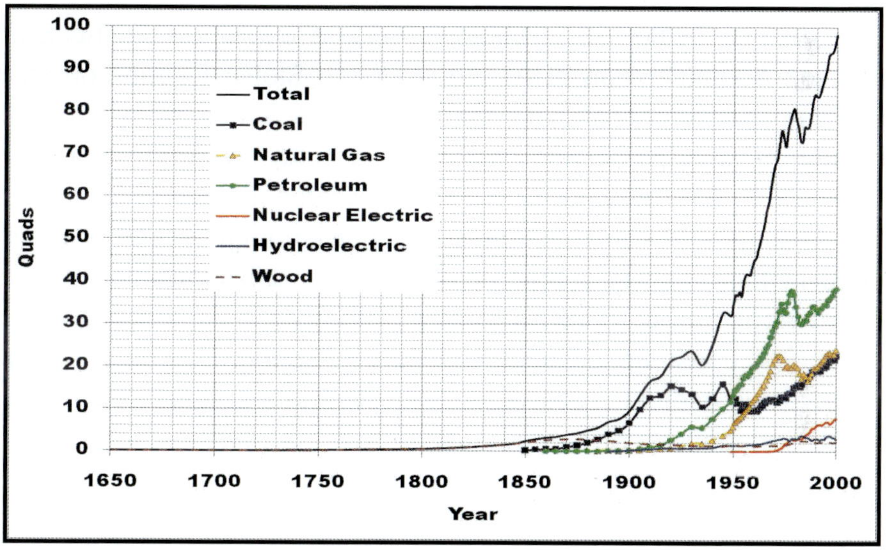

Figure 1-2. Historical Energy Consumption in the United States [US EIA AER, 2001]

Table 1-3 shows the production and consumption of energy in the United States. The unit of energy is the quad, which is one quadrillion BTU. The three dominant energy sources are fossil fuels, nuclear electric power, and renewable energy. The table shows that approximately 75 quads of energy were produced and 97 quads of energy were consumed by the United States in 2010. The energy that was not produced in the United States was imported.

Table 1-3
2010 United States Energy Production and Consumption
[US EIA AER, 2015]

Energy Source	Production (quads)	Consumption (quads)
Total	74.76	97.39
Fossil Fuels	58.22	80.89
Nuclear Electric Power	8.43	8.43
Renewable Energy	8.11	8.07

The data in Table 1-3 is taken from data published by the Energy Information Administration (EIA) of the United States Department of Energy. Additional historical data for the United States from the EIA is presented in Appendix B, and analogous data for the world is presented in Appendix C. Energy data for the United States is used to illustrate energy production and consumption for an individual nation. All of the data reported in this book is subject to revision as data collection agencies make revisions and corrections to data from institutions that monitor and report energy consumption and production. Although the data is subject to revision, it is useful to view actual reported data so that we can get an idea of the magnitude of energy needed and the relative contributions of different energy sources to the energy mix.

Table 1-4 presents the relative contributions of different energy sources to 2010 United States energy production and consumption as percent of total values shown in Table 1-3. Fossil fuels are the dominant contributor to the United States energy mix. Data covering a period of decades (see Appendix B) show that the relative contribution of energy sources to the United States energy mix is changing. The contribution of fossil fuels is declining while the contribution of renewable energy sources is increasing. The increasing role of renewable energy sources is also occurring on a global scale, as discussed in Section 1.5.

Table 1-4
2010 United States Energy Production and Consumption as % of Annual Amount [US EIA AER, 2015]

Energy Source	Production (%)	Consumption (%)
Total	100.0	100.0
Fossil Fuels	77.9	83.1
Nuclear Electric Power	11.3	8.7
Renewable Energy	10.9	8.3

Figure 1-3 shows how energy consumed in the United States was distributed between end-use energy sectors from 2004 to 2014. The transportation sector is the largest consumer of energy, followed by the industrial, residential and commercial sectors. The transportation sector includes vehicles that transport people or goods. Specific definitions of the composition of each energy sector are provided by the United States EIA.

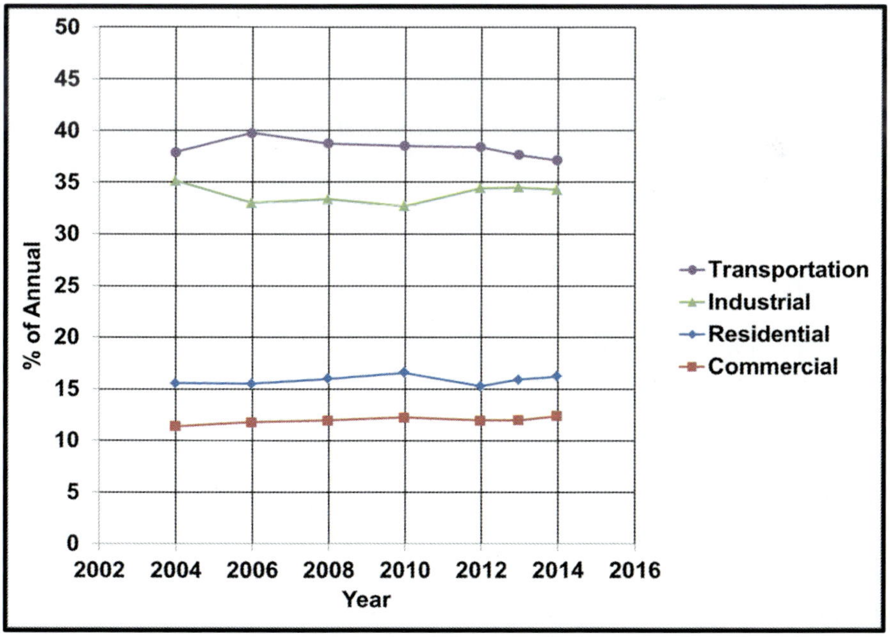

Figure 1-3. United States Energy Consumption by End-Use Sector [LLNL Energy Flow Charts, 2015]

1.3 Energy Consumption and the Quality of Life

The per capita energy consumption discussed above gives us an idea of how much energy is consumed today, but it does not tell us how much energy should be consumed. One way to estimate the amount of energy each person should use is to examine the relationship between energy consumption and quality of life.

Quality of life is a subjective concept that can be quantified in several ways. The United Nations calculates a quantity called the Human Development Index (HDI) to provide a quantitative measure of the quality of life. The HDI is a fraction that varies from zero to one. A value of HDI that approaches zero is considered a relatively low quality of life, while a value of HDI that approaches one is considered a high quality of life. The HDI measures human development in a country using three basic factors: health, education, and standard of living. Table 1-5 presents measures for each factor.

Table 1-5
Basic Factors of the United Nations Human Development Index

Factor	Measure
Health	Life expectancy at birth
Education	Combination of mean years of schooling and expected years of schooling
Standard of Living	Gross National Income per capita

Standard of Living is quantified using Gross National Income (GNI) per capita. GNI is Gross Domestic Product (GDP) plus income received from other countries less payments made to other countries. Gross Domestic Product (GDP) is the market value of the goods and services produced by a country. GDP is a measure of the economic growth of a nation.

A plot of HDI versus annual per capita electricity consumption is shown in Figure 1-4. Annual per capita electricity consumption is the total amount of electricity consumed by the nation in a year divided by the population of the nation. The amount of electricity is expressed in kilo-watt-hours (kWh) in the figure. Per capita primary energy consumption is an estimate of the annual amount of primary energy consumed by each individual in the nation. The calculation of per capita primary energy consumption establishes a common basis for comparing the consumption of primary energy between nations with large populations and nations with small populations.

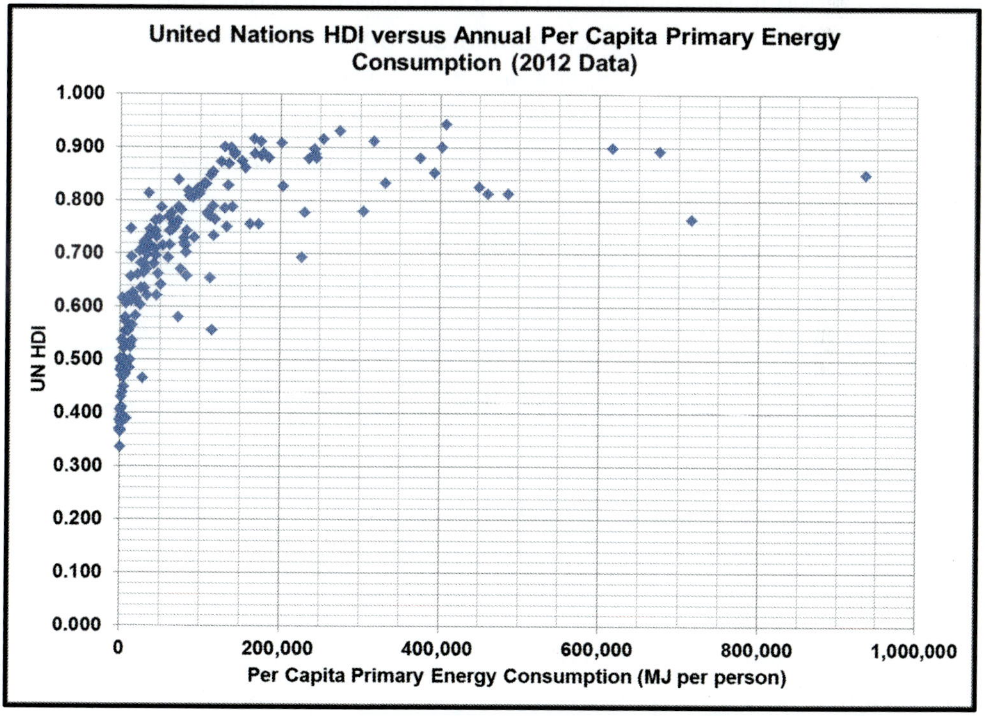

Figure 1-4. Human Development and Annual Energy Consumption

The HDI data in Figure 1-4 are 2012 data from the United Nations Human Development Index website [UN HDI, 2015], and annual per capita primary energy consumption data are 2012 data reported by the Energy Information Administration of the United States Department of Energy [US EIA TPEC, 2015]. The figure shows that quality of life increases as annual per capita primary energy consumption increases. The increase is not linear; the improvement in quality of life begins to level off when per capita energy consumption rises to about 200,000 megajoules per person.

The HDI is a quantitative measure that can be used to compare the quality of life from one nation to another. Nevertheless, it should be noted that the HDI is subject to modification. The factors that go into the HDI were redefined in November, 2010 by the United Nations with the changes taking effect in 2011. When compared to the first two editions of this text, the changes in the definition of the Education (previously titled "Knowledge") and Standard of Living categories were significant enough to cause a change in the order of the countries listed in Table 1-6.

Table 1-6
10 Countries with the Largest UN HDI
(2012 data)

Country	UN HDI	Per Capita Primary Energy Consumption (MJ/person)
Norway	0.943	408574
Australia	0.931	275878
Switzerland	0.916	167572
Netherlands	0.915	254779
United States	0.912	318572
Germany	0.911	176524
New Zealand	0.908	202824
Ireland	0.901	131441
Canada	0.901	404038
Denmark	0.900	139116

Table 1-6 lists per capita consumption of primary energy for the 10 countries with the largest UN HDI. The numbers are subject to change when databases are updated or corrected, but the data in the table illustrates the relative magnitude of per capita primary energy consumption for each country.

Point to Ponder: How can we use quality of life to forecast energy consumption?

The data used to prepare Figure 1-4 can also be used to make a quick forecast of energy demand. Suppose we assume that the world population will stabilize at approximately 8 billion people in the 21st century and that all people will want the quality of life represented by an HDI value of 0.9. In this scenario, the annual per capita energy demand from Figure 1-4 is approximately 200,000 megajoules per person, or 1.6×10^{15} MJ $\approx$ 1500 quads for 8 billion people. The world population of approximately 6.9 billion people in 2012 consumed approximately 524 quads of energy. According to this scenario, worldwide energy demand would need to triple by the end of the 21st century when compared to worldwide energy consumption in 2012. This calculation illustrates the types of assumptions that must be made to prepare forecasts of energy demand. At the very least, a forecast of demand for energy at the end of the 21st century needs to provide an estimate of the size of the population and the per capita demand for energy at that time. [Fanchi, 2004, Chapter 1]

1.4 The Path to Sustainability

Coal was the first fossil fuel to be used on a large scale. J.U. Nef [1977] described 16th century Britain as the first major economy in the world that relied on coal. Britain was dependent on wood before it switched to coal. The transition from wood to coal during the period from about 1550 A.D. to 1700 A.D. was made necessary by the excessive consumption of wood that was leading to the deforestation of Britain. Coal was a combustible fuel that could be used as an alternative to wood.

Coal was the fuel of choice during the Industrial Revolution. It was used to boil steam for steam turbines and steam engines. Coal was used in transportation to provide a combustible fuel for steam engines on trains and ships. The introduction of the internal combustion engine made it possible for oil to replace coal as a fuel for transportation. Coal is used today to provide fuel for many coal-fired power plants.

People have used oil for thousands of years [Yergin, 1992, Chapter 1]. Civilizations in the Middle East, such as Egypt and Mesopotamia, collected oil in small amounts from surface seepages as early as 3000-2000 B.C. During that period, oil was used in building construction, waterproofing boats and other structures, setting jewels, and mummification. Arabs began using oil to create incendiary weapons as early as 600 A.D. By the 1700's, small volumes of oil from sources such as surface seepages and mine shafts were being used in Europe for medicinal purposes and in kerosene lamps. Larger volumes of oil could have been used, but Europe lacked adequate drilling technology.

Pulitzer Prize winner Daniel Yergin [1992, page 20] chose George Bissell of the United States as the person most responsible for creating the modern oil industry. American demand for nighttime illumination in the middle of the 19th century exceeded the volume of whale oil produced each year by refineries in New England and New York. Bissell realized in 1854 that rock oil – as oil was called in the 19th century to differentiate it from vegetable oil and animal fat – could be used as an illuminant in lamps. He gathered a group of investors together in the mid-1850's. The group formed the Pennsylvania Rock Oil Company of Connecticut and selected James M. Townsend to be its president.

Bissell and Townsend knew that oil was sometimes produced along with water from water wells. They believed that rock oil could be found below the surface of the earth by drilling for oil in the same way that water wells were drilled. Townsend commissioned Edwin L. Drake to drill a well in Oil Creek, near Titusville, PA. The location had several oil seepages. The project began in 1857 and encountered many problems. A letter from Townsend was en route to Drake to inform him that funds were to be cut off [van Dyke, 1997] when Drake struck oil on Aug. 27, 1859.

Drake's well began a transformation of the American energy mix. Oil could be refined for use in lighting and cooking. The substitution of rock oil for whale oil, which was growing scarce and expensive, reduced the need to hunt whales for fuel to burn in lamps. The importance of Drake's well is shown in an 1861 cartoon from Vanity Fair (Figure 1-5). The cartoon shows whales celebrating the discovery of rock oil as a replacement for whale oil. It is fair to say that, at the time, oil saved the whales.

Grand Ball given by the Whales in honor of the discovery of the Oil Wells in Pennsylvania.

Figure 1-5. Vanity Fair, 1861: "Grand ball given by the whales in honor of the discovery of the oil wells in Pennsylvania"

Within fifteen months of Drake's strike, Pennsylvania was producing 450,000 barrels of oil per year from seventy-five wells. By 1862, three million barrels of oil were being produced and the price of oil dropped to ten cents a barrel [Kraushaar and Ristinen, 1993].

The Pennsylvania oil fields provided a relatively small amount of oil to meet demand. In 1882, the invention of the electric light bulb caused a drop in the demand for kerosene which was a fuel derived from rock oil. The drop in demand for rock oil was short lived, however. The quickly expanding automobile industry needed oil for fuel and lubrication. New sources of oil were discovered in the early 20th century. Oil was found in Ohio and Indiana, and later in the San Fernando Valley in California and near Beaumont, Texas.

The world's first gusher, a well that produced as much as 75,000 barrels of oil per day, was drilled at Spindletop Hill near Beaumont (Figure 1-6). The well was named Lucas No. 1 after Anthony F. Lucas, its driller and an immigrant from the Dalmatian coast (now Croatia). The well was notable because it was drilled using rotary drilling, a modern drilling technique discussed in more detail later, and it produced so much oil at such a high rate that it demonstrated oil could be used as an energy source on a global scale.

Figure 1-6. Spindletop, Gladys City Boomtown Museum, Beaumont, Texas (© Fanchi, 2003)

Industrialist John D. Rockefeller began Standard Oil in 1870 and by 1879 the company held a virtual monopoly over oil refining and transportation in the United States. Rockefeller's control of the oil business made him rich and famous, but it also reduced competition in the marketplace. The Sherman Antitrust Act of 1890 was used by the United States government to break Rockefeller's grip on the oil industry. Standard Oil was found guilty of restraining trade and a Federal court ordered the dissolution of Standard Oil in 1909. The ruling was upheld by the United States Supreme Court in 1911 [Yergin, 1992, Ch. 5].

By 1909, the United States produced more oil than all other countries combined, producing half a million barrels per day. Up until 1950, the United States produced more than half of the world's oil supply. Discoveries of large oil deposits in Central America, South America and the Middle East reduced global dependence on United States production. Production in the United States peaked in 1970 while oil demand in the United States and elsewhere in the world continued to grow. Oil production rate in 2014 and 2015 was comparable to the 1970 peak because of the addition of shale oil production. The United States has imported oil to satisfy domestic demand since 1948.

Until 1973, oil prices were influenced by market demand and the supply of oil that was provided in large part by a group of oil companies called the "Seven Sisters." This group included Exxon, Royal Dutch/Shell, British Petroleum (BP), Texaco, Mobil, Standard Oil of California (which became Chevron), and Gulf Oil. In 1960, Iran, Iraq, Kuwait, Saudi Arabia and Venezuela founded the Organization of Petroleum Exporting Countries, commonly known as OPEC, in Baghdad, Iraq. In 1973 OPEC became a major player in the oil business by raising prices on oil exported by its members. This rise in price contributed to the "first oil crisis" as prices for consumers in many countries increased significantly. OPEC members in 2015 were Algeria, Angola, Ecuador, Iran, Iraq, Kuwait, Libya, Nigeria, Qatar, Saudi Arabia, United Arab Emirates, and Venezuela [OPEC, 2015].

Today, fossil fuels are still the primary fuels for generating electrical power, but society is becoming increasingly concerned about the global dependence on finite resources and the environmental impact of fossil fuel combustion. As a result, society is in the process of changing the global energy mix from an energy portfolio that is heavily dependent on fossil fuels to an energy portfolio that depends on several energy sources. The transition process began in the latter half of the 20th century.

Figure 1-7 shows total energy consumption in quads from 1940 to 2000. In 1940, the world relied on firewood, coal and oil. Natural gas, energy from water, especially hydropower from dams, and nuclear energy joined firewood, coal and oil as important contributors to the energy mix by the end of the 20th century. Other energy sources – identified as wind, solar, biomass, geothermal and waste in Figure 1-7 – were beginning to make an appearance in the global energy mix at the beginning of the 21st century. They do not appear in the figure because their impact was negligible in the last half of the 20th century.

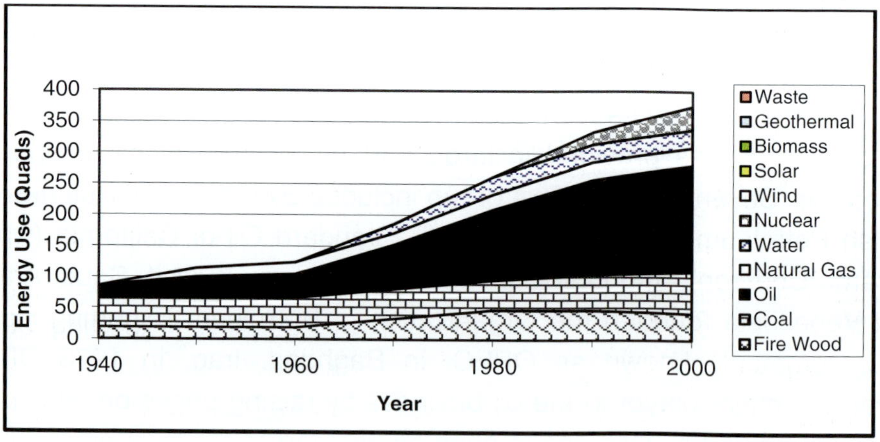

Figure 1-7. World Historical Energy Consumption

One of the factors that supported the selection of fossil fuels and nuclear energy as fuels of choice is energy density. Fossil fuels have relatively large energy densities and have been preferentially chosen as the raw fuel for power plants. Oil, coal, natural gas, and uranium are raw fuels that are present in nature and can be used to provide primary energy.

The dominance of fossil fuels in the energy mix at the end of the 20th century is being replaced by a move toward sustainable energy. Sustainable energy is the mix of energy sources that will allow society to meet its present energy needs while preserving the ability of future generations to meet their needs. This definition is a variation of the concept of sustainable development introduced in 1987 in a report prepared by the United Nations' World Commission on Environment and Development. The Commission, known as the Brundtland Commission after chairwoman Gro Harlem Brundtland of Norway, said that society should adopt a policy of sustainable development that allows society to meet its present needs while preserving the ability of future generations to meet their own needs [WCED, 1987].

The fossil fuels coal, oil and natural gas are carbon-based fuels. The dependence of the global economy on the extensive consumption of fossil fuels has raised concerns about the reliability and extent of the supply of fossil fuels, the impact of fossil fuel consumption on the environment, and the use of fossil fuels as geopolitical weapons. The desire to minimize the use of carbon-based fuels is discussed in more detail in Chapter 4.

Point to Ponder: Why should I care about the global distribution of energy?

Suppose a country with a population of 20 million people wants to provide enough energy to sustain a quality of life that consumes approximately 200,000 megajoules of energy per person each year. This corresponds to approximately 127 power plants with 1000 megawatts capacity each [Fanchi, 2004, Exercise 1-10]. Where will this energy come from?

Today, energy on a national scale comes primarily from fossil fuels such as oil, gas and coal. In a few countries such as France, it is provided by nuclear fission. If the country does not have significant reserves of fossil fuels or uranium – a material needed for most nuclear fission reactors – it will have to import the materials it needs. In this case the country is a "have not" country that is dependent on countries that have the resources and technology it needs. This creates an opportunity for "have" countries to manipulate "have not" countries. On the other hand, it creates an incentive for "have not" countries to use their human resources to take what is needed. For example, the "have not" country could maintain a large standing army or sponsor acts of violence to influence "have" countries. The global distribution of energy influences relationships between nations and can affect geopolitical stability.

1.5 Activities

True-False

Specify if each of the following statements is True (T) or False (F).

1. The steam engine transformed solar energy to mechanical energy.
2. The substitution of rock oil for whale oil reduced the need to hunt whales.
3. Per capita energy consumption is the amount of energy consumed per person.
4. The Sherman Antitrust Act of 1890 was used to break up Standard Oil in 1909.
5. The use of rock oil reduced the need to hunt whales for fuel.
6. A quad is a unit of energy and refers to a quadrillion BTU.
7. Per capita energy consumption is the amount of energy consumed per day.

8. Energy efficiency has a value between 0% and 100%.
9. Fossil fuels are preferred as raw energy sources because of their small energy densities.
10. "Sustainable Development" seeks to reach a balance between meeting current needs and preserving the ability of future generations to meet their needs.

Questions

1. What is the maximum value of the Human Development Index that a country can reach (the highest possible number)?
2. Which of the following companies was *not* one of the "Seven Sisters": Exxon, Royal Dutch/Shell, British Petroleum, Marathon, Texaco, Mobil, or Gulf Oil?
3. Name three non-fossil energy sources included in the global energy mix.
4. What are the three components of the United Nations Human Development Index?
5. Why did Britain switch to coal from wood as its primary energy source in the 16th century?
6. Use the following table to estimate the percent of world energy consumption that was due to fossil fuels in 2010.

Primary Energy Type	**Total World Energy Consumption [Source: US EIA IES, 2015]**
Oil	34.6 %
Natural Gas	23.0 %
Coal	29.0 %
Hydroelectric	6.5 %
Nuclear	5.4 %
Geothermal, Solar, Wind, Wood, etc	1.5 %

7. A measure of the energy that is available for doing useful work is called ______.
8. Which measure of energy is larger: one BTU or one megajoule?
9. What is Gross Domestic Product?
10. Why might a "have not" country with a large standing army be dangerous?

CHAPTER 2

FOSSIL ENERGY – COAL

After reading this chapter, you should be able to...

- Describe the geologic history of the earth
- Explain the origin of coal and identify the types of coal
- Define and differentiate between the concepts of resources and reserves
- Describe the Resource Pyramid
- Describe the distribution and production of coal
- Describe the combustion of fossil fuels
    - *P2P: Why is fossil fuel combustion a problem?*
- Explain how a coal-fired power plant works
- Define coal gas and describe how coal gas is produced

Fossil fuels are the dominant energy source in the modern global economy. They include the carbon-based fuels coal, oil, and natural gas. Coal is a combustible rock that is composed primarily of carbon-rich organic (carbonaceous) material. Coal contributed about 29% to the total energy consumed in the world in 2010 (Appendix C). The formation and distribution of fossil fuels depends on the geologic history of the earth, which is where we begin our discussion of coal.

2.1 Geologic History of the Earth

Most scientists believe that the earth first began to form 4 to 5 billion years ago. The earth was formed from hot, gaseous matter that whirled through space for several hundred million years. This body of matter slowly cooled and condensed. A solid crust gradually formed around the molten interior. The cross-section in Figure 2-1 bisects the earth and shows that the interior of the earth consists of an iron core wrapped inside a mantle of rock with a thin crust at the surface. Molten rock beneath the surface of the earth is called magma. If magma is expelled in a molten state by a volcano, it is called lava.

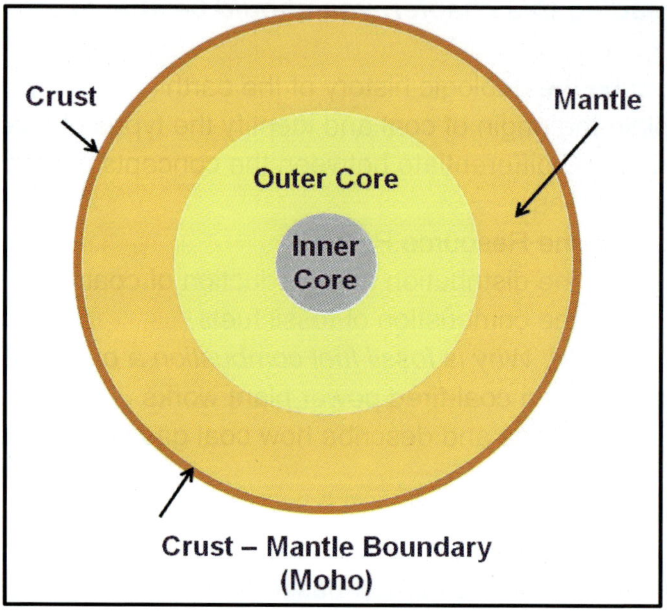

Figure 2-1. Schematic Cross-Section of the earth (not to scale)

The thickness of the crust is small compared to the diameter of the earth. The crust consists of oceanic crust and continental crust (Figure 2-2). The mobile part of the upper mantle and crust is called the lithosphere. Lithospheric plates drift on a denser layer of semi-molten basalt called the asthenosphere. Basalt is composed of magnesium and iron silicates. The semi-molten asthenosphere behaves as a ductile material; it can be deformed without breaking.

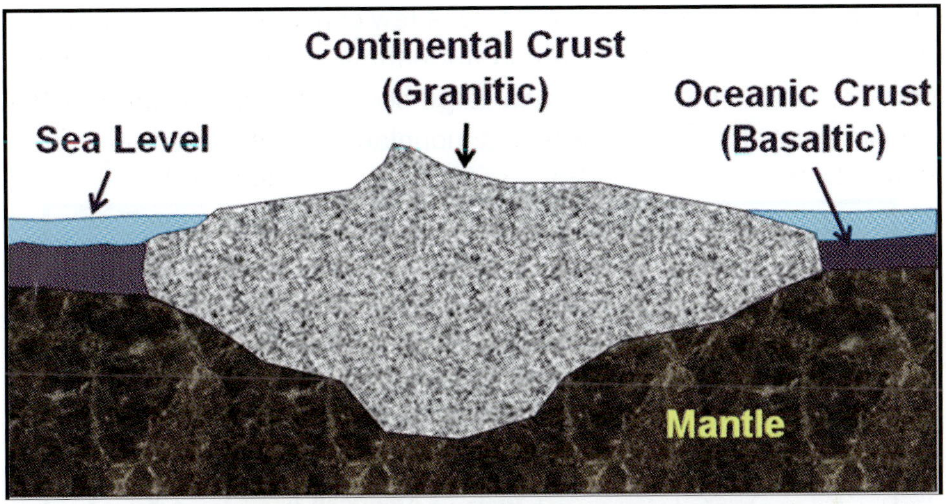

Figure 2-2. Cross-Section of the earth

As the earth cooled from its original state, the crust of the earth was subjected to forces that caused great changes in its topography, including the formation of continents and the uplift of mountain ranges. Pressure from the earth's interior could crack the sea floor and allow less dense molten material to flow onto the sea floor to form subsea ridges. Satellite measurements of the earth's gravitational field have identified sea floor ridges and boundaries between continents. The shapes of the boundaries between continents are suggestive of vast plates, as depicted in Figure 2-3. Only the largest of the known plates are depicted in the figure. These plates are referred to as tectonic plates. Sea floor ridges are long, subsea mountain ranges that can serve as plate boundaries.

In the theory of plate tectonics, the entire crust of the earth is considered a giant, ever-shifting jigsaw puzzle. Figure 2-4 shows the hypothesized movement of tectonic plates during the past 225 million years. It begins at the time that all surface land masses were thought to be coalesced into a single land mass known as Pangaea. Geoscientists believe that Pangaea was formed by the movement of tectonic plates, and the continued movement of plates led to the breakup of the single land mass into the surface features we see today. Tectonic plates move in relation to one another at the rate of a few centimeters per year and are driven by forces which originate in the earth's interior. As the plates pull apart or collide, they can cause large-scale geologic events such as volcanic eruptions, earthquakes and mountain range formation.

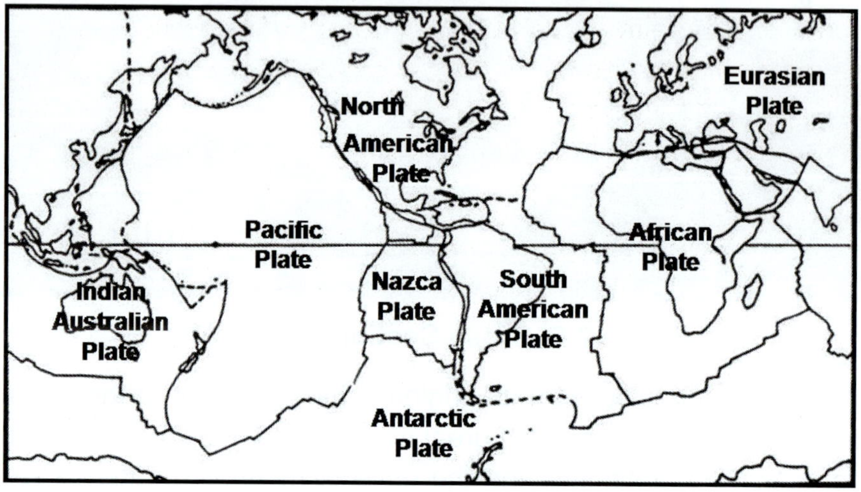

Figure 2-3. Tectonic Plates

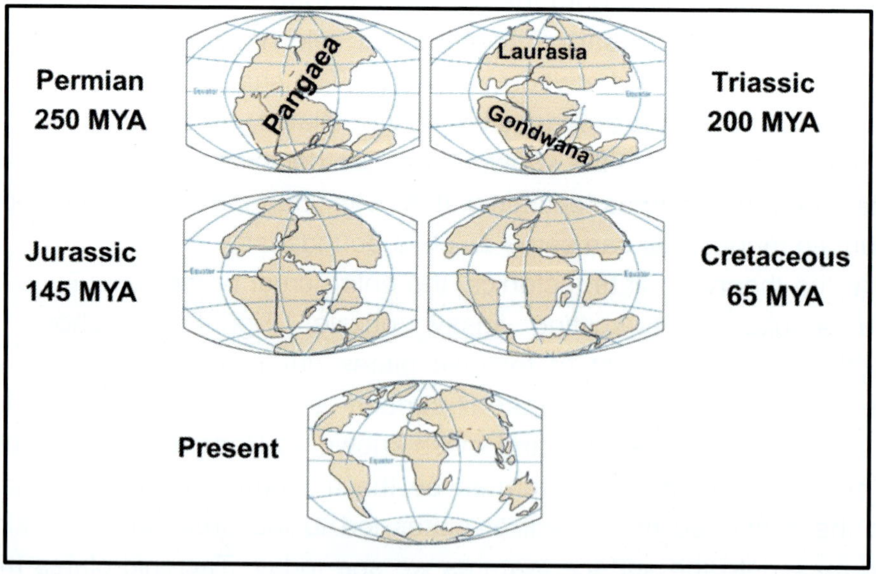

Figure 2-4. Tectonic Plate Movement [after USGS Historical, 2015]

The movement of tectonic plates across the surface of the globe generated forces powerful enough to cause rocks to form and change shape. Geologists recognize three primary types of rocks: igneous, sedimentary, and metamorphic. The process of transforming rock from one type to another is known as the rock cycle, and is illustrated in Figure 2-5.

Igneous rocks are formed by the cooling and solidification of molten rock. Sedimentary rocks form when materials on the surface of the earth are weathered, transported, deposited, and cemented together. Metamorphic rocks form when buried sedimentary rocks are subjected to heat and pressure.

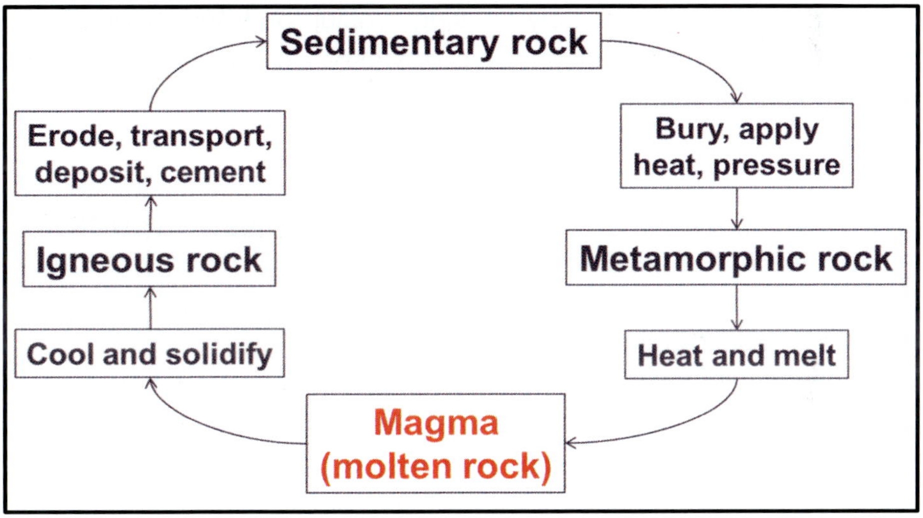

Figure 2-5. Rock Cycle

Porosity and permeability quantify the storage and flow characteristics, respectively, of a porous material like rock. Pores in a rock are the void space between the grains. The bulk volume of the rock is the sum of pore volume and grain volume. Porosity in a porous material is the ratio of pore volume to bulk volume and is typically expressed as a fraction. The continuity of void space from pore to pore is expressed in terms of permeability. Permeability is expressed in oilfield units as the Darcy, which is named after 19th century French engineer Henri Darcy. One Darcy is equal to 0.986923 × 10^{-12} m^2 and may be thought of as a cross-sectional area for fluid flow through pores. The permeability of a traditional oil reservoir is on the order of 0.1 Darcy, or 100 millidarcies.

Sediment can be lithified, or made rock-like, by the movement of minerals into sedimentary pore spaces. A mineral is a naturally occurring, inorganic solid with a specific chemical and crystalline structure. In a process called cementation, minerals can form cement that binds grains of sediments together into a rock-like structure that has less porosity than the original sediment. Compaction occurs when pressure compresses the rock-like structure.

Slowly, over millions of years, tectonic plates move up and down relative to sea level, alternately causing erosion and deposition. Erosion occurs when chemical and physical processes cause exposed rock to break into smaller and smaller fragments. Wind and water transport these fragments from their source location. The fragments become finer as they collide with other objects during the transport process. A fragment will be deposited along with other fragments when the energy of the wind or water dissipates to the point where there is not enough energy to transport the fragment. The accumulation of fragments thickens as fragments are deposited in a specific location.

Deposition can range from thousands of feet of sediment in an area to none at all. Erosion can carve canyons, level once-jagged mountains, or remove all traces of a formation that was once hundreds of feet thick. High pressure and temperature can cause rocks to change character in a process called metamorphism. Fragments may become fused together to form considerably larger objects. Given enough time, pressure and heat, rocks will melt and start the cycle again.

The formation of new rock can occur at plate boundaries and over "hot spots" within the earth's mantle. As plates collide, pressure and heat may cause part of the plate to melt, and result in molten rock being thrust to the surface. After cooling, surface rock is subjected to atmospheric phenomena.

Tectonic plate movement is considered the cause of extensive changes on the surface of the earth. Tectonic plate movement can lead to changes in sea level and the biosphere. Lower sea level can result in a period of vast erosion, while an increase in sea level can result in a period of flooding and deposition. The biosphere is also affected by these changes. Plants and animals may thrive in one set of conditions, and readily become extinct when the conditions change. Plate movement also provides a mechanism to help explain the geographic distribution of organisms around the world. For example, the creation of a land bridge allows migration while the destruction of the land bridge can impose geographic isolation.

An ecosystem is composed of physical and biological components. Geologists use significant historical changes to the global ecosystem as markers on a geologic timeline. Table 2-1 shows an abridged version of the geologic time scale since the earth's formation. The geologic timeline is divided into several durations. Eons are subdivided into eras; eras are subdivided into periods; and periods are subdivided into epochs. The acronym MYBP in Table 2-1 stands for millions of years before the present. The geologic time scale is from the Geological Society of America [Cohen, et al., 2015; and Gradstein, et al., 2012]. Our understanding of the chronology of the earth is approximate.

The oldest terrestrial rocks formed during the Precambrian Eon. Microscopic organisms are believed to have originated during this eon. Organisms with cellular nuclei (eukaryotes) appeared during the Proterozoic Era of the Precambrian.

The Cambrian Period in the Paleozoic Era of the Phanerozoic Eon is the era when life began to blossom. The demise of trilobites during the Permo-Triassic extinction marked the end of the Permian Period at the end of the Paleozoic Era. The Mesozoic Era includes the age of dinosaurs, and its end was marked by the Cretaceous-Tertiary (K-T) extinction. Cenozoic rocks are relatively new, being less than 70 million years old. Mammals began to flourish during the Cenozoic. The variations in life forms give rise to variations in the fossil record that can help geoscientists characterize the stratigraphic column, which shows the ordering of rock units as we drill down into the crust of the earth.

Table 2-1
Geologic Time Scale

Eon	Era	Period	Epoch	Approximate End of Interval (MYBP)
Phanerozoic	Cenozoic	Quaternary	Holocene (Recent)	0.01
			Pleistocene	2.6
		Neogene	Pliocene	5.3
			Miocene	23.0
		Paleogene	Oligocene	33.9
			Eocene	56.0
			Paleocene	66.0
	Mesozoic	Cretaceous		145
		Jurassic		201
		Triassic		252
	Paleozoic	Permian		299
		Carboniferous	Pennsylvanian	323
			Mississippian	359
		Devonian		419
		Silurian		444
		Ordovician		485
		Cambrian		541
Precambrian	Proterozoic			2500
	Archean			4000
	Hadean			4600

2.2 ORIGIN OF COAL

Coal was formed from organic debris by a process known as coalification. When some types of organic materials are heated and compressed over time, they can form water, gas, and coal. In some cases, a high-molecular weight, waxy oil is also formed. For example, bog or swamp vegetation may be buried under anaerobic conditions and become peat. In biology, oxygen is available for consumption by organisms in an aerobic system, while oxygen is not available for consumption in an anaerobic system. Consequently, aerobic organisms consume oxygen, while anaerobic organisms do not. Peat is a soft, spongy material formed by the decay of vegetation under anaerobic conditions is a coal precursor. It is an unconsolidated deposit of partially carbonized vegetable matter in a water-saturated environment. If peat is buried by rock in a depositional environment, it is subjected to increasing temperature and pressure. Volatile products and water migrate away from the formation. The resulting carbon-rich, compressed residue is coal. Figure 2-6 illustrates the coal formation process.

Organisms that form coal when subjected to coalification include algae, phytoplankton, and zooplankton. Coal can also be formed by the bacterial decay of plants and, to a lesser extent, animals. Organic debris is composed primarily of carbon, hydrogen, and oxygen. It may also contain minor amounts of other elements such as nitrogen and sulfur. The organic origin of coal provides an explanation for the elemental composition of coal, which ranges from almost pure carbon to compounds of carbon with other elements, notably hydrogen, oxygen, and sulfur.

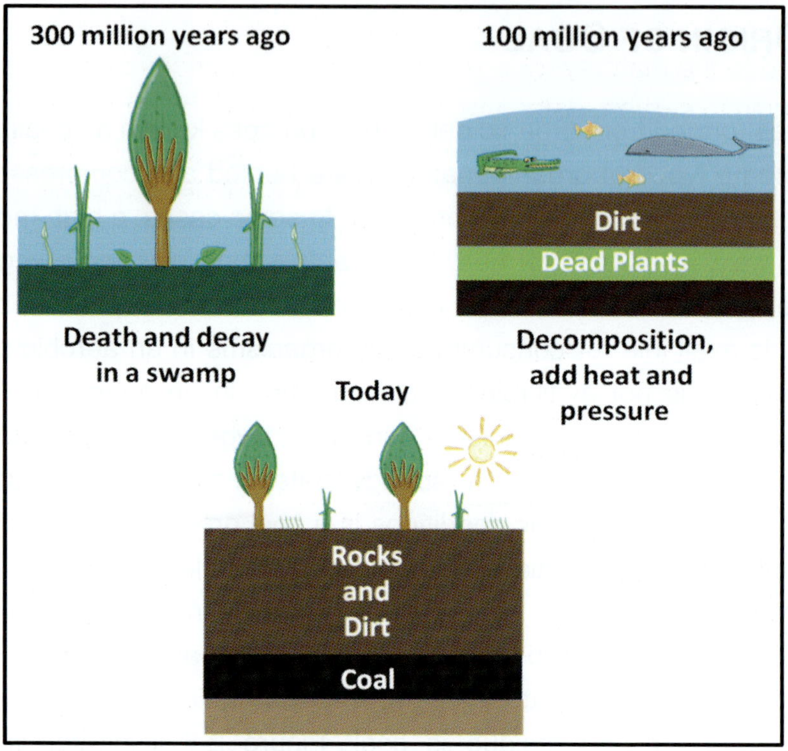

Figure 2-6. A Schematic View of the Formation of Coal [US EIA NEED, 2015]

2.3 COAL

Coals are classified by rank. Rank is a measure of the degree of coalification or maturation of carbonaceous material. The lowest rank coal is lignite, followed in order by sub-bituminous coal, bituminous coal, anthracite and graphite. Table 2-2 summarizes the relative properties of coal for four coal ranks that are used as energy sources. The moisture content of lignite is high compared to the moisture content of anthracite. High moisture content is associated with low heating value, while low moisture content is associated with high heating value.

Coal becomes darker and denser with increasing rank. Coal rank is correlated to the maturity, or age, of the coal. As a coal matures, the ratio of hydrogen to carbon atoms and the ratio of oxygen to carbon atoms decrease. The composition of the highest rank coal, graphite, has carbon content in excess of 98%. Graphite is difficult to ignite and is not typically used as a fuel, therefore it is not included in Table 2-2.

Table 2-2
Relative Coal Properties
[US EIA NEED, 2015]

Coal	Rank	Carbon Content (%)	Moisture Content	Heating Value
Lignite	Low	25 – 35	High	Low
Sub-bituminous		35 – 45		
Bituminous		45 – 86		
Anthracite	High	86 – 97	Low	High

Coals burn better if they are relatively rich in hydrogen; this includes lower rank coals with higher hydrogen to carbon ratios. The percentage of volatile materials in the coal decreases as coal matures. Volatile materials include water, carbon dioxide and methane. Coal gas is primarily methane with lesser amounts of ethane and carbon dioxide. It is absorbed in the coal structure or coal matrix, and adsorbed on the surface of the coal. Absorption is the penetration of one substance into another substance. By contrast, adsorption is the accumulation of one substance on the surface of another.

The amount of gas that is present in the coal matrix depends on rank. As rank increases, the amount of methane in the coal matrix increases because the molecular structure of higher rank coals forms a structure that has a greater capacity to absorb gas and can therefore contain more gas.

Coals are organic, sedimentary rocks. Figure 2-7 shows an idealized representation of the physical structure of a coal seam. A coal seam is the stratum or bed of coal. It is a collection of coal matrix blocks bounded by natural fractures. The fracture network in coalbeds consists of micro-fractures called "cleats." An inter-connected network of cleats allows coal gas to flow from the coal matrix blocks when the pressure in the fracture declines. This is an important mechanism for producing coal gas.

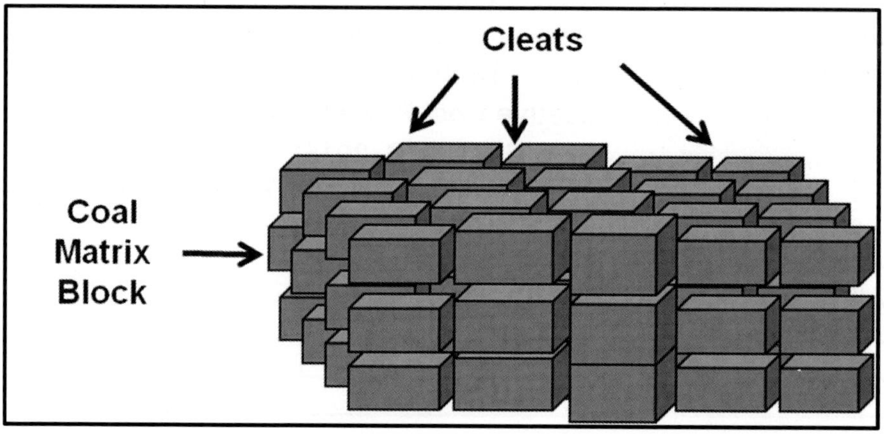

Figure 2-7. Schematic of a Typical Coal Seam

2.4 Distribution and Production of Coal

Coal is a nonrenewable resource. The amount of coal is finite and takes more time to form than it does to consume. Consequently, the consumption of coal results in a reduction of the known supply. Since coal is a nonrenewable resource, the amount of coal is expressed in terms of resources and reserves.

2.4.1 Resource Pyramid

The value of a resource can be displayed using the resource pyramid (Figure 2-8). A resource container (such as a coal seam, oil reservoir, gas reservoir, uranium ore, gold ore, etc.) is placed at the top of the pyramid if it contains a relatively large concentration of resource that is relatively inexpensive to extract. By contrast, a resource container is placed lower in the pyramid if the concentration of the resource decreases or the extraction becomes more difficult or expensive. If the resource container has a relatively small concentration of resource and is difficult to extract, it will be placed near the base of the pyramid. As an illustration, a large, shallow coal seam in an easily accessible part of the earth would be at the top of the pyramid, while a small, deep coal seam in an extreme environment such as the Arctic would be at the base of the pyramid. A pyramid is used because the amount of resource that is both easy to extract and has a high concentration is expected to be small, while the amount of resource that is found in a low concentration container and is difficult to extract is expected to be large.

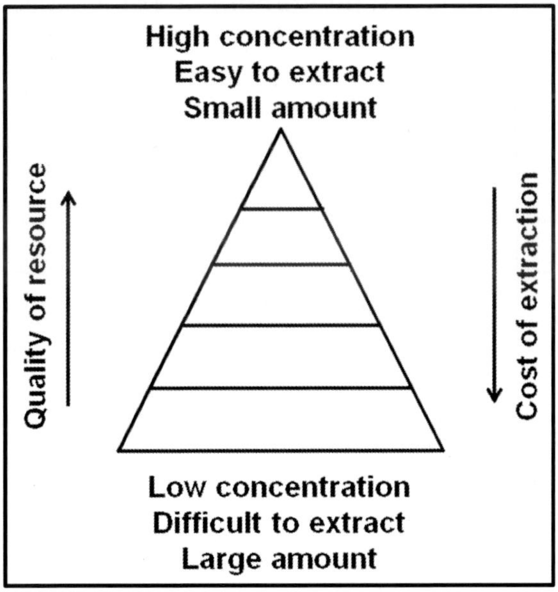

Figure 2-8. Illustration of a Resource Pyramid

Coal resources include all coal that has been identified. Coal reserves refer to the subset of coal resources that is economically producible when they were classified. Improvements in technology and an increase in the price of coal can lead to the reclassification of coal from a resource to a reserve. The reserve concept applies to other nonrenewable resources, including oil, gas and uranium.

2.4.2 Coal Distribution

The five countries with the most recoverable coal in 2011 are shown in Table 2-3. Total recoverable coal for each country in Table 2-3 is expressed as a percent of total global recoverable coal. Many countries have coal reserves of 1% or less. The United States, Russia, and China contain the largest percentage of recoverable coal.

Table 2-3
National Recoverable Coal as Percent of Total Global Recoverable Coal in 2011* [US EIA Coal, 2015]

Country	Recoverable Coal (Billion short tons)	% of Total Global Recoverable Coal
United States	258.6	26.4
Russia	173.0	17.7
China	126.2	12.9
Australia	84.2	8.6
India	66.8	6.8
* Total global recoverable coal in 2011 was approximately 980 billion short tons (1 short ton = 2000 lbs)		

Major coal regions in the United States are illustrated in Figure 2-9. Lignite and anthracite are found predominantly in the Gulf Coast (Texas) and Northern Great Plains (North Dakota, South Dakota, and eastern Montana) regions. Sub-bituminous coal is located in the Rocky Mountain region. Bituminous coal is located in central Texas, central Colorado and Utah, the Interior region of Illinois, Missouri, and Iowa, and the Eastern region stretching from Alabama to Pennsylvania.

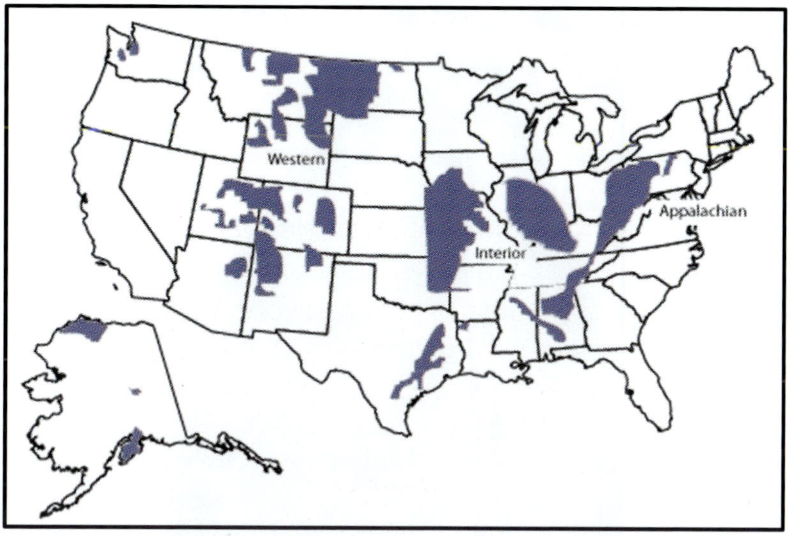

Figure 2-9. Major Coal Regions in the United States as of January 1, 2008 [US EIA Coal, 2009]

2.4.3 Coal Production

Coal is usually produced by extraction from coal beds. Mining is the most common extraction method. There are several types of mining techniques. Some of the more important coal mining techniques are strip mining, drift mining, deep mining, and longwall mining.

Coal seams relatively close to the surface of the earth are strip mined. Consequently, strip mining is also known as surface mining. Any overlying material, such as topsoil and vegetation, is removed and coal is extracted by scraping. Drift mines are used to extract coal from coal seams that are exposed by the slope of a mountain. Drift mines typically have a horizontal tunnel entrance into the coal seam. Deep mining extracts coal from beneath the surface of the earth. In the case of deep mining, coal is extracted by mining the coal seam and leaving the bounding overlying layers and underlying layers undisturbed. Longwall mining is used to mine a longwall panel of coal. The panel is a block of coal that can be 1000 feet wide and 2 to 3 miles long.

Figure 2-10. Coal Trains in Canada (© Fanchi, 2002)

Coal is transported to consumers by ground transportation such as trains (Figure 2-10) and, to a lesser extent, ships. A relatively inexpensive means of transporting coal is the coal slurry pipeline. Coal slurry is a mixture of water and finely crushed coal. Coal slurry pipelines are not widely used because it is often difficult to obtain rights of way for coal slurry pipelines that extend over long distances, particularly in areas where a coal slurry pipeline would compete with an existing railroad right of way.

2.5 Fossil Energy and Combustion

Coal is a carbon-based fossil fuel. The chemical energy in fossil fuels is released by the process of combustion. The complete combustion of fossil fuel with oxygen in the absence of nitrogen can be represented by the chemical reaction

$$Fuel + Oxygen \rightarrow heat + carbon\ dioxide + water$$

or

$$C_xH_y + \left(x + \frac{y}{4}\right)O_2 \rightarrow heat + xCO_2 + \frac{y}{2}H_2O$$

where x and y are stoichiometric coefficients. The stoichiometric coefficients for pure carbon are $x = 1,\ y = 0$ and the stoichiometric coefficients for methane are $x = 1,\ y = 4$.

Our air contains both nitrogen and oxygen. If a carbon-based fuel is burned in air, the carbon can react with oxygen and nitrogen to produce carbon dioxide, carbon monoxide, and nitrogen oxides. Nitrogen oxides produced by combustion of fossil fuels in air are abbreviated NOx, and include nitrogen-oxygen molecules with a single nitrogen atom, namely nitric oxide (NO) and nitrogen dioxide (NO_2). Water is also formed when hydrogen is present in the carbon-based fuel because hydrogen reacts with oxygen to form water.

Chemical reactions involved in fossil fuel combustion are exothermic reactions, that is, reactions that have a net release of energy. An exothermic reaction between two reactants *A*, *B* has the form

$$A + B \rightarrow products + energy$$

An exothermic reaction forms products with a net release of energy. By contrast, an endothermic chemical reaction consumes more energy than it releases. The energy that must be added to a system to initiate a reaction is called activation energy.

Fossil fuels release a relatively large amount of energy during the combustion process, but they also release gases that are considered pollutants. Table 2-4 presents a comparison of air pollutants emitted by fossil fuel power plants. Coal emits almost twice as much carbon dioxide as natural gas to provide the same amount of energy. The environmental impact of the emission of combustion byproducts is a factor that must be taken into account when considering the inclusion of fossil fuels in an energy mix.

Table 2-4
Pounds of Air Pollutants Produced per Billion BTU of Energy [US EIA Emissions, 1998]

Pollutant	Natural Gas	Oil	Coal
Carbon Dioxide	117,000	164,000	208,000
Carbon Monoxide	40.00	33.00	208.0
Nitrogen Oxides (NOx)	92.00	448.00	457.0
Sulfur Dioxide	0.6	1,122	2,591
Particulates	7.0	84.00	2744.00
Formaldehyde	0.8	0.22	0.22
Mercury	0	0.007	0.016

Clean coal technology is being developed to minimize the environmental impact of carbon-based fuel combustion by-products. For example, a coal scrubber can be attached to a coal fired power plant. Flue gas from the power plant boiler is passed through the coal scrubber. The coal scrubber is designed to remove particulates and sulfur from the flue gas. The remaining flue gas is then cooled and passed through a carbon dioxide absorber and a carbon dioxide stripper to remove carbon dioxide. In principle, the recovered carbon dioxide can then be sequestered. The adoption of clean coal technology can reduce the emission of carbon dioxide into the atmosphere, but increases the cost of the power plant.

Point to Ponder: Why is fossil fuel combustion a problem?
Fossil fuel combustion provided a new source of energy that helped prevent deforestation and now supports a relatively high quality of life in industrialized nations. Unfortunately, fossil fuel combustion also releases nitrogen oxides (NOx), carbon monoxide, and carbon dioxide.

NOx can react in the presence of sunlight with volatile organic compounds (VOCs) to form smog. VOCs are organic chemicals that can exist as gases in the atmosphere. Examples of VOCs include low molecular weight hydrocarbons such as methane and ethane, and formaldehyde. The smog can adversely affect a person's ability to breathe. Figure 2-11 is a satellite image of smog over eastern China.

Carbon monoxide and carbon dioxide are examples of greenhouse gases. The presence of greenhouse gases in the atmosphere can trap heat energy. Many scientists believe that the additional heat is increasing the temperature of the atmosphere and is causing climate change. The impact of fossil fuel combustion on the climate is discussed in more detail in Chapter 4.

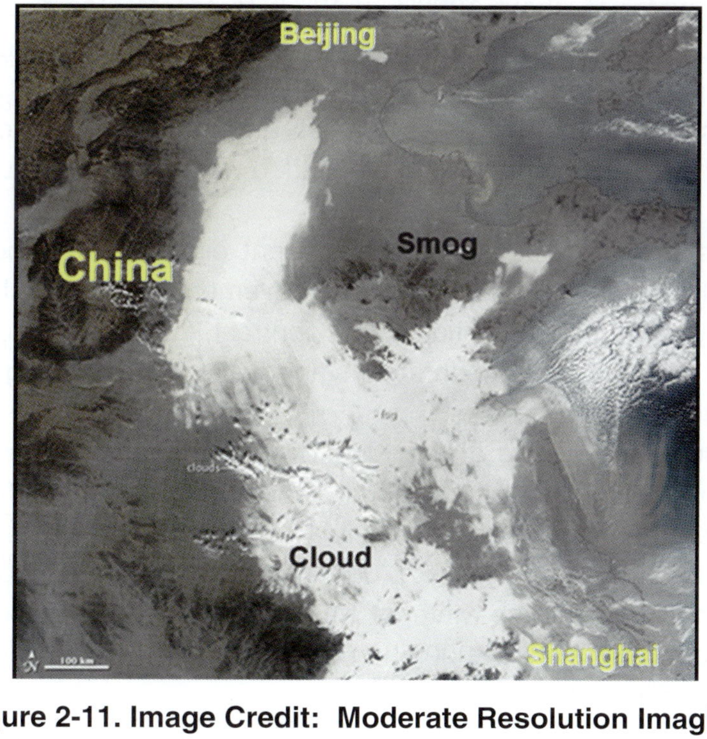

Figure 2-11. Image Credit: Moderate Resolution Imaging Spectroradiometer (MODIS), NASA Terra satellite; image acquired December 7, 2013

2.6 Case Study: A Coal-Fired Power Plant

Coal-fired power plants generate electricity by burning coal to produce steam from water. Figure 2-12 illustrates the key systems of a coal-fired power plant. Flowing steam spins a turbine that turns a magnet in a generator to produce electricity. In a closed system, the steam is cooled, condenses into water, and the water is recycled to the boiler so that the process can be repeated. Generated electricity is transmitted through power lines to consumers. Electricity generation and distribution is discussed in more detail in Chapter 11.

An example of a coal-fired power plant is the Kingston Fossil Plant near Knoxville, Tennessee. The plant is managed by the Tennessee Valley Authority (TVA). According to the TVA [TVA, 2015], boilers at the plant heat water to about 1000°F (540°C) to create steam. The steam flows through pipes at pressures in excess of 1800 psi into turbines. The turbines spin at approximately 3600 revolutions per minute and drive generators. The steam exits the turbines and is cooled by a condenser using river water. The steam is condensed to water before it returns to the boiler.

The generators make alternating current electricity at a potential of 20,000 volts. Alternating current is the movement of electric charge that periodically reverses direction. By contrast, direct current is the movement of electric charge in a constant direction. Alternating current is widely used in homes.

Nine coal-fired boilers at the Kingston Fossil Plant burn a total of 5 million short tons of low-sulfur coal per year when operating at full capacity. The plant generates approximately 10 billion kilowatt-hours a year of electricity to supply energy to 670,000 homes. The net dependable generating capacity during the winter is 1450 megawatts. The amount of coal burned is approximately 14,000 short tons per day, which is enough to fill 140 railroad cars.

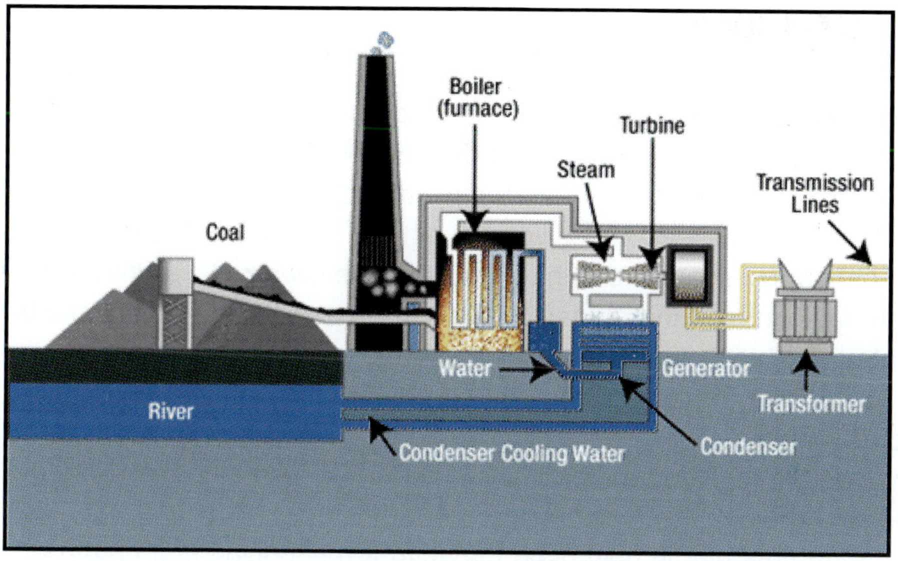

Figure 2-12. Tennessee Valley Authority [TVA, 2015]

The TVA has added equipment to control gas emissions from the plant. The emissions control equipment includes scrubbers to reduce sulfur dioxide (SO_2) emissions, and catalytic reduction systems to reduce NOx emissions. The addition of equipment to control gas emissions is an example of clean coal technology and is designed to reduce the environmental impact of energy generation as a result of coal combustion.

In addition to gas emissions, coal-fired power plants produce ash, which is the residue left over from burning coal. A coal ash retention facility at the Kingston Fossil Plant spilled a large volume of sludge into the Emory River and surrounding countryside on December 22, 2008. According to knoxnews.com, the spill prompted several lawsuits for financial payments from the TVA to cover damages such as destroyed homes [see Barker, 2009a], a call by a state of Tennessee advisory board for tougher regulations of ash impoundments [Barker, 2009b], and a restructuring of management [Marcum, 2009]. TVA cleaned up the coal ash spill with the goals of satisfying government requirements, regain the public trust, and leave the impacted area in a better condition than it had been before the spill [TVA, 2015; TVA Spill, 2015].

2.7 Coal Gas

Coal gas is a gas obtained from coal and is an example of a biogenic gas. Biogenic gases form when buried organic matter decays in lower temperature environments on or near the subsurface. These gases are primarily methane with lesser compositions of ethane and propane. Biogenic gases are ordinarily associated with compost heaps, swamps, landfills and coalbeds. By contrast, thermogenic gases are formed when organic matter is heated in high temperature, subsurface environments. Thermogenic gases typically have between one carbon atom (methane) and six carbon atoms (hexane).

The term coalbed methane was historically used to describe coal gas because methane is typically the largest component of the gas mixture. The term coal gas is now used because it recognizes that coal gas contains methane and smaller amounts of other components such as ethane, propane, carbon dioxide, nitrogen and hydrogen. Coal is an abundant source of methane.

Coal gas is a product of coalification and exists as a monomolecular layer (a layer that is one molecule thick) on the internal surface of the coal matrix. Coal gas is able to diffuse into the natural fracture network when a pressure gradient exists between the matrix and the fracture network.

Gas recovery from coal begins with desorption, or escape, of gas from the internal surface to the micropores of the coal matrix. The gas then diffuses through the micropores into the cleats. Finally, gas flows through the cleats to the production well. The flow rate depends, in part, on the pressure gradient in the cleats and the density and distribution of cleats. The controlling mechanisms for gas production from coal are the rate of desorption of gas from the coal surface to the micropores of the coal matrix, the rate of diffusion from the coal matrix to the cleats, and the rate of flow of gas through the cleats.

Figure 2-13 is a schematic of a coal gas well. Coal gas production often requires the production of large volumes of water before significant volumes of coal gas will flow. Additional energy may be needed to lift the water to the surface. This additional energy is provided by a submersible pump shown in Figure 2-13.

Coal gas recovery can be increased by injecting carbon dioxide (CO_2) into the coal seam. Carbon dioxide is generated in large quantities by a variety of processes. For example, CO_2 is produced by biological, industrial and transportation processes. In particular, we exhale CO_2 with every breath, and CO_2 is a by-product of fossil fuel combustion in power plants. Coal absorbs more CO_2 than methane so the injection of CO_2 into a coalbed can accomplish two objectives: increase methane production from coal, and store CO_2 in coal. Storage of CO_2 in coal is an example of geological sequestration, or the long-term storage of a gas in a geologic formation. Geological sequestration reduces the amount of CO_2 that might otherwise be emitted into the atmosphere. The presence of CO_2 in the atmosphere can have an adverse environmental impact.

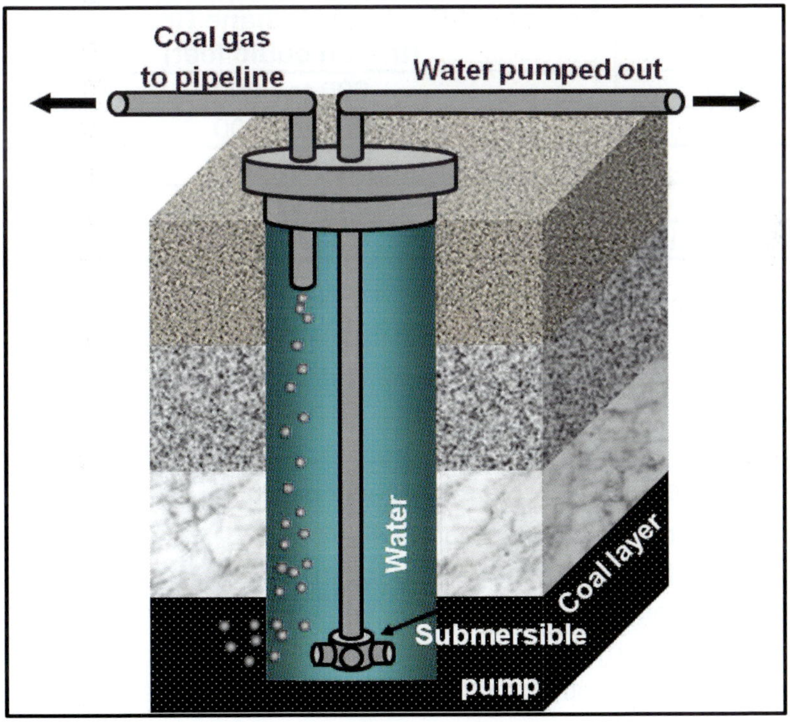

Figure 2-13. Schematic of Coal Gas Well

2.7.1 Global Distribution of Coal Gas

The out-gassing of gas from coal is well known to coal miners as a safety hazard, and occurs when the pressure in the cleat system declines. The gas in the microscopic pore structure of the coal is also a source of natural gas. The World Coal Association (WCA) reports that the largest coal gas resources are in the coal seams of Canada, Russia, China, Australia and the United States. Table 2-5 lists the range of estimated coalbed methane resources for these countries. Much of the coal gas resource is underdeveloped.

Table 2-5
Nations with Largest Coal Gas Resources in 2006
[WCA, 2012]

Country	Estimated Coalbed Methane Resource Base (trillion cubic feet)
Canada	600 to 3200
Russia	600 to 2800
China	1000 to 1200
Australia	280 to 490
United States	140 to 390

2.8 Activities

True-False

Specify if each of the following statements is True (T) or False (F).

1. An exothermic reaction releases energy.
2. Porosity is the fraction of void space in rock.
3. An endothermic chemical reaction forms products with a net release of energy.
4. Molten rock beneath the surface of the earth is called lava.

5. Molten rock cools and solidifies into igneous rock as part of the rock cycle.
6. Tectonic plate movement can affect sea levels, rock creation, and the biosphere.
7. The dinosaurs became extinct during the Permo-Triassic extinction.
8. Coals are organic, metamorphic rocks.
9. Strip mining, drift mining, deep mining and longwall mining are common types of coal mining.
10. Coal gas is primarily methane.

Questions

1. How are coals classified?
2. What is the primary negative byproduct of combustion?
3. What is a cleat in a coal seam?
4. What hydrocarbon gas can be found in coalbeds in abundant amounts?
5. What are the three primary rock types?
6. List the four stages of the rock cycle.
7. What is coalification?
8. What is the difference between a coal resource and a coal reserve?
9. How do coal-fired power plants generate electricity?
10. How is CO_2 used in coal gas production?

CHAPTER 3

FOSSIL ENERGY – OIL AND GAS

After reading this chapter, you should be able to...

- Define petroleum in relation to oil and gas
  - *P2P: What is petroleum?*
- Describe the geology of oil and gas reservoirs
- Explain the origin of oil and gas in terms of Biogenic and Abiogenic Theories
  - *P2P: Why does it matter which theory is right?*
- Apply the classifications of oil and gas
- Describe historical and modern methods of producing oil and gas
- Explain modern drilling methods, and how they have changed access to resources
- Describe how oil and gas move from the reservoir to the consumer
- Define shale oil, tar sands and extra heavy oil
- Define unconventional gas
- Describe the global distribution of oil and gas

Oil and gas are terms that refer to mixtures of hydrocarbon molecules in the liquid phase and gas phase, respectively. Hydrocarbon molecules are compounds of carbon and hydrogen. Table 3-1 lists the most common elements found in crude oil and natural gas. Crude oil is "a mixture of hydrocarbons that exists in liquid phase in natural underground reservoirs and remains liquid at atmospheric pressure" after passing through facilities on the surface that separate gas and liquid [US EIA Glossary, 2015].

Table 3-1
Elemental Composition of Crude Oil and Natural Gas

Element	Composition (% by mass)
Carbon	84% – 87%
Hydrogen	11 % – 14%
Sulfur	0.6% – 8 %
Nitrogen	0.02% – 1.7%
Oxygen	0.08% – 1.8%
Metals	0% – 0.14%

The actual elemental composition depends on factors such as molecular composition of the source, reservoir temperature and reservoir pressure. Hydrocarbon molecules are compounds of carbon and hydrogen. They are organic molecules, that is, molecules that consist of carbon atoms attached to other atoms such as hydrogen, oxygen and nitrogen. Inorganic molecules such as nitrogen, carbon dioxide, and hydrogen sulfide can also be found in oil and natural gas.

Petroleum is another term for a hydrocarbon mixture found in nature. Petroleum liquid typically refers to crude oil and natural gas liquids. Natural gas contains methane with lesser amounts of heavier hydrocarbon molecules like ethane and propane. If the natural gas contains liquids dissolved in the gas phase at high temperatures and pressures, the liquids can condense out of the gas phase when the temperature and pressure of the gas are reduced. The change in temperature and pressure typically occurs when gas is produced from a subsurface formation to a facility on the surface. The liquids obtained from the gas are referred to as condensates or natural gas liquids. Table 3-2 displays the typical composition of natural gas products including liquefied natural gas (LNG), liquefied petroleum gas (LPG), and natural gas liquids (NGL). Non-hydrocarbon gases include the inorganic molecules carbon dioxide (CO_2), water (H_2O), hydrogen sulfide (H_2S), and nitrogen (N_2).

Table 3-2
Typical Composition of Natural Gas Products

Component	# of C atoms	Natural Gas	LNG	Residential Gas	LPG	NGL
Methane	1	X	X	X		
Ethane	2	X	X	X		X
Propane	3	X	X		X	X
Butanes	4	X			X	X
Pentanes	5	X				X
Natural gasoline	5+	X				X
Non-hydrocarbon gases (CO_2, H_2O, H_2S, N_2, etc.)		X				
Water		X				

The total energy consumed in the world in 2010 included approximately 35% petroleum liquid and 23% natural gas (Appendix C). Including coal, global energy from fossil fuel consumption exceeded 80% in 2010. Large segments of modern economies are designed to use oil and gas even though oil and gas are considered finite, nonrenewable resources. Most crude oil is refined for use in transportation, while gas is used to generate electric power and as a fuel in industrial, commercial, and residential sectors of the economy.

The objective of this chapter is to discuss how oil and gas resources are formed, distributed, produced and consumed throughout the world. We begin by describing the geology of oil and gas reservoirs.

> **Point to Ponder: What is petroleum?**
> The term petroleum is widely used but inconsistently defined. The Society of Petroleum Engineers (SPE) defined petroleum as "naturally occurring liquids and gases which are predominately comprised of hydrocarbon compounds" [SPE Definitions, 2009]. More recently, the SPE-sponsored glossary defines petroleum in terms of its Latin meaning of 'rock oil' [SPE Glossary, 2015]. By contrast, the United States Energy Information Administration defines petroleum as "a broadly defined class of liquid hydrocarbon mixtures. Included are crude oil, lease condensate, unfinished oils, refined products obtained from the processing of crude oil, and natural gas plant liquids." [US EIA Glossary, 2015]. Both SPE and EIA define petroleum as naturally occurring mixtures of hydrocarbon molecules. The SPE defines petroleum in terms of liquids and gases, while EIA defines petroleum in terms of liquids and keeps gas separate. It is important to recognize context to understand the meaning of the term petroleum.

3.1 Geology of Oil and Gas Reservoirs

Naturally occurring hydrocarbon mixtures are usually found in the pore space of sedimentary rocks. The collection of physical and chemical changes that occur during the conversion of sediments to rock is called diagenesis. According to the rock cycle, sedimentary rocks are formed by erosion, transport, deposition, and cementing of igneous rock. Unlike sedimentary rocks, igneous and metamorphic rocks originated under conditions of high pressure and temperature that did not favor the formation or retention of petroleum. Any oil and gas that might have occupied the pores of a metamorphic or igneous rock would ordinarily be cooked away by heat and pressure. Consequently, we focus our attention on sedimentary rock formations as geologic environments for containing oil and gas.

3.1.1 Rock Attributes

The key attributes used to classify sedimentary rock are mineral composition, grain size, color, and structure. Each of these attributes is considered here.

Mineral content is an important characteristic of a rock. The minerals that comprise a rock help determine the chemical and physical properties of the rock. For example, quartz may be able to withstand multiple cycles of erosion and deposition since quartz is much less reactive than other minerals. As another example, the presence of clays in the pore space can cause reduced productivity if a change in the salinity of the formation causes clay swelling. Fluids introduced into the reservoir through the wellbore can react with clays to swell and plug clay-bearing formations.

Grains which form sedimentary rocks are created by weathering processes at the surface of the earth. Weathering creates fragments that can be practically any size, shape, or composition. A glacier may create and transport a fragment the size of a house, while a desert wind might create a uniform bed of very fine sand. The fragments, also known as sediments, are transported to the site of deposition. Mineral transport is often achieved by water-based processes such as rivers, floods, and tides. When fragments are transported over long distances, only the most durable fragments survive the transport. Grains which started out as angular chunks of rock slowly become smaller and more rounded as they roll and bump along the transport pathway.

The shape of rock grain edges tells us something about the history of a rock. If rock edges are sharp, the rock probably did not get transported very far because the sharp edges did not get smoothed by collisions with other objects. Rounded edges indicate a longer period of transport. Rocks made up of rounded edges may have better permeability than rocks that are flat or have sharp edges.

Sorting refers to the uniformity of grain size. Fluids typically flow better through a well-sorted rock than a poorly-sorted rock. The ability to flow is related to permeability. Fluid flows faster through a rock with large permeability than it does through a rock with low permeability. Permeability is an important property that is a measure of flow capacity, and has a direct impact on the economic viability of producing fluids from a formation.

3.1.2 Depositional Environment

The environment under which a rock forms is called the environment of deposition. The depositional environment can vary from one location to another and it can change locations. A changing depositional environment can leave a laterally continuous progression of rock which is distinctive in character. For example, the rise or fall of sea level relative to a continental plate can change shorelines and their associated environments of deposition. If the progression of rock is large enough to be mapped, it can be called a formation.

Formations are the basic descriptive unit for a sequence of sediments. The formation represents a recognizable, mappable rock unit that was deposited under a uniform set of conditions at one time. Formations can extend for miles and vary in thickness from a few feet to hundreds of feet. The thickness is related to the length of time an environment was in a particular location, and how much subsidence was occurring during that period. An example of a stratified rock column is shown in Figure 3-1. The figure shows layers of sedimentary rock (labelled strata), a coal seam, and a fault separating the coal seam from the rock layers.

Figure 3-1. Example of Stratified Rock (© Fanchi, 2006)

Each environment of deposition causes a different type of rock sequence to be deposited. The Mississippi River is an example of a fluvial (river) environment of deposition, and the mouth of the Mississippi River is a deltaic environment. Sandstone can be deposited in both a fluvial environment and a deltaic environment. The deposited sandstone is different in each environment. A fluvial system contains rocks that meander along, while rocks in a delta tend to stay in one place after settling. A large volume of sediment can be deposited in a deltaic system. In a fluvial system, the grain size of deposited sand is finer at shallower depths. In addition, there is a coarser deposit at the base of the formation that distinguishes the fluvial system from a deltaic deposit. Lying above the sandstones of a meandering river are typically mudstones that may have traces of roots and desiccation cracks that can occur when mud dries. The sandstones are interpreted to be formed in the river channel, while the mudstones are considered floodplain deposits.

The term facies refers to those characteristics of rocks that can be used to identify their depositional environment. In the fluvial deposition example, sandstone and mudstone are distinct facies.

3.2 Origin of Oil and Gas

Sedimentary rock formations are the most common geologic environments for storing oil and gas, but that does not mean that sedimentary rock formations are the geologic environments where oil and gas originated. Two theories of the origin of fossil fuels are considered here: the biogenic theory, and the abiogenic theory.

3.2.1 Biogenic Theory

The biogenic theory is the mainstream scientific view of the origin of fossil fuels. In the biogenic theory, fossil fuels are formed by a type of biochemical precipitation called organic sedimentation. We saw in Chapter 2 that a carbon rich organic material called peat can be formed when vegetation dies and decays in aqueous environments such as swamps. If peat is buried by subsequent geological activity, the buried peat is subjected to increasing temperature and pressure, and peat can eventually be transformed into coal by the process of coalification. An analogous process is used in the biogenic theory of the origin of oil and gas.

Hydrocarbon mixtures can exist in solid, liquid or gas phase in the reservoir. The phase depends on the composition of the mixture and the temperature and pressure of the reservoir. The elemental mass content of naturally occurring hydrocarbon mixtures ranges from approximately 84% to 87% carbon and 11% to 14% hydrogen, which is comparable to the carbon and hydrogen content of life. This is one piece of evidence for the origin of petroleum from biological sources.

The biochemical process for the formation of oil and gas is illustrated in Figure 3-2. It begins with the death of microscopic organisms such as algae and plankton. The remains of the organisms settle into the sediments at the base of an aqueous environment as organic debris. Lakebeds and seabeds are examples of favorable sedimentary environments. Subsequent sedimentation buries the organic debris.

The rise and fall of sea level, as well as other geologic processes, continue the process of burying the organic debris. As burial continues, the

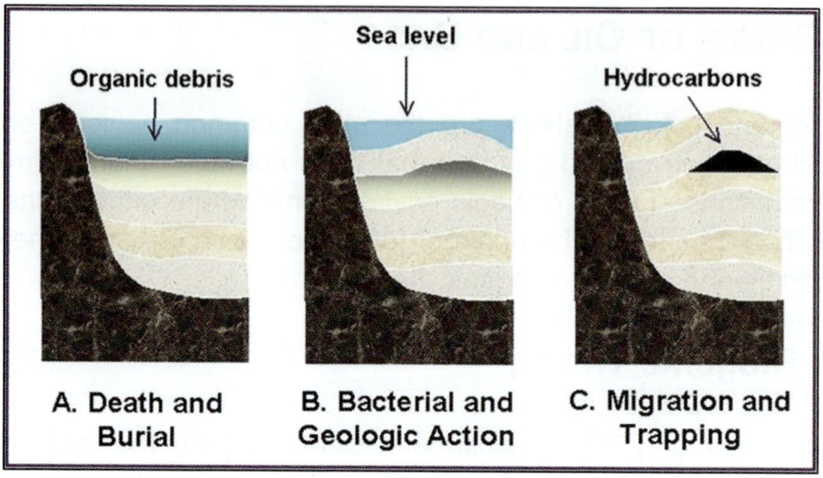

Figure 3-2. Biogenic Origin of Oil and Gas

organic material is subjected to increasing temperature and pressure, and is transformed by bacterial action into oil and gas. In many cases the organic material is contained in sediment that originally had a muddy composition, such as silts and clays. The transformation of organic material into oil and gas often occurs in shale, which can be formed by the application of heat and pressure to muddy sediment. Consequently, shale is a common source rock for oil and gas. Petroleum fluids are usually less dense than water and will migrate upwards until they encounter impermeable barriers and are collected in geologic traps. The accumulation of a hydrocarbon mixture in a geologic trap forms an oil and gas reservoir.

3.2.2 Abiogenic Theory

In the biogenic theory, the origin of oil and gas begins with the death of organisms that live on or near the surface of the earth. An alternative hypothesis called the abiogenic theory says that processes deep inside the earth, in the earth's mantle, form petroleum fluids. Thomas Gold, an advocate for the abiogenic theory, argued that simple inorganic and organic molecules in the interior of the forming earth were subjected to increasing heat and pressure, and eventually formed more complex molecules.

Gold [1999] presented several pieces of evidence in support of the abiogenic theory, and the possible existence of a biological community deep inside the earth. Some of Gold's evidence includes the existence of microbial populations that can thrive in extreme heat. These single-celled microbes, notably bacteria and archaea, grow at hot, deep ocean vents and can feed on hydrogen, hydrogen sulfide, and methane. Gold considered life forms at deep ocean vents transitional life forms that exist at the interface between two biospheres.

One biosphere is the surface biosphere and includes life that lives on the continents and in the seas on the crust of the earth. Gold postulated that a second biosphere exists in the mantle of the earth. He called the second biosphere the deep biosphere. The surface biosphere uses chemical energy extracted from solar energy, while the deep biosphere feeds directly on chemical energy. Oxygen is a requirement in both biospheres. Gold's deep biosphere is the source of life that eventually forms hydrocarbon mixtures in the earth's mantle. Crustal oil and gas reservoirs are formed by the upward migration of petroleum fluid until the fluid is stopped by impermeable barriers and accumulates in geological traps. The abiogenic theory is illustrated in Figure 3-3.

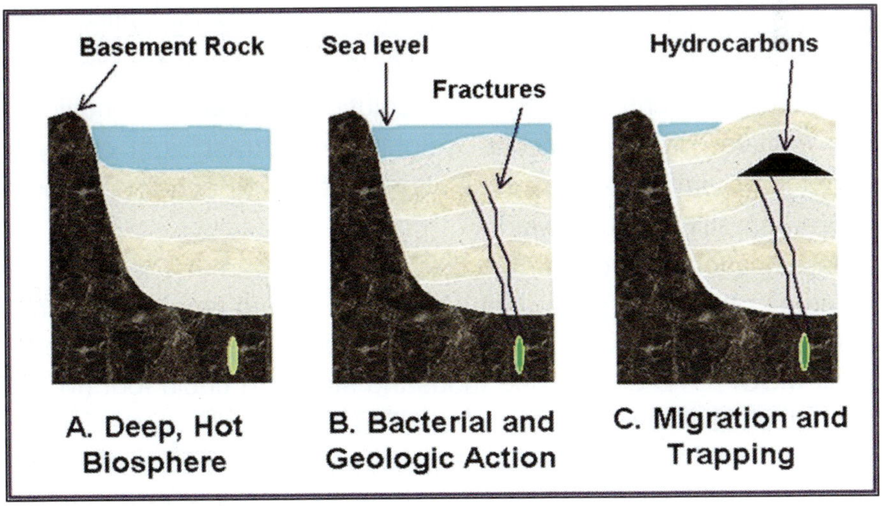

Figure 3-3. Abiogenic Origin of Oil and Gas

Point to Ponder: Why does it matter whether the biogenic theory or the abiogenic theory is right?

Two of the arguments driving a transition from fossil energy to other forms of energy are the belief that the earth contains a finite amount of fossil fuels, and that fossil fuels are not produced by natural processes fast enough to allow fossil fuels to be used as an inexhaustible source of energy. These arguments are based on the assumption that the biogenic theory is correct. If the abiogenic theory is correct, existing estimates of the amount of petroleum fluid and the rate at which it is renewed could be significantly understated. There are still other arguments for reducing our dependence on fossil fuels. They are considered in more detail later.

3.2.3 Classification of Oil and Gas

Hydrocarbon fluids that are one phase at the temperature and pressure in the reservoir often become two-phase fluids when they are produced to the surface where the temperature and pressure are lower. Natural gas is a hydrocarbon mixture in the gaseous state at surface conditions. Crude oil is a hydrocarbon mixture in the liquid state at surface conditions. Heavy oils do not contain much gas in solution at reservoir conditions and have a relatively large molecular weight. By contrast, light oils typically contain a large amount of gas in solution at reservoir conditions and have a relatively small molecular weight.

A classification of hydrocarbon fluid types is given in Table 3-3. Separator gas-oil ratio (GOR) is a useful indicator of fluid type. It is the ratio of volume of gas produced at the surface to the volume of liquid produced. The unit MSCF/STB equals one thousand standard cubic foot (MSCF) of gas per stock tank barrel (STB) of oil. The gas volume in SCF is at surface temperature and pressure and the oil volume in STB is at stock tank temperature and pressure. The stock tank is a storage tank on the surface.

Table 3-3
Rules of Thumb for Classifying Hydrocarbon Fluid Types

Fluid Type	Separator GOR (MSCF/STB)	Behavior when Pressure Drops
Dry gas	No surface liquids	Remains gas
Wet gas	> 50	Remains gas
Condensate	3.3 - 50	Gas with liquid drop out
Volatile oil	2.0 – 3.3	Liquid with significant gas
Black oil	< 2.0	Liquid with some gas
Heavy oil	≈ 0	Negligible gas formation

Another way to classify hydrocarbon liquids is to compare the properties of the hydrocarbon liquid to water. Two key properties are viscosity and density. Viscosity is a measure of the ability to flow, and density is the amount of material in a given volume. Water viscosity is 1 cp (centipoise) and water density is 1 gram per cubic centimeter (1 g/cc) at 60°F. A liquid with smaller viscosity than water flows more easily than water. Gas viscosity is much less than water viscosity. Tar, on the other hand, has a very high viscosity relative to water.

A comparison of hydrocarbon liquid density with water density is made using API gravity, where API refers to the American Petroleum Institute. The ratio of hydrocarbon liquid density (HC liquid density) to water density is the specific gravity of the hydrocarbon liquid. Specific gravity of a hydrocarbon liquid is calculated as

$$\text{HC liquid density} = (\text{water density})/(\text{specific gravity})$$

API gravity is calculated using specific gravity:

$$\text{API} = (141.5/\text{specific gravity}) - 131.5$$

The API gravity of water is 10°API, which is read as 10 degrees API.

Table 3-4 shows a hydrocarbon liquid classification scheme using API gravity and viscosity. Water properties are included in the table for comparison. Bitumen is a hydrocarbon mixture with large molecules and high viscosity. Light oil, medium oil, and heavy oil are different types of crude oil that are less dense than water. Extra heavy oil and bitumen are denser than water. In general, lower density oil will float on water, while extra heavy oil and bitumen will sink in water.

Table 3-4
Classifying Hydrocarbon Liquid Types Using API Gravity and Viscosity

Liquid Type	API Gravity (degrees API)	Viscosity (centipoises)
Light oil	> 31.1	
Medium oil	22.3 to 31.1	
Heavy oil	10 to 22.3	
Water	10	1 cp
Extra heavy oil	4 to 10	< 10,000 cp
Bitumen	4 to 10	> 10,000 cp

Oil and gas resources can also be classified as conventional or unconventional. The United States Energy Information Administration defines conventional oil and natural gas production as "Crude oil and natural gas that is produced by a well drilled into a geologic formation in which the reservoir and fluid characteristics permit the oil and natural gas to readily flow to the wellbore" [US EIA Glossary, 2015]. Unconventional oil refers to hydrocarbon production from low permeability shale (shale oil) and tar sands. Unconventional gas refers to gas production from coal (coal gas), low permeability sands (tight gas), and low permeability shale (shale gas).

The primary difference between conventional and unconventional oil and gas is the ability of the fluid to flow through rock.

3.2.4 Oil and Gas Reservoirs

Several key factors must be present to allow the development of a conventional hydrocarbon reservoir:

1. A source rock for the hydrocarbon must be present. For example, oil and gas can be formed by the decay of organisms like plankton. Shale formed by the heating and compression of silts and clays is a good source rock. Hydrocarbon mixtures can form when the remains of an organism are subjected to increasing pressure and temperature.
2. A flow path must exist between the source rock and reservoir rock.
3. Once hydrocarbon fluid has migrated to a suitable reservoir rock, a trapping mechanism must exist. If the hydrocarbon fluid is not stopped from migrating, buoyancy and other forces will cause it to move towards the surface.
4. Overriding all of these factors is timing. A source rock can provide large volumes of oil or gas to a reservoir, but a trap must exist at the time oil or gas enters the reservoir so that fluid can accumulate.

If all of these conditions are met, a conventional hydrocarbon bearing reservoir can form.

Unconventional reservoirs can be considered reservoirs that cannot be produced at economic rates or volumes without stimulation treatments or special recovery processes and technologies. For example, shale oil and shale gas reservoirs are unconventional reservoirs with very low permeabilities. What distinguishes shale reservoirs from conventional reservoirs is that shales are often source rock for forming conventional reservoirs. Shale reservoirs are sometimes called source rock reservoirs.

3.3 Oil and Gas Production Operations

The oil and gas industry consists of three sectors: upstream, midstream, and downstream. The upstream sector refers to operations designed to find and produce oil and gas. The downstream sector includes natural gas processing, crude oil refining, and marketing and distribution of products obtained from crude oil and natural gas. The midstream sector refers to operations that transfer hydrocarbons from upstream production operations to downstream refining and processing operations. We discuss the upstream sector in this section, and the midstream and downstream sectors in the next section.

3.3.1 Drilling Methods

Seismic surveys are typically used to explore for structures that can trap oil and gas. A seismic survey is based on vibrations that are sent into the earth. Part of the vibrational energy is reflected back to the surface when the vibrations encounter changes from one type of rock formation to another in the subsurface. The reflected energy can be detected and used to infer a picture of the geology of the subsurface.

When a prospect is found, wells are drilled to see if hydrocarbon is actually present. A well is a "dry hole" if it does not encounter hydrocarbons. The well that first encounters hydrocarbon in a new reservoir is the discovery well. Delineation wells are drilled to establish the size of the reservoir, but can also be used for production or injection later in the life of the reservoir. Reservoir boundaries are established by seismic surveys and delineation wells.

Production can begin immediately after the discovery well is drilled, or years later after several delineation wells have been drilled. The number of wells used to develop the field, the location of the wells, and their flow characteristics are among the many issues that must be addressed by reservoir management.

The first method of drilling a well for oil in the modern era was introduced by Edwin Drake in the 1850's and is known as cable-tool drilling. In this method, a rope connected to a wood beam had a drill bit attached to the end. The beam was rocked back and forth so that it lifted and then dropped the bit into the ground. Cable-tool drilling does not work in soft-rock formations where the sides of the hole can collapse. Cable-tool drilling has been largely replaced by rotary drilling.

Developed in France in the 1860's, rotary drilling was first used in the United States in the 1880's because it could drill into the soft-rock formations of the Corsicana oilfield in Texas. Rotary drilling uses a rotating drill bit to penetrate the earth. Drilling mud is circulated down through the center of cylindrical drill pipe and exits through nozzles in the drill bit. The mud flows up the annulus between the drill pipe and the wellbore wall. Drilling mud is designed to lubricate and cool the drill bit, and carry rock cuttings up the wellbore to the surface.

Rotary drilling gained great popularity after Captain Anthony F. Lucas drilled the Lucas 1 well at Spindletop, near Beaumont, Texas. The Lucas 1 well was a discovery well. So much oil flowed up the well that it engulfed the drilling derrick and it appeared as if the earth was gushing oil. Instead of flowing at the expected 50 barrels of oil per day, the well produced up to 75,000 barrels per day. The Lucas gusher began the Texas oil boom [Yergin, 1992, pages 83-85]. Since then, rotary drilling has become the primary means of drilling.

A modern drilling rig is shown in Figure 3-4. The rotary mechanism that is used to rotate the drill pipe with the attached drill bit is set in the drill floor. The traveling block is used to raise and lower the pipe. A cable is attached to the traveling block, wraps over the crown block, and extends down the side of the derrick to a drawworks at the base of the derrick. The drawworks is used to wind and unwind a spool of cable. An engine provides power to all of the systems that require power.

Once a hole has been drilled, it is necessary to complete the well. A well is said to be completed when it is prepared for production. The first

Figure 3-4. Illustration of a Drilling Rig (© Fanchi, 2009)

completed well of the modern era was completed in 1808 when two American brothers, David and Joseph Ruffner, used wooden casings to prevent low-concentration salt water from diluting the high-concentration salt water they were extracting from deeper in their saltwater well [van Dyke, 1997, pages 145-146].

A modern wellbore is sketched in Figure 3-5. Steel casing and cement are typically used to isolate shallow formations from fluid in the wellbore. Cement is injected through the drill pipe and into the annulus between the drill pipe and the borehole wall. Cement is used to form a barrier between the borehole and adjacent formations. The steel casing and cement protect formations with drinking water. Leaks in the cement can result in fluid movement from the producing interval to other formations. This can affect production and injection of fluids. Cement bond log measurements can be used to check the integrity of the cement.

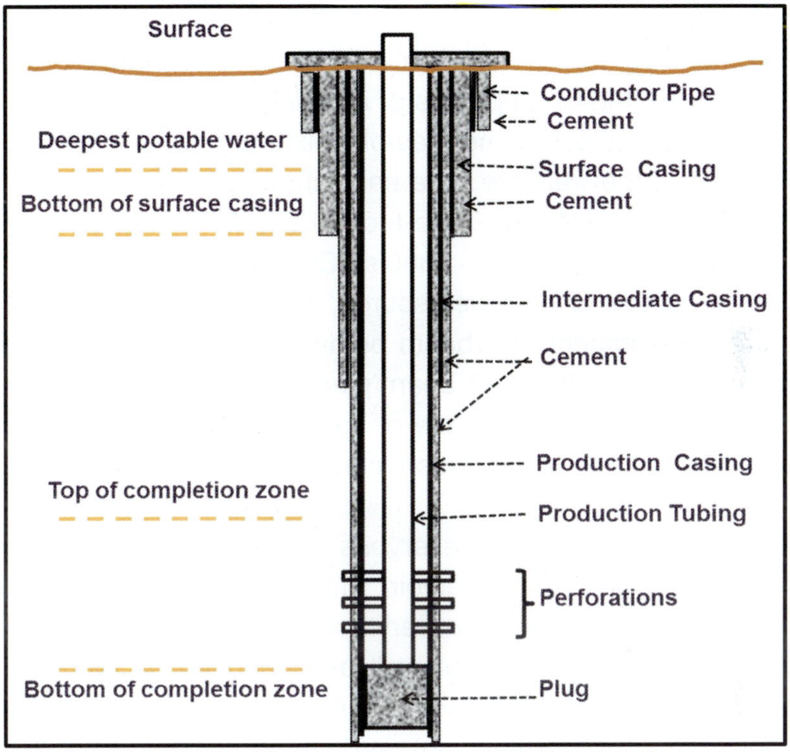

Figure 3-5. Illustration of a Wellbore Diagram for a Vertical Well

Wellbores may be completed using a variety of techniques. A common completion technique is to place tubing adjacent to the productive interval and then use a perforation gun to shoot perforations through the tubing walls, as illustrated in Figure 3-5. Reservoir fluid can flow into the tubing through the perforation holes. A plug is set at the bottom of the completion zone in Figure 3-5 to prevent fluid from flowing into the borehole below the completion zone. In open hole completions, tubing is set above the productive interval and fluid is allowed to fill the open wellbore until it can flow into the tubing and up the wellbore. A wire screen can be placed in the wellbore adjacent to the productive interval in reservoirs where rock fragments or grains can move into the wellbore. This can happen, for example, when the productive interval is an unconsolidated formation, that is, a formation of rock that is not bound or only loosely bound together.

It is sometimes necessary to provide energy to extract oil from reservoirs. Oil can be lifted using pumps or injecting gas into the well stream to increase the buoyancy of the gas-oil mixture that is being produced through tubing in the wellbore. The earliest pumps used the same wooden beams that were used for cable-tool drilling. Oil companies developed central pumping power in the 1880's. Central pumping power used a prime mover, which is a power source, to pump several wells. In the 1920's, the development of a beam pumping system made it feasible to allocate a pump to each well. A beam pumping system is a self-contained unit that is mounted at the surface of each well and operates a pump in the hole. More modern techniques include gas-lift and electric submersible pumps.

Energy can be provided to reservoirs by injecting fluids like water, air, or carbon dioxide into the reservoir. Improved recovery technology includes traditional secondary recovery processes such as water flooding and immiscible gas injection, as well as enhanced oil recovery (EOR) processes. EOR processes include techniques that inject high molecular-weight polymers (polymer flooding), surfactants (chemical flooding), gases that mix with oil (miscible flooding), heat or steam (thermal flooding), and microscopic organisms (microbial flooding).

3.3.2 Modern Drilling Methods

Advances in drilling technology are extending the options available for prudently managing subsurface reservoirs and producing fossil fuels, especially oil and gas. Modern drilling technologies discussed here include horizontal wells, multilateral wells, and infill drilling.

A well is a string of connected, concentric pipes. The path followed by the string of pipes is called the trajectory of the well. Historically, wells were drilled vertically into the ground and the well trajectory was essentially a straight, vertical line. Today, wells can be drilled so that the well trajectory is curved. A curved wellbore trajectory is possible because the length of each straight pipe that makes up the well is small compared to the total well length. The length of a typical section of pipe in a well is 30 ft to 40 ft, while wells can be thousands of feet long. Wells with one or more horizontal trajectories are sketched in Figure 3-6.

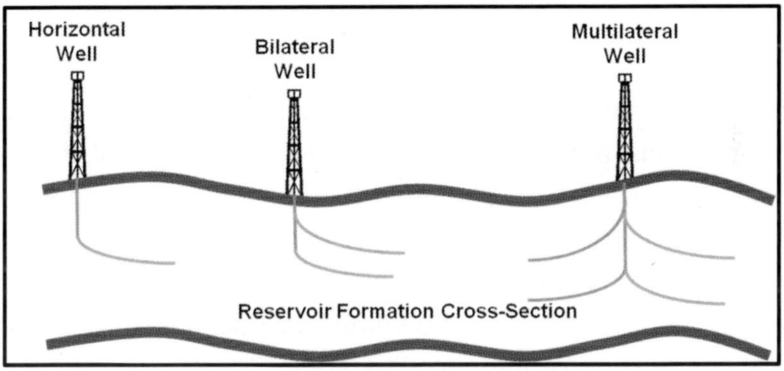

Figure 3-6. Multilateral Wells

Wells with more than one hole can be drilled. Each hole is called a lateral or branch and the well is called a multilateral well. For example, a bilateral well is a well with two branches. Figure 3-6 shows examples of modern multilateral well trajectories.

A well can begin as a vertical well, and then be modified to a horizontal or multilateral well. The vertical section of the well is called the main (mother) bore or trunk. The point where the main bore and a lateral meet is called a junction. When the vertical segment of the well reaches a specified depth called the kick-off point (KOP), mechanical wedges (whipstocks), or other downhole tools are used to change the direction of the drill bit and alter the well path. The beginning of the horizontal segment near the vertical wellbore is the heel and the end of the horizontal segment is the toe. The distance, or reach, of a well from the drilling rig to the final bottomhole location can exceed 6 miles. Wells with unusually long reach are called extended reach wells.

Multilateral wells make it possible to connect multiple well paths to a common wellbore. Multilateral wells have many applications. For example, multilateral wells are used in offshore environments where the number of well slots is limited by the amount of space available on a platform. They are also used to produce fluids from reservoirs that have many compartments. A compartment in a reservoir is a volume that is isolated from other parts of the reservoir by barriers to fluid flow such as sealing faults.

Horizontal, extended reach, and multilateral wellbores that follow subsurface formations are providing access to more parts of the reservoir from fewer well locations. This provides a means of minimizing environmental impact associated with drilling and production facilities, either on land or at sea. Extended reach wells make it possible to extract petroleum from beneath environmentally or commercially sensitive areas by drilling from locations outside of the sensitive areas. In addition, offshore fields can be produced from onshore drilling locations, and reduce the environmental impact of drilling by reducing the number of surface drilling locations.

Horizontal drilling provided significant advantages needed to make unconventional gas production economical. For example, shale formations can be a few hundred feet thick and extend laterally for miles. The Barnett Shale in the United States is up to 450 feet thick and covers an area of over 5,000 square miles. A vertical well, which may pass through 400 feet of shale, can extract gas from a much smaller volume of the shale than a horizontal well that could run through two miles or more of shale. Figure 3-7 shows that a horizontal well can intersect more of the reservoir when the areal extent of the reservoir is substantially larger than the thickness of the reservoir. The orientation of manmade fractures associated with hydraulic fracturing depends in part on the orientation of the well, and also on the rock properties of the reservoir.

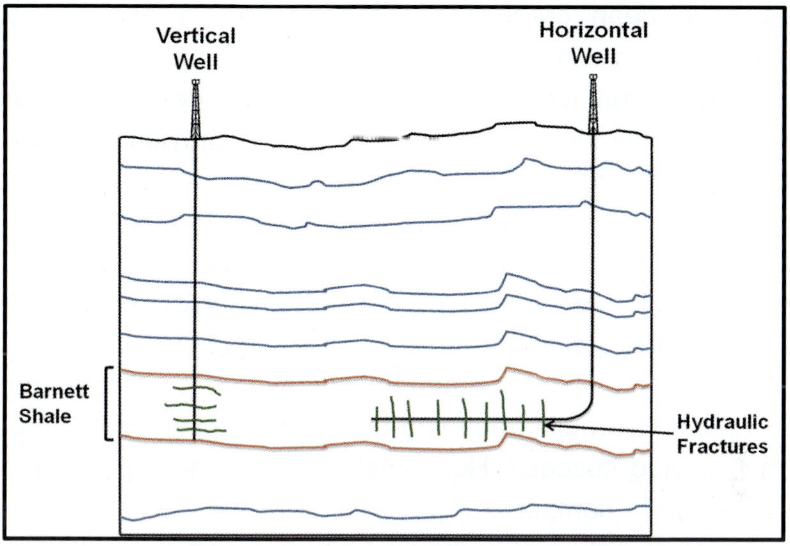

Figure 3-7. Schematic of Barnett Shale Cross Section

Multiple wells drilled from a single well pad decrease the need for large numbers of potentially noisy and aesthetically unappealing surface well sites. With multi-well drilling sites, the number of locations that must have pipelines for transporting fluids and that must be operated and maintained is minimized, saving time and money for operators. Horizontal drilling makes it possible to locate a well site outside of a highly populated area and drill underneath the population without being seen or heard. Horizontal drilling increases public support for urban drilling.

3.3.3 Hydraulic Fracturing

Wells drilled in low permeability rock such as shale often need to be stimulated. One key stimulation technology is hydraulic fracturing. In the 1980's and 1990's, developments in stimulation technology made it possible to create cracks in rock using a process called hydraulic fracturing (or "fracking"). The hydraulic fracturing process involves the injection of liquid slurry at a pressure that is high enough to break the rock. Once fractures have been created in the formation, a proppant such as manmade pellets or coarse grain sand is injected into the fracture to prevent the fracture from closing, or healing, when injection is discontinued. The proppant keeps the fracture open enough to provide a higher permeability flow path for fluid to flow to the production well. Today, fracking is used to improve oil and gas flow rates from low permeability rock. Modern applications include oil shale, gas shale, and tight gas sands.

The combination of hydraulic fracturing and horizontal drilling technologies catalyzed the growth of the shale oil and gas production industry. In 2002, Mitchell Energy implemented horizontal drilling in the Barnett Shale in North Texas. Hydraulic fracturing had been in place for years, but the results had limited success. Horizontal drilling made it possible for horizontal wellbores to contact a larger volume of the reservoir than vertical drilling. The combination of hydraulic fracturing and horizontal drilling increased the rate that fluid could be produced from targets like the gas-rich Barnett shale in Texas or the oil-rich Bakken shale in North Dakota. Many horizontal wells can be drilled from a single well pad in an urban area. This reduces the surface area needed for operating drilling rigs and installing surface facilities such as wellheads, storage tanks and compressors.

3.3.4 Production System

A production system can be thought of as a collection of the subsystems illustrated in Figure 3-8. Drilling rigs are used to drill wells. An example of a drilling rig on an offshore platform in dry dock is shown in Figure 3-9. The performance of the well depends on the properties of the reservoir rock, the interaction between the rock and fluids in the reservoir, and properties of the fluids in the reservoir. Reservoir fluids include the fluids originally contained in the reservoir as well as fluids that may be introduced as part of the reservoir management process. Well performance also depends on the properties of the well itself, such as its cross-section, length, trajectory, and completion. The completion of the well establishes the connection between the well and the reservoir.

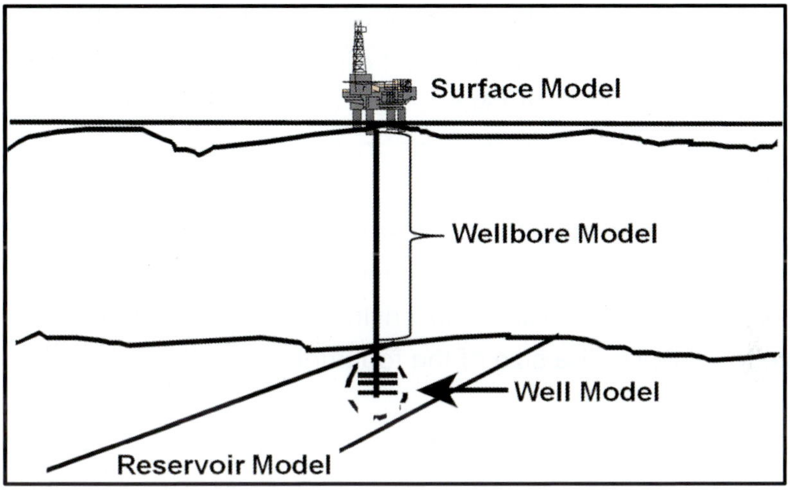

Figure 3-8. Production System

Figure 3-9. Offshore Platform in Dry Dock, Galveston, Texas (© Fanchi, 2003)

Surface facilities such as drilling rigs, storage tanks, and compressor stations are needed to drill, complete and operate wells. Drilling rigs may be moved from one location to another on trucks, ships, or offshore platforms; or drilling rigs may be permanently installed at specified locations. The facilities may be located in desert climates in the Middle East, stormy offshore environments in the North Sea, arctic climates in Alaska and Siberia, and deepwater environments in the Gulf of Mexico and off the coast of West Africa.

The surface area required for installing all of the facilities needed to develop a resource is called the footprint. The size of the footprint has an impact on project economics and environmental impact. As a rule, it is desirable to minimize the size of the footprint.

3.3.5 Oil and Gas Reserves

The amount of recoverable fluid in hydrocarbon-bearing reservoirs is known as reserves. The definition of reserves adopted by two influential organizations is summarized in Table 3-5 [SPE-PRMS, 2007]. The acronyms SPE and WPC refer to the Society of Petroleum Engineers and the World Petroleum Congress. The definitions are from the Petroleum Resources Management System (PRMS). Reserve estimates have many commercial applications. For example, reserve estimates are used to compare projects, establish the value of a resource, and support applications for project financing.

Oil and gas reserves are not directly measured. They must be estimated from a limited set of observations. The resulting value of reserves is an estimate that should include our lack of knowledge, or uncertainty, about important factors such as size of resource. The uncertainty may be expressed in terms of a probability of actually attaining the estimate of reserves. We illustrate this concept by associating a normal probability distribution with the reserves definitions.

Table 3-5
SPE/WPC Reserves Definitions

Proved Reserves	• Those quantities of petroleum, which by analysis of geoscience and engineering data, can be estimated with reasonable certainty to be commercially recoverable, from a given date forward, from known reservoirs and under defined economic conditions, operating methods, and government regulations. • There should be at least a 90% probability (P_{90}) that the quantities actually recovered will equal or exceed the estimate.
Probable Reserves	• Those additional reserves which analysis of geoscience and engineering data indicate are less likely to be recovered than Proved Reserves but more certain to be recovered than Possible Reserves. • There should be at least a 50% probability (P_{50}) that the quantities actually recovered will equal or exceed the estimate.
Possible Reserves	• Those additional reserves which analysis of geoscience and engineering data suggests are less likely to be recoverable than Probable Reserves. • There should be at least a 10% probability (P_{10}) that the quantities actually recovered will equal or exceed the estimate.

For a normal distribution with a mean and standard deviation (abbreviated std dev), the reserves definitions follow:

$$\text{Proved Reserves} = P_{90} = \text{mean} - 1.28 \times (\text{std dev})$$

$$\text{Probable Reserves} = P_{50} = \text{mean}$$

$$\text{Possible Reserves} = P_{10} = \text{mean} + 1.28 \times (\text{std dev})$$

An estimate of reserves is needed to compare the commercial values of different reservoirs. Uncertainty for a normal distribution is captured by the value of the standard deviation. A large standard deviation is associated with a large uncertainty in reserves estimation.

Figure 3-10 shows a normal distribution for a mean of 200 million barrels of oil (MMSTBO) and three different standard deviations (20 MMSTBO, 40 MMSTBO, and 60 MMSTBO). The difference between possible and proved reserves gets smaller as the standard deviation decreases. The unit MMSTBO signifies a million stock tank barrels of oil. A stock tank barrel is a barrel of oil stored in a stock tank.

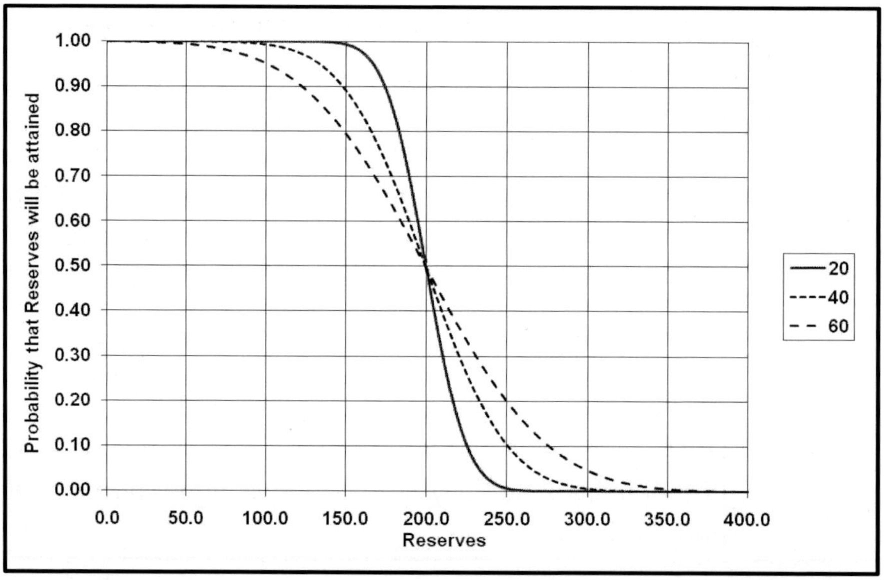

Figure 3-10. Illustration of a Reserves Distribution

3.4 Oil and Gas Transport and Processing

Produced fluids must be recovered, transported to storage facilities, and eventually to the consumer. Many modes of transportation are used to transport oil and gas. They include pipelines, tanker trucks, double hulled tankers, and ships capable of carrying liquefied natural gas. The mode of transport depends on such factors as safety, distance, and state of the fluid. The transport distance can be a few miles from field to processing facility, or thousands of miles from refinery to consumer.

Processing can begin at the well site where the produced wellstream is separated into oil, water and gas phases. Further processing at refineries separates the hydrocarbon fluid into marketable products (Figure 3-11). A typical barrel of crude oil is used to produce gasoline, diesel, jet fuel, liquefied petroleum gas, heavy fuel oil, and other products.

Figure 3-11. South Texas Refinery (© Fanchi, 2002)

3.4.1 Gas-Fired Power Plants

Fossil fuels were used throughout the 20th century to drive turbines in electricity generating power plants. Combustion of fossil fuels, especially coal, emits undesirable pollutants (see Section 2.5). One way to reduce the emissions problem is to use cleaner natural gas in power plants.

A conventional gas-fired power plant is the combined cycle power plant. It uses heat from combustion in more than one electricity generating cycle. The primary cycle uses hot gases from combustion to drive a gas turbine. Waste heat, typically in the form of exhaust, is captured and used to create steam, which is then used to drive a steam turbine. Both the gas turbine and the steam turbine generate electricity.

An alternative configuration to the combined cycle configuration is obtained by using waste heat from the gas turbine for direct heating. The use of waste heat for space heating or water heating is an example of cogeneration. Cogeneration is the simultaneous production and application of two or more forms of energy. Both the combined cycle configuration and the cogeneration configuration improve the efficiency of energy use.

Gas-fired power plants are cleaner than coal-fired power plants, but they still emit some greenhouse gases, notably carbon dioxide. Another variation of the gas-fired combined cycle power plant is to attach a carbon capture and sequestration (CCS) system to the power plant. An example of a CCS system is a system that captures carbon dioxide from the gas combustion process and injects the captured carbon dioxide into a geological reservoir. Gas storage in a geological reservoir is known as geological sequestration.

3.5 Global Distribution of Oil and Gas

The global distribution of oil and gas is illustrated by presenting the size of a nation's reserves. Table 3-6 lists fifteen countries with the largest crude oil and natural gas proved reserves in 2013. National reserves are found by summing the reserves for all of the reservoirs in the nation.

Table 3-6
Nations with Largest Crude Oil and Natural Gas Proved Reserves in 2013 [US EIA Petroleum, 2015]

Crude Oil Reserves (billion barrels)		Natural Gas Reserves (trillion cubic feet)	
World	1649	World	6845
Venezuela	298	Russia	1688
Saudi Arabia	268	Iran	1187
Canada	173	Qatar	890
Iran	155	United States	308
Iraq	141	Saudi Arabia	288
Kuwait	104	Turkmenistan	265
United Arab Emirates	98	United Arab Emirates	215
Russia	80	Venezuela	195
Libya	48	Nigeria	182
Nigeria	37	Algeria	159
United States	33	China	141
Kazakhstan	30	Iraq	112
Qatar	25	Indonesia	108
China	24	Kazakhstan	85
Brazil	13	Malaysia	83

Table 3-7 shows a regional distribution of crude oil and natural gas reserves. Eurasian reserves are primarily Russian reserves. The table shows that the Middle East has the largest reserves in both categories and Europe has the smallest reserves.

Table 3-7
Regional Distribution of Crude Oil and Natural Gas Proved Reserves in 2013
[US EIA Petroleum, 2015]

Crude Oil Reserves (billion barrels)		Natural Gas Reserves (trillion cubic feet)	
World	1649	World	6845
Middle East	802	Middle East	2823
Central & South America	326	Eurasia	2178
North America	217	Asia & Oceania	521
Africa	128	Africa	515
Eurasia	119	North America	393
Asia & Oceania	45	Central & South America	269
Europe	12	Europe	146

The geographic distribution of crude oil and natural gas shown in Table 3-7 can change. Discovery of major new fields can alter the amount of a resource in one area relative to another. Offshore West Africa and deepwater Gulf of Mexico are examples of regions where significant new fields are being discovered. Other factors such as advances in technology and changes in price can affect the distribution of reserves. The development of unconventional shale oil and gas in North America has substantially increased recoverable oil and gas resources, but price volatility can influence the economics of resource recovery.

3.5.1 Shale Oil and Shale Gas

Shale oil is solid bituminous material contained in low permeability shale. An oily liquid is obtained when the material is heated. The global distribution of shale oil is shown in Table 3-8. The table shows the 10 nations with the largest technically recoverable shale oil resource. The volume of technically recoverable resource is the volume of resource that can be recovered using existing exploration and production technology.

Table 3-8
Nations with Largest Technically Recoverable Shale Oil and Shale Gas Resources [US EIA ARI Shale, 2013, pg. 6]

Shale Gas Resource (trillion cubic feet)		Shale Oil Resource (billion barrels)	
World	7795	World	335
United States	1161	Russia	75
China	1115	United States	48
Argentina	802	China	32
Algeria	707	Argentina	27
Canada	573	Libya	26
Mexico	545	Australia	18
Australia	437	Venezuela	13
South Africa	390	Mexico	13
Russia	285	Pakistan	9
Brazil	245	Canada	9

Gas shale is a low permeability source rock found throughout the world. It has enough porosity and volume to hold large amounts of gas but permeability on the order of nanodarcies makes it difficult to economically produce hydrocarbons.

Gas production from shale has flow mechanisms that are similar to those required for gas production from coal. Shale gas must diffuse through the shale until it reaches a permeable path that allows it to flow to a well. Permeable pathways are typically natural or induced fractures.

Gas recovery from tight gas sands and shale depends on our ability to produce gas at an economical rate. Advances in drilling and well completion technologies have made oil and natural gas production from low permeability sands and shale increasingly cost-effective. Two technologies have greatly improved the economics: horizontal drilling and hydraulic fracturing. Price volatility can have a significant impact on the economics of producing very low permeability resources. If hydrocarbon price is too low, resources that are costly to develop may be delayed or abandoned.

Some communities do not allow hydraulic fracturing in their jurisdiction because of concerns about environmental impact. Others do not believe the environmental impact is significant and have embraced the technology. Environmental concerns are among the factors motivating a change from fossil fuels to an energy supply that is clean. Clean energy refers to energy that has little or no detrimental impact on the environment. Natural gas is a source of relatively clean energy that can be obtained from such sources as conventional oil and gas fields, unconventional gas resources, landfill gas, and municipal solid waste gas (MSW gas). Gas from landfills and municipal solid waste (MSW) gas are obtained from the decay of organic waste. Landfill gas and MSW gas are not fossil fuels; they are renewable energy sources.

3.5.2 Tar Sands and Extra Heavy Oil

Tar sands, also known as natural bitumen, are a combination of clay, sand, water, and bitumen. The World Energy Council [WEC, 2010] estimated that the total original oil in place of natural bitumen (tar sand) and extra heavy oil at the end of 2008 were approximately 3.3 trillion barrels and 2.1 trillion barrels, respectively. Nations with the largest volumes of natural bitumen and extra heavy oil are listed in Table 3-9.

Table 3-9
Nations with Largest Volumes of Natural Bitumen and Extra Heavy Oil at the end of 2008 [WEC, 2010, Tables 4-1 and 4-2]

Natural Bitumen (Tar Sands) Total Original Oil in Place (billion barrels)		Extra Heavy Oil Total Original Oil in Place (billion barrels)	
Canada	2434	Venezuela	2112
Kazakhstan	421	United Kingdom	11.9
Russia	347	Azerbaijan	8.9
United States	53	China	8.8
Nigeria	38	Italy	2.7

Total original oil in place listed in Table 3-9 is an estimate of the size of the resource. It does not represent the amount of resource that will be developed. We can use an estimate of resource size as one factor in prioritizing the importance of the resource.

Although difficult to produce, the relatively large volume of low API gravity hydrocarbon resource has motivated efforts to improve production techniques. Shale oil, tar sands and extra heavy oil can be extracted by mining when they are close enough to the surface. Tar pits have been found around the world and have been the source of many fossilized dinosaur bones. In locations where oil shale, tar sands, and extra heavy oil are too deep to mine, it is necessary to increase the mobility of the hydrocarbon.

An increase in permeability or a decrease in viscosity can increase fluid mobility. Increasing the temperature of shale oil, tar or asphalt can significantly reduce viscosity. If there is enough permeability to allow injection, steam or hot water can be used to increase formation temperature and reduce hydrocarbon viscosity. Steam Assisted Gravity Drainage (SAGD) is used to increase recovery of oil from tar sands in Canada. In many cases, however, permeability is too low to allow significant injection of a heated fluid. As alternatives to fluid injection, radio frequency heating has been used in Canada, and electromagnetic heating techniques are being developed for other parts of the world.

3.5.3 Gas Hydrates and Tight Gas Sands

Unconventional gas resources include gas hydrates, tight gas sands, coal gas, and shale gas. Coal gas and shale gas were discussed previously. Gas hydrates and tight gas sands are discussed here.

Gas Hydrates

Gas hydrates have traditionally been considered a problem for oil and gas field operations, but their potential commercial value as a clean energy resource is changing industry perception. Gas hydrates are naturally occurring, crystalline complexes that are formed when one type of molecule completely encloses another type of molecule in a lattice. In the case of gas hydrates, hydrogen-bonded water molecules form a cage-like structure in which molecules of gas (such as methane, ethane and propane) are absorbed or bound. A methane molecule in the center of the cage-like structure is shown in Figure 3-12 for methane hydrate. The figure shows burning methane hydrate, which is sometimes referred to as burning ice because it can sustain a flame.

Methane hydrates contain a relatively large volume of methane in the hydrate complex. The hydrate complex contains about 85 mole percent water and approximately 15 mole percent guests, where a guest molecule is methane or some other relatively low molecular weight hydrocarbon.

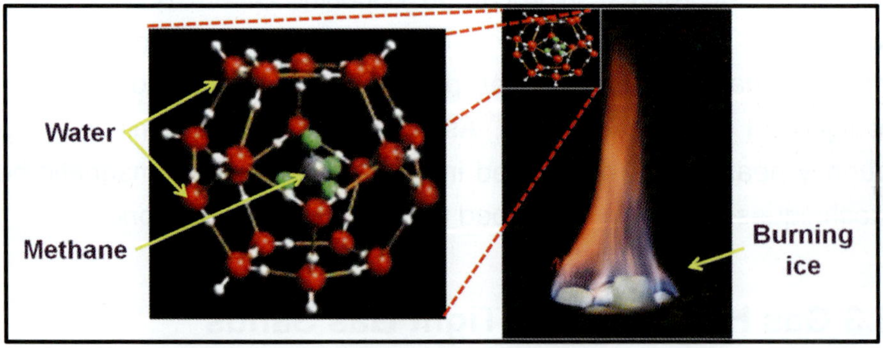

Figure 3-12. Illustration of Methane Hydrate [after USGS Hydrate, 2009]

Methane hydrates are found around the world. They exist on land in sub-Arctic sediments and on seabeds where the water is near freezing. They can be found in Arctic sands, marine sands, fractured muds, and shale. Recognizing that the estimate of global hydrate resource size is very uncertain, Boswell [2009] reported that gas hydrates may contain on the order of 680,000 trillion cubic feet of methane. For comparison, natural gas consumption in the United States during 2014 was approximately 26 trillion cubic feet [US EIA NG, 2015]. The size of the gas hydrate resource is a huge target for future development. Difficulties in cost-effective production of methane hydrates have hampered the production of methane from hydrates.

Tight Gas Sands

Coal gas, tight gas sands, and shale gas are characterized by low permeabilities. Conventional gas reservoirs have permeabilities on the order of millidarcies. Tight gas sands have permeabilities on the order of microdarcies.

The geographic distribution of tight gas sands and gas shale is better understood in the United States than it is elsewhere in the world. Table 3-10 compares the volumes of unconventional gas resources for the United States at the beginning of 2009. These estimates were compiled by the Natural Gas Supply Association (NGSA) from data provided by the United States EIA, the United States Federal Energy Regulatory Commission (FERC), and the United States Potential Gas Committee (PGC). The former two organizations (EIA and FERC) are government agencies and the latter organization (PGC) is a non-profit organization.

Table 3-10
Size of Unconventional Gas Resources in the United States circa January 2009 [NGSA Unconventional Gas, 2009]

Resource	Trillion Cubic Feet	Source
Tight gas	309	US EIA
Shale gas	742	US FERC
Coal gas	163	US PGC

3.6 Global Oil and Gas Production and Consumption

The global distribution of oil and gas production and consumption is illustrated by presenting the leading nations in production and consumption categories. The nations and values change from year to year, so this data should be viewed as a snapshot of the distribution of oil and gas production and consumption. Table 3-11 lists the five countries with the largest production and consumption of oil in 2013. Table 3-12 lists the five countries with the largest production and consumption of dry natural gas in 2013. If a country consumes more oil than it produces, the additional oil must be imported from countries that produce more oil than they consume. The United States, for example, had to import oil in 2013. By contrast, Saudi Arabia produced 11.7 million barrels per day and consumed 3.0 million barrels per day in 2013, so it was able to export the difference. Similar comments apply to dry natural gas imports and exports.

Table 3-11
Top 5 Oil Producing and Consuming Nations in 2013
[US EIA Petroleum, 2015]

Producing Nations (million barrels per day)		Consuming Nations (million barrels per day)	
World	91.0	World	91.2
United States	12.3	United States	19.0
Saudi Arabia	11.7	China	10.5
Russia	10.8	Japan	4.6
China	4.6	India	3.7
Canada	4.1	Russia	3.5

Table 3-12
Top 5 Dry Natural Gas Producing and Consuming Nations in 2013 [US EIA Petroleum, 2015]

Producing Nations (billion cubic feet per day)		Consuming Nations (billion cubic feet per day)	
World	332.3	World	332.5
United States	66.7	United States	71.7
Russia	60.7	Russia	42.7
Iran	15.6	China	15.8
Qatar	15.3	Iran	15.2
Canada	14.1	Japan	12.3

3.7 Activities

True-False

Specify if each of the following statements is True (T) or False (F).

1. A hydrocarbon reservoir must be able to trap fluids.
2. The biogenic theory of the origin of fossil fuels depends on the death and decay of life from the surface biosphere.
3. Wells with unusually long extent are called extended reach wells.
4. Gas wells have to be drilled vertically.
5. Seismic measurements use vibrations to provide images of the subsurface.
6. "Easy" oil is found at the bottom of the resource pyramid.
7. Rotary drilling has been largely replaced by cable-tool drilling.
8. The amount of oil and gas reserves in the world does not change.
9. The Middle East has only oil reserves and no natural gas reserves.
10. SPE includes gases in its definition of petroleum.

Questions

1. How many joints of pipe are needed in a vertical well to reach a target depth of 6000 ft? Assume the average length of a joint of pipe is 30 ft.

2. What role does cement play in the completion of a wellbore?
3. What are the two primary elemental constituents of petroleum fluids?
4. Which of the following factors are needed to form a hydrocarbon reservoir?

a.	The biogenic theory must be correct
b.	A hydrocarbon source must exist
c.	A flow path must exist between hydrocarbon source and reservoir
d.	A trap must exist for capturing fluids
e.	The timing must be right, that is, the fluid trap must exist at the time that fluid is moving from the hydrocarbon source to the reservoir

5. Which of the following theories is considered the mainstream theory of the origin of fossil fuels: biogenic or abiogenic?
6. What role does injected sand play in hydraulic fracturing?
7. Calculate the SPE-PRMS reserves for a mean of 200 MMSTBO and a standard deviation of 40 MMSTBO.
8. How does volatile oil behave when pressure drops?
9. Hydrocarbon liquids can be classified using ____ and viscosity.
10. What is SAGD and where is it used?

CHAPTER 4

ENERGY IN TRANSITION

After reading this chapter, you should be able to...

- Explain Hubbert's oil supply forecast
- Discuss alternative oil supply forecasts
- Define "decarbonization" and its impact on the energy mix
    - *P2P: Why would an oil rich country develop nuclear and renewable energy?*
- Describe environmental issues associated with oil and gas production
    - *P2P: Should society encourage investment in unconventional oil and gas?*
- Discuss climate change and its possible causes
    - *P2P: Is geologic sequestration a desirable policy?*
- Summarize international efforts to mitigate emissions
    - *Point to Ponder: Does the COP21 agreement require global cooperation?*
- Is the climate change debate settled?
    - *Point to Ponder: Is it true that 97% of scientists believe that anthropogenic climate change (ACC) is occurring?*
- Introduce energy security

Fossil fuels were the fuel of choice to meet global energy needs at the end of the 20th century. Many experts believe that the global dependence on fossil fuels is not sustainable and needs to be changed. The most common concerns are the following: the supply of fossil fuels is finite because fossil fuels are nonrenewable, the consumption of fossil fuels by combustion is having an adverse effect on the environment, and the distribution of fossil fuels contributes to geopolitical instability because fossil fuels can be used as a weapon to influence national foreign policies. Each of these concerns is discussed below.

4.1 Peak Oil

The transition to a new energy mix is motivated in part by the concern that the supply of nonrenewable fossil fuels is limited. As a limited resource, there will come a time when fossil fuel production is unable to meet growing energy demand. The justification for this concern and the associated concept of peak oil are discussed here.

4.1.1 Hubbert's Oil Supply Forecast

M. King Hubbert [1956] studied the production of conventional oil in a mature oil producing region and found that oil production in this limited geographic region could be modeled as a function of time. The annual production of oil increased steadily until a maximum was reached, and then began to decline as it became more difficult to find and produce. The maximum annual oil production is considered a peak. Hubbert used his method to predict peak annual oil production in the contiguous United States, which excludes Alaska and Hawaii. He predicted that the peak would occur between 1965 and 1970. Hubbert then used the methodology that he developed for the contiguous United States to predict the peak of global annual oil production. He predicted that global oil production would peak around 2000 at a peak rate of 12 to 13 billion barrels per year or approximately 33 to 36 million barrels per day (Figure 4-1).

A comparison of Hubbert's predictions with actual production data for the contiguous United States provided support for Hubbert's method. Average annual crude oil production rate in the contiguous United States peaked at 9.4 million barrels per day in 1970. A second peak for the United States occurred in 1988 when Alaskan oil production rate peaked at

2.0 million barrels per day. The second peak is not considered the correct peak to compare to Hubbert's prediction because Hubbert restricted his analysis to the production from the contiguous United States. Many modern experts considered the 1970 oil peak to be a validation of Hubbert's methodology, and have tried to apply the methodology to global oil production. Analyses of historical data using Hubbert's methodology typically predict that global oil production should peak in the first quarter of the 21st century.

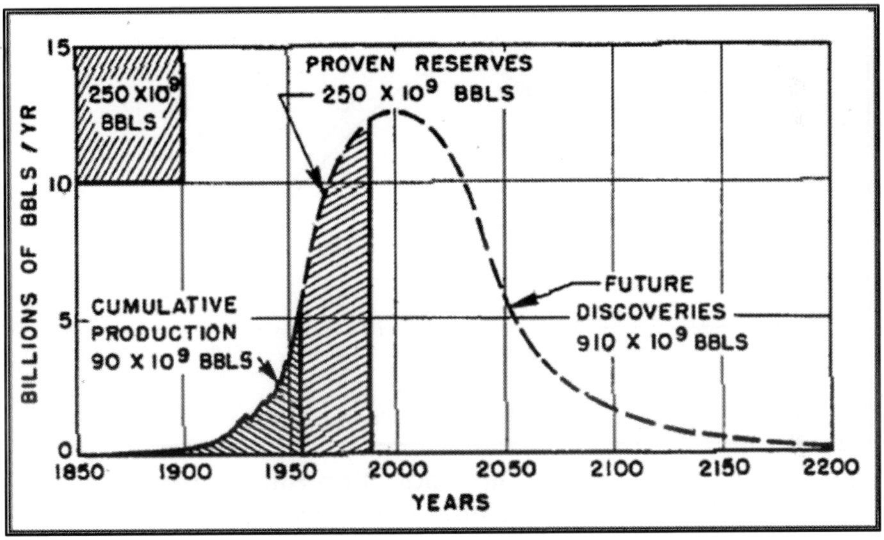

Figure 4-1. Hubbert's Global Oil Prediction [Hubbert, 1956, Figure 20]

4.1.2 World Oil Production Rate Peak

Forecasts based on an analytical fit to historical data can be readily prepared using publicly available data. Figure 4-2 shows a fit of world oil production rate (in millions of barrels per day) from the United States EIA database. Three fits are displayed. The fit labeled "Gaussian Fit 2000" uses data through year 2000, the fit labeled "Gaussian Fit 2008" uses data through year 2008, and the fit labeled "Gaussian Fit 2014" uses data through year 2014. Each fit was designed to match the most recent part of the production curve most accurately. The 2000 and 2008 fits give oil production rate peaks between 2010 and 2015, while the 2014 fit gives an oil production rate peak of approximately 2030.

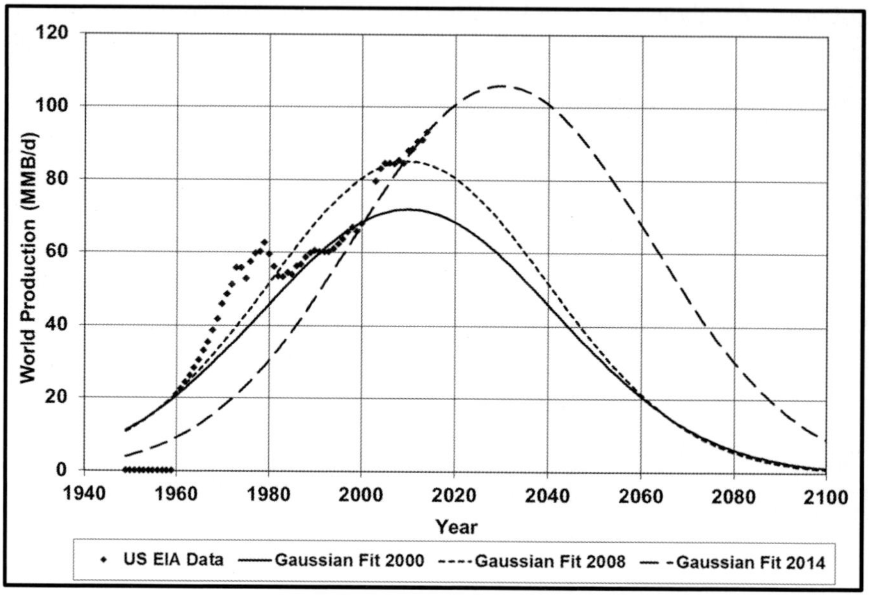

Figure 4-2. World Oil Production Rate Forecast Using Gaussian Curve

There are a couple of details worth highlighting in the figure. The significant increase in actual world oil production rate between 2000 and 2010 was due to a change in infrastructure capacity in Saudi Arabia. This period coincided with a significant increase in oil price per barrel, which justified an increase in facilities needed to produce, collect, and transport an additional one to two million barrels per day of oil. The significant increase in actual world oil production rate after 2010 was primarily due to the commercial development of shale oil in the United States. Hubbert's method was applied to conventional oil production. It did not take into account the development of unconventional oil resources.

The actual data displayed in Figure 4-2 does not show the beginning of a decline in global oil production rate in 2015 as the result of an oversupply of oil. The oversupply in 2015 was the result of a Saudi Arabian policy to protect Saudi Arabian market share by making it very difficult for high cost oil producers, like shale oil producers, to compete at lower oil prices.

4.1.3 World per Capita Oil Production Rate Peak

We saw in Section 4.1.2 that the evidence for a peak in world oil production rate is inconclusive. On the other hand, suppose we consider world per capita oil production rate, which is annual world oil production rate divided by world population for that year. Figure 4-3 shows world per capita oil production rate (in barrels of oil produced per day per person) for the period from 1960 through 2014.

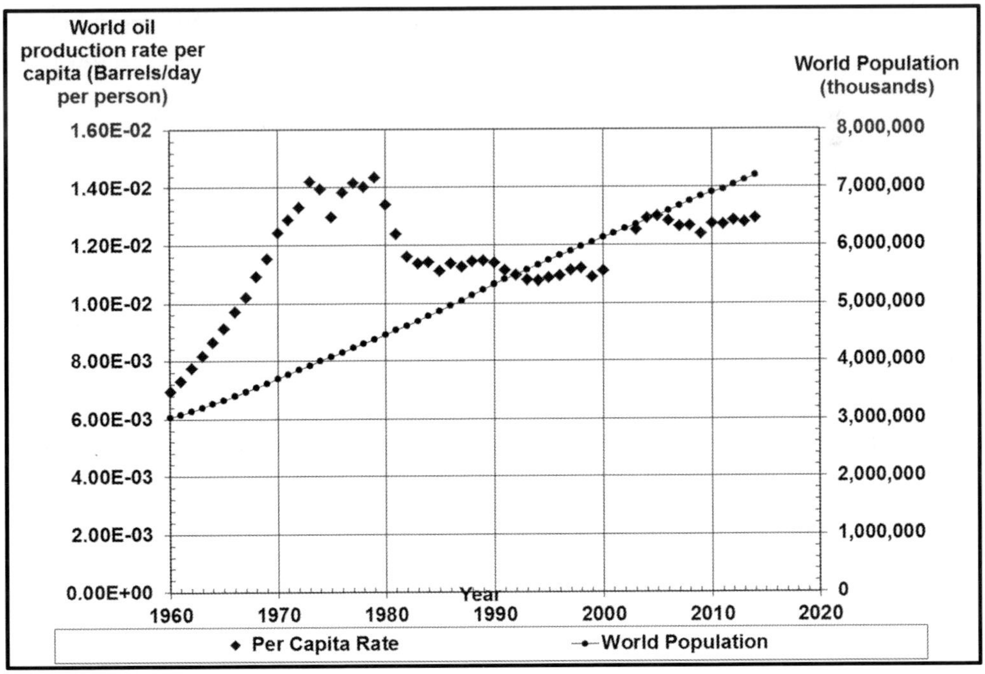

Figure 4-3. World per Capita Oil Production Rate

Figure 4-3 shows two peaks in the 1970's. The first peak was at the time of the first oil crisis, and the second peak occurred when Prudhoe Bay, Alaska oil production came online. World per capita oil production rate was relatively constant from the end of the 1970's to 2000. The increase in Saudi Arabian oil production and shale oil production led to the increase in world per capita oil production rate from 2000 to the present. Despite these increases, the peak in world per capita oil production rate occurred in the 1970's. We do not know if another, higher peak can be achieved because of the continuing growth in world population.

4.2 "DECARBONIZATION"

The peak oil concept says that the annual supply of oil will eventually peak and then decline. The decline in produced oil will occur despite an increasing demand for energy. The combination of increased demand and reduced supply will lead to a significant increase in the price of oil. According to this scenario, other sources of energy will begin to replace the increasingly expensive oil.

Table 4-1 presents historical data for global energy production reported by the United States Energy Information Administration for the period from 1970 to 2010 and a forecast for 2020. The table shows historical energy production as a percent contribution to the total energy mix. The row of data labeled "Fossil Fuels" includes coal, petroleum and natural gas. Petroleum refers to hydrocarbon liquids such as crude oil and natural gas plant liquids. The row of data labeled "Renewable Energy" includes hydroelectric energy, geothermal energy, wind energy, solar energy, and bioenergy. The row of data labeled "Total (quads)" shows the total energy in quads produced for the specified year.

Table 4-1
World Primary Energy Production by Source, 1970 – 2020
[Appendix C]

Primary Energy	Primary Energy Production (as % of total)					
	1970	1980	1990	2000	2010	2020*
Total	100.0	100.0	100.0	100.0	100.0	100.0
Fossil Fuels	93.2	90.1	86.7	85.4	86.5	79.1
Nuclear Electric Power	0.4	2.6	5.8	6.5	5.4	6.0
Renewable Energy	6.4	7.2	7.5	8.1	8.1	14.9
Total (quads)	215.4	287.6	349.9	395.7	505.0	592.1

* Estimate [IEA WEO, 2015]

A forecast for 2020 is shown in Table 4-1. The forecast is representative of the expectation by many energy agencies that fossil fuel consumption will become a smaller contributor to the global energy mix as renewable energy consumption increases. This does not necessarily mean that fossil fuel consumption will decline in terms of total amount consumed since growth in energy demand could require more energy must be provided to the market. The data in Table 4-1 are graphically displayed in Figure 4-4 and show that fossil fuels continue to be the major contributors to the global energy mix.

The data presented in Table 4-1 shows a trend toward "decarbonization," that is, a move away from fuels with many carbon atoms to fuels with few or no carbon atoms. H.J. Ausubel [2000, page 18] defined decarbonization as "the progressive reduction in the amount of carbon used to produce a given amount of energy." Figure 4-5 illustrates how the carbon to hydrogen ratio (C:H) declines as the fuel changes from carbon-rich coal to carbon-free hydrogen. The result of decarbonization will be a low carbon economy, that is, an economy that minimizes the consumption of carbon-based fuels.

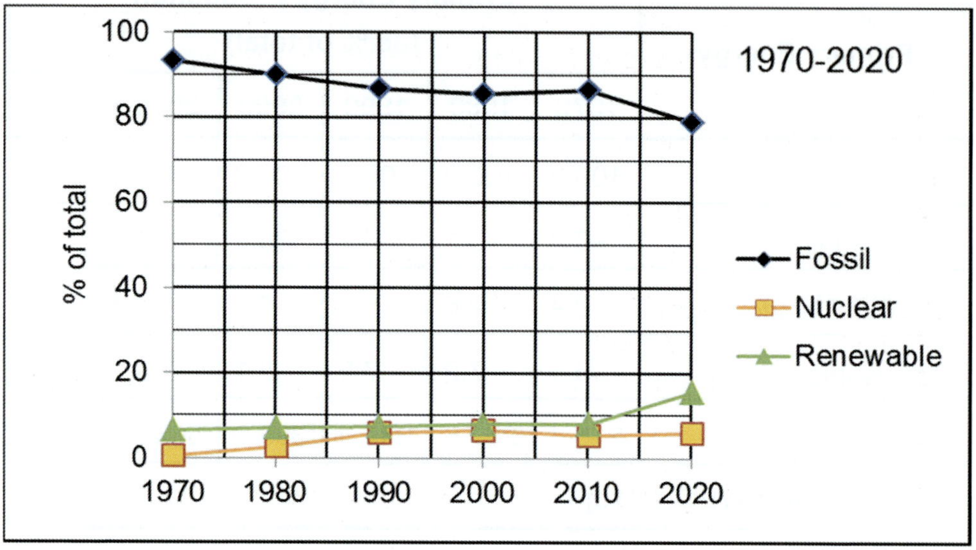

Figure 4-4. Percent Contribution of Different Energy Types to Historical Energy Production for the World from 1970 to 2010

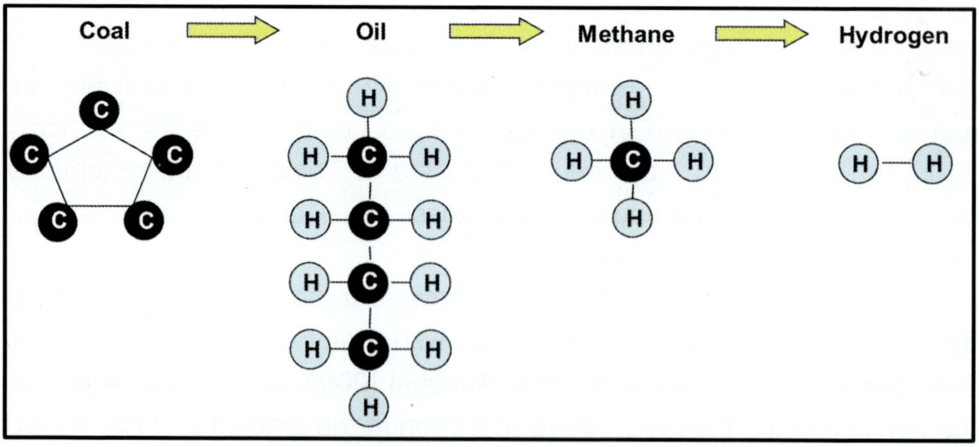

Figure 4-5. Decarbonization of Fuels

Point to Ponder: Why would an oil rich country develop nuclear and renewable energy?

The trend toward decarbonization can help explain why a country like Iran that is rich in fossil fuels would seek to develop nuclear energy and renewable energy sources. Even if the country is able to export fossil fuels such as oil and gas for decades to come at its current rate of export, the country can look at forecasts of energy consumption and see that the market is changing. They can use revenue from the sale of fossil fuels to help them transition to new energy sources. The remaining fossil fuels, especially oil, can be used for other applications besides transportation fuel. For example, oil is used in the manufacture of refined products such as lubricants and plastics. We see in Chapter 5 that the level of enrichment of uranium, a nuclear fission fuel, can indicate whether the development of a nuclear program is intended for peaceful or military purposes.

4.3 Environmental Issues

Fossil fuel producers recognize that oil and gas field management should protect the environment. This is due to factors such as a desire by producers to be good stewards of the earth, a strategy to recognize the economic interests of a broad group of stakeholders, and government regulation. Environmental issues associated with the production of conventional fossil fuels include handling naturally occurring radioactive materials (NORM), minimizing emission of pollutants into the environment, and preventing spills. The following examples illustrate environmental issues that are not part of the climate change discussion. Climate change is considered in Section 4.4.

One example of a facility that is designed to protect the environment is the Louisiana Offshore Oil Production (LOOP) facility. LOOP is designed to keep the transfer of hydrocarbons between pipelines and tankers away from sensitive coastal areas. Periodic water sampling of surface and produced waters can assure that fresh water sources are not contaminated. In addition, periodic testing for the excavation or production of NORM helps assure environmental compliance and makes it possible for operators to collect and properly dispose of NORM.

Failure to develop an environmental management plan that accounts for the interests of all stakeholders, including the public, can lead to unexpected consequences. For example, in 1995 Shell UK reached an agreement with the British government to dispose an oil storage platform called the Brent Spar in the deep waters of the Atlantic. The environmental group Greenpeace led a public challenge to the plan by occupying the platform and supporting demonstrations that, in some cases, became violent. Greenpeace and its allies were concerned that oil left in the platform would leak into the Atlantic. Shell UK abandoned the plan and eventually found another use for the structure as a ferry quay. Governments throughout Europe subsequently changed their rules regulating disposal of offshore facilities.

Operating a field with prudent consideration of environmental issues can pay economic dividends. In addition to improved public relations, sensitivity to environmental issues can minimize adverse environmental effects that may require costly remediation and financial penalties. Remediation often takes the form of expensive cleanup operations such as those required after the oil spill from the grounding of the 1989 Exxon-Valdez oil tanker in Alaska or the 2010 explosion and sinking of the BP Deepwater Horizon offshore platform in the Gulf of Mexico. The operator, BP, agreed to pay approximately $18.7 billion to government agencies and injured parties in 2015 following the Deepwater Horizon disaster.

Technologies are being developed to improve our ability to clean up environmental pollutants. For example, bioremediation using living microorganisms or their enzymes can accelerate the rate of degradation of environmental pollutants. Other methods of handling spills include skimming to collect the surface oil, burning the oil, or allowing some types of spilled oil to evaporate naturally.

Environmental issues associated with oil and gas production from shale include meeting the demand for water to conduct hydraulic fracture treatments, disposing of produced water containing pollutants, controlling the emanation of toxic chemicals from storage tanks, and preventing the occurrence of seismic events when disposing of produced water. Urban operations require controlling truck traffic, acquiring access to rights-of-way, mitigating noise and visual pollution from operations, obtaining well pad locations and reclaiming the environment when the well site is abandoned. King [2012] has provided an extensive review of hydraulic fracturing issues associated with oil and gas production from shale.

As noted earlier, fossil fuels are commonly consumed by burning. A key environmental issue associated with combustion is the emission of pollutants into the atmosphere. Efforts are underway to minimize the release of gases that are the byproduct of hydrocarbon combustion because of their impact on climate change. This is discussed in more detail later.

Point to Ponder: Should society encourage investment in unconventional oil and gas?

The ability to economically produce unconventional sources of oil and gas has increased the commercial supply of fossil fuels. Production of unconventional oil and gas and replacing oil-based vehicles with vehicles that operate using cleaner-burning natural gas or electricity from nuclear or clean renewable sources can give society more time to make the transition from fossil fuels to other sources of energy. On the other hand, the use of unconventional oil and gas could delay the transition to energy sources that do not emit greenhouse gases. If you accept the premise that climate change resulting from greenhouse gas emissions associated with human activity is a major problem that must be addressed immediately, then development of unconventional oil and gas would exacerbate existing environmental problems.

4.4 CLIMATE CHANGE

Many people believe that our current reliance on the combustion of fossil fuels is having an adverse impact on climate. The term "climate change" is widely used to characterize this impact. Our purpose here is to review and evaluate the evidence for climate change. We then consider what might happen if the byproducts of combustion are altering the climate. A plan for making the transition to a sustainable energy portfolio that provides cost-effective energy while protecting the environment is discussed in Chapter 14.

4.4.1 What is Climate Change?

We have all experienced short-term weather changes associated with storms, droughts and changing seasons. Climate is weather averaged over time scales on the order of decades or longer and may be defined as the long-term weather pattern. For example, scientific evidence shows that events like ice ages have occurred in the past. Ice ages affect the climate for millennia and are examples of changes in the long-term weather pattern. The modern debate about climate change is focused on the impact of human activity on changes to the long-term weather pattern. Here we use the term anthropogenic climate change to describe climate change due to human activity. Advocates of anthropogenic climate change (ACC) believe that human activity is altering the temperature of the atmosphere, increasing the rate of glacier melting, causing sea levels to rise, and increasing the acidity of ocean water. We discuss these issues below.

What can affect the atmosphere?

Life has been known to affect the atmosphere. We can see this by examining how the content of oxygen in the earth's atmosphere has changed from the origin of the earth approximately 4.5 billion years ago to the present. Scientists believe that oxygen was not present in the atmosphere when the earth formed. Holland [2006] said that oxygen first appeared in the atmosphere approximately 2.5 billion years ago when cyanobacteria appeared. The amount of oxygen began to increase as life evolved from relatively simple cyanobacteria into more complex eukaryotes and algae. Today, air typically consists of approximately 21 volume % oxygen and 78 volume % nitrogen. The remaining 1 volume % of air includes argon, carbon dioxide, methane, and water vapor.

One source of atmospheric oxygen is plant photosynthesis. Carbon dioxide and water react in the presence of sunlight and chlorophyll to form a carbohydrate and oxygen. Plant photosynthesis converts carbon dioxide into oxygen in the atmosphere. Another photosynthetic process, bacterial photosynthesis, forms a carbohydrate and water.

Human breathing is the source of a significant amount of carbon dioxide. We inhale air containing 78 volume percent nitrogen and 21 volume percent oxygen. When we exhale, we exhale 78 volume percent nitrogen, 14 to 16 volume percent oxygen, and 4 to 5 volume percent carbon dioxide. We rely on plants to convert exhaled carbon dioxide into oxygen.

The observation that humans are a source of atmospheric carbon dioxide gains in importance when we look at the global population trend. Actual global population reached 7 billion people in 2012. UN Global population forecasts (Figure 4-6) show that global population may increase to as many as 16 billion people by 2100 AD. The recent history of population growth in some Western nations suggests that it is possible to reduce population by educating people about birth control. If global population increases, which is likely in the near-term, the associated human exhalation of carbon dioxide will also increase.

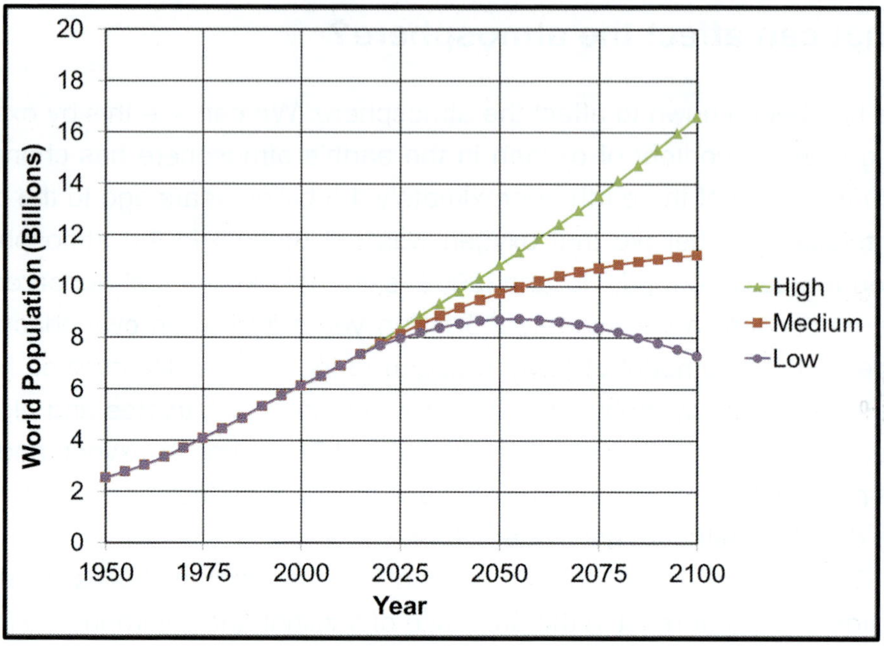

Figure 4-6. UN Global Population Forecasts [UN Population, 2015]

4.4.2 Are we changing the atmosphere?

One concern about climate change is that the atmosphere is warming. According to the National Oceanic and Atmospheric Administration [NOAA Temperature, 2015], a global average temperature is calculated from temperature measurements taken at several locations around the world. The difference between the observed temperature and long-term average temperature for each location and date is calculated using measured temperatures and is called temperature anomaly. Measurements of global temperature anomalies are shown in Figure 4-7 from 1850 to 2015. Temperature increased by approximately 1.2°C during that period. The increase in atmospheric temperature since 1850 coincides with the use of fossil fuels during the industrial age as well as an increase in global human population from approximately 1 billion people to over 7 billion people.

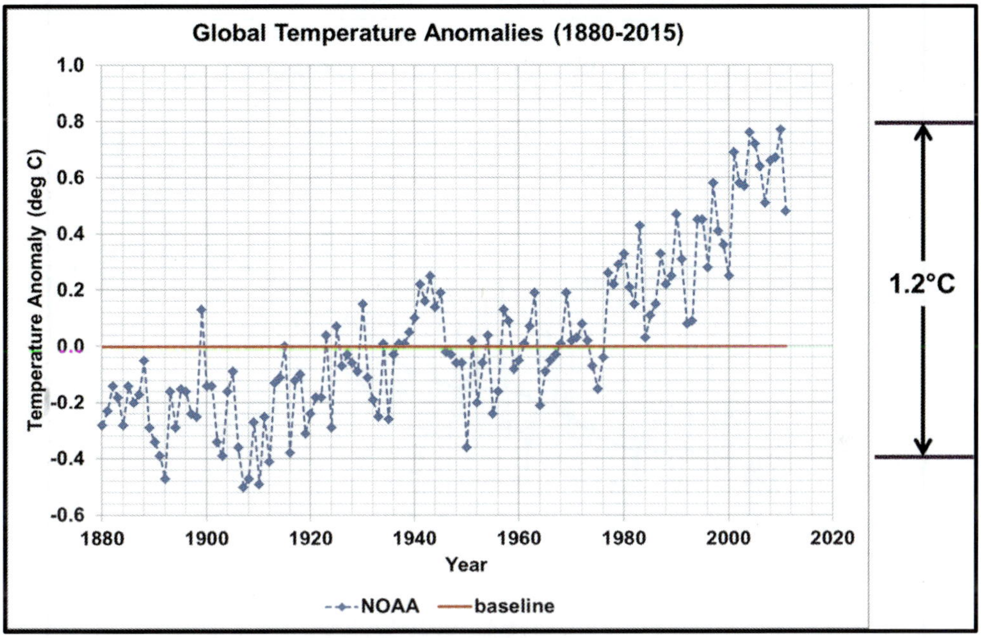

Figure 4-7. Global Temperature Anomalies [NOAA Temperature, 2015]

One possible explanation for the observed increase in atmospheric temperature is that the sun caused the temperature increase. Figure 4-8 compares global surface temperature with solar insolation between 1978 and 2009. Solar insolation is the solar power arriving at the top of the atmosphere per square meter of surface area of atmosphere. It is labelled "Sun's Energy Output" in the figure. We see that the energy arriving at the top of the atmosphere from the sun is following its normal 11-year cycle while global surface temperature has been increasing during the same period. It appears that the sun has not caused the change in global surface temperature.

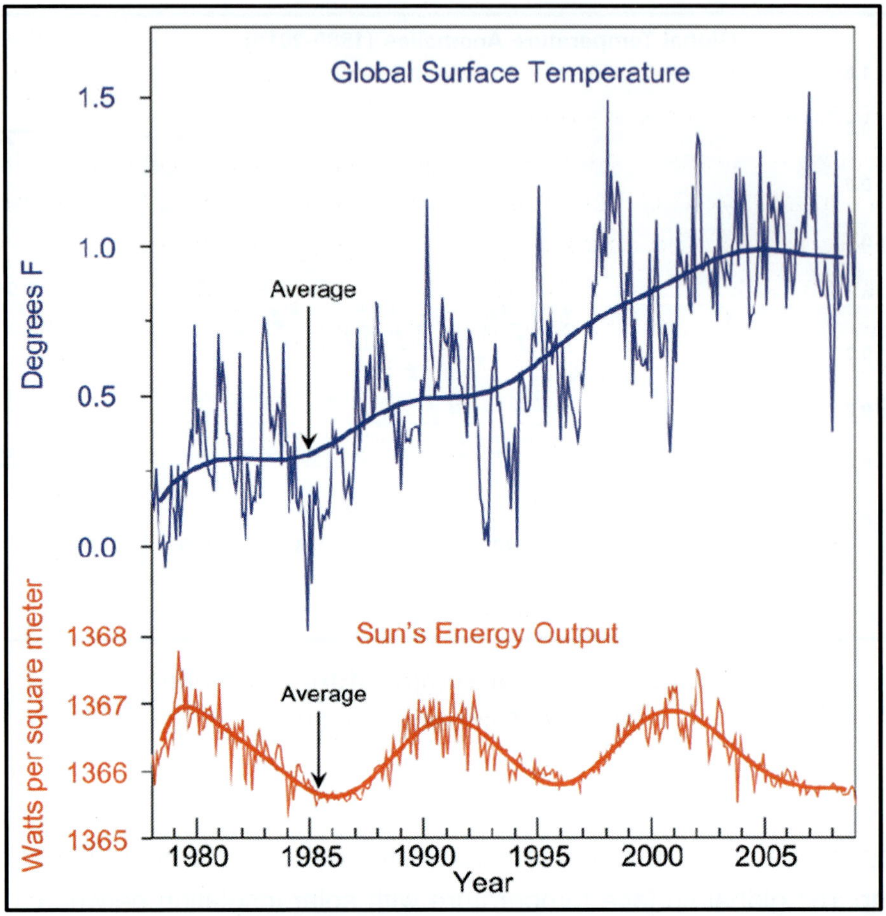

Figure 4-8. Did the Sun cause the Temperature Increase? [NOAA Solar Impact, 2015]

The increase in global temperature anomalies has been associated with the combustion of fossil fuels. When a carbon-based fuel burns, carbon can react with oxygen and nitrogen in the atmosphere to produce carbon dioxide, carbon monoxide, and nitrogen oxides (NOx). The combustion byproducts, including water vapor, are emitted into the atmosphere in gaseous form. Some of the gaseous byproducts are called greenhouse gases because the gases capture energy from sunlight that is absorbed and then emitted as infrared radiation from the earth's surface. Greenhouse gases include water vapor, carbon dioxide, methane, and nitrous oxide, as well as other gases such as volatile organic compounds and hydrofluorocarbons.

Figure 4-9 shows the increase in greenhouse gas concentrations from 0 A.D. to 2005 A.D. Concentrations are in parts per million (ppm) or parts per billion (ppb). The concentration refers to the number of molecules of the greenhouse gas per million (ppm) or billion (ppb) air molecules in an atmospheric sample. The shape of the curves in Figure 4-8 look like hockey sticks with the blades pointing up.

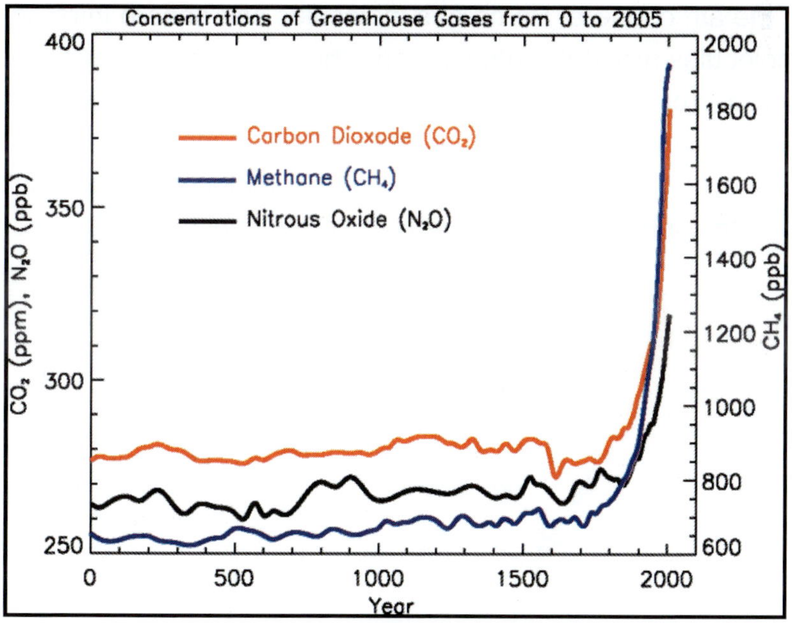

Figure 4-9. Greenhouse Gas Concentrations in the Atmosphere [IPCC, 2007]

Mann, et al. [1999] were the first to publish a hockey stick shaped graph of temperature anomaly in the northern hemisphere versus time for the past 1000 years. The Intergovernmental Panel on Climate Change (IPCC) of the United Nations featured the "hockey stick" graph in its 2001 report on climate. The graph appeared to demonstrate that the increase in temperature was associated with the use of fossil fuels in an increasingly industrialized society.

The Greenhouse Effect is illustrated in Figure 4-10. Sunlight is energy traveling from the sun to the earth. Some sunlight gets reflected back into space when it is reflected by the atmosphere and the rest passes through the atmosphere to the surface of the earth. A portion of the sunlight that arrives at the earth's surface is converted into heat energy. The sunlight can be transformed into thermal energy and re-radiated back toward space. When that happens, the frequency of incident sunlight can change from visible or ultraviolet to infrared. Re-radiated infrared light has to pass through the atmosphere. Some gas molecules can absorb infrared light. If these molecules are in the atmosphere, they behave as greenhouse gases. When a greenhouse gas (GHG) molecule absorbs infrared light, the increase in kinetic energy of the gas molecule increases the average temperature of the gas in the region of the molecule. The resulting increase in atmospheric temperature is the greenhouse effect.

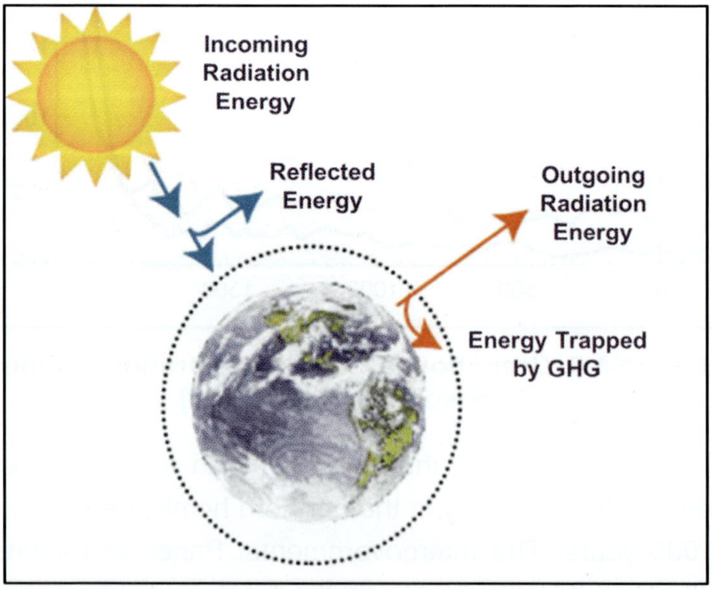

Figure 4-10. The Greenhouse Effect

A study of the behavior of sunlight reaching the earth shows that much of the solar energy arriving at the top of the atmosphere is not making it to the surface of the earth. Many of the gaps in the frequencies of sunlight reaching the surface are associated with light absorption by a greenhouse gas molecule. The most common greenhouse gases are water vapor, carbon dioxide, methane, and nitrous oxide. Other greenhouses gases include volatile organic compounds such as hydrofluorocarbons.

An important method for measuring the composition of the earth's atmosphere in the distant past is ice coring in Antarctica. Ice cores have been obtained at Lake Vostok, Antarctica by drilling into the ice with a core drilling bit and extracting cores of ice from different depths. The formation of ice by compaction of snow or freezing water would capture samples of air that was present at the time the ice formed. The depth of the ice sample represents a point of time in the earth's past. Deeper ice samples are older than shallow ice samples. When a sample of ice is brought to the surface and allowed to melt, the gas released from pores in the ice core represents air present at the time the ice was formed. A sample of the released gas is captured and analyzed. The composition of the gas that was trapped in the ice provides information about the composition of the atmosphere at the time the ice was formed.

Figure 4-11 shows data for Vostok, Antarctica ice core that is dated back over 400,000 years. The data shows change in temperature (A), methane and carbon dioxide concentrations (B), and dust concentration (C). The different graphs are on the same time scale so you can see the relationship between temperature change and concentrations of methane, carbon dioxide and dust. Higher dust levels are believed to be caused by cold, dry periods.

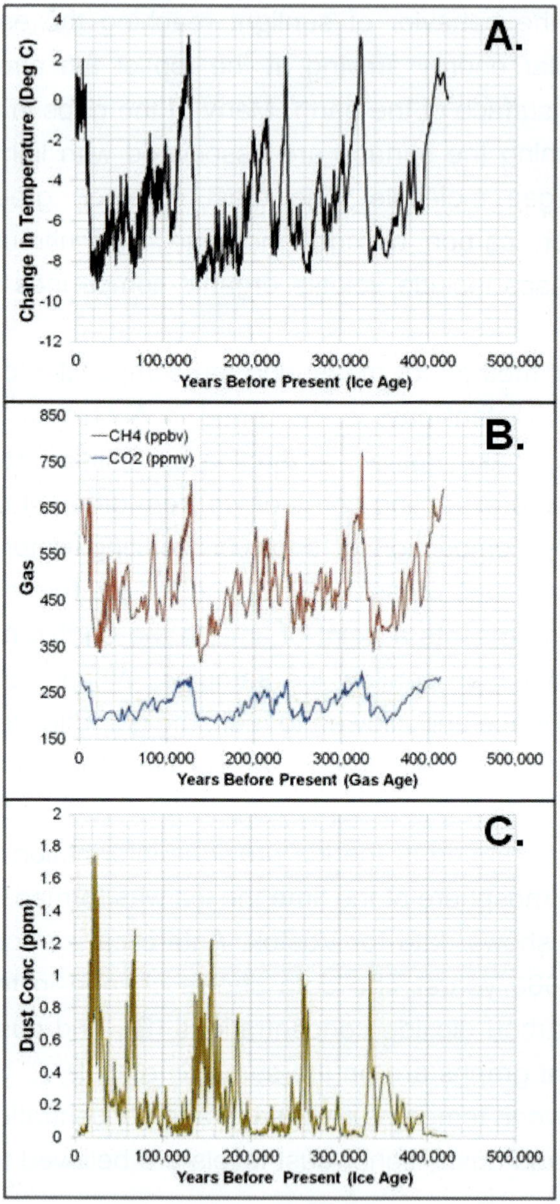

Figure 4-11. Data from Vostok ice core include A. change in temperature, B. methane and carbon dioxide concentrations, and C. dust concentration [Petit, et al., 2001]

Vostok ice core measurements show that CO_2 concentration has historically varied from 150 ppm to 300 ppm. Measurements of CO_2 concentration in the past two centuries show that CO_2 concentration is greater than 300 ppm and continuing to increase.

One way to measure the concentration of carbon dioxide in the atmosphere is represented by the Keeling Curve. Charles David Keeling began measuring carbon dioxide concentration in the atmosphere in 1958. He made his measurements at the Mauna Loa Observatory on the Big Island of Hawaii. The top of the Mauna Loa volcano is a good location for measuring the composition of gases in the atmosphere because it is above typical pollution sources closer to sea level. Figure 4-12 shows that there has been a steady increase in carbon dioxide concentration since measurements began in 1958. The available data in Figure 4-12 begin in 1974. The saw tooth structure of the curve shows an annual cycle. The carbon dioxide concentration was originally measured at a little over 310 parts per million (ppm). Today it is approximately 400 parts per million. These measurements show that carbon dioxide concentration in the atmosphere has been increasing since the beginning of the Keeling curve.

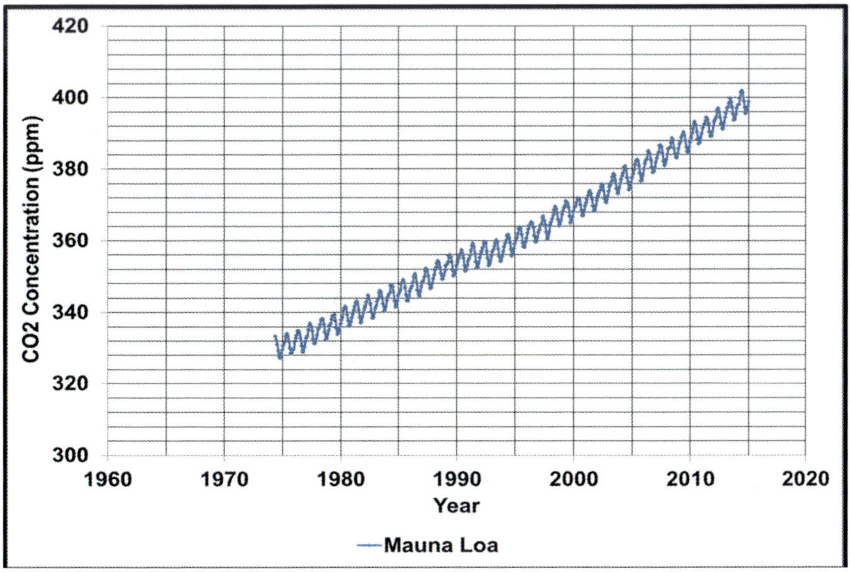

Figure 4-12. The Keeling Curve; the CO_2 reading on 12-27-15 is 401.76 [Scripps Keeling, 2015; NOAA Keeling, 2015]

The Keeling method has been used at locations around the world to see if carbon dioxide concentrations are following the same pattern as the Mauna Loa Keeling curve. The United States National Oceanographic and Atmospheric Administration (NOAA) has made the data publicly available [NOAA Keeling, 2015]. The data cover a shorter period of time than the Mauna Loa data, but plots of the data at locations around the world show annual variations of CO_2 concentration and a steadily increasing average similar to the pattern exhibited by the Mauna Loa Keeling curve. Figure 4-13 presents an overlay of CO_2 concentration data from relatively isolated monitoring stations at Mauna Loa, Hawaii; Barrow, Alaska; Summit, Greenland; Tutuila, American Samoa; and South Pole, Antarctica.

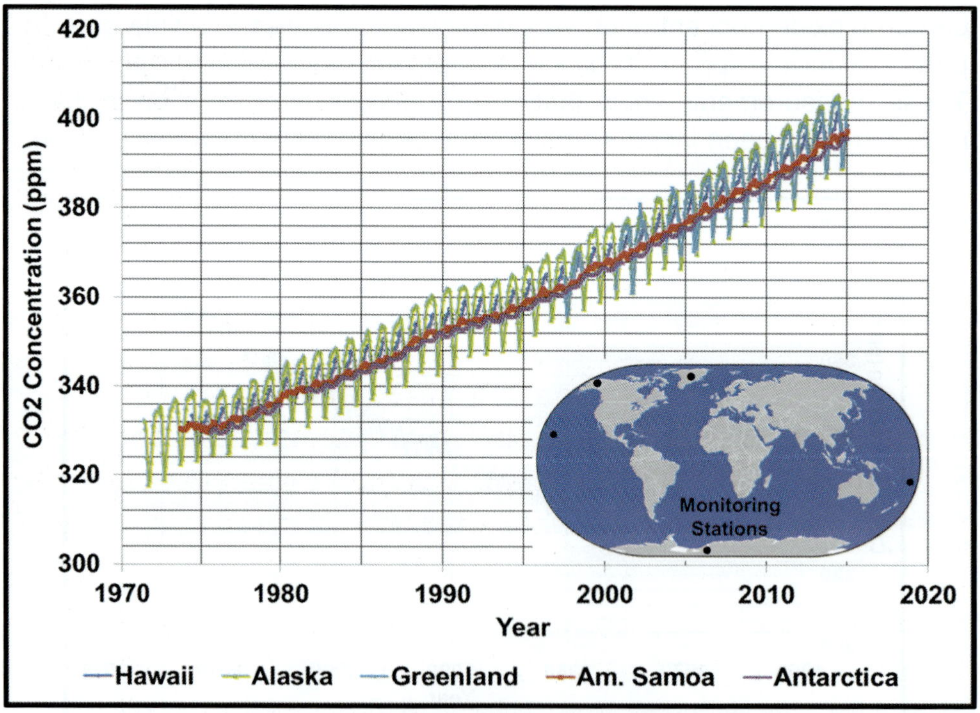

Figure 4-13. Overlay of CO_2 concentration from different parts of the world [NOAA Keeling, 2015]

The Mauna Loa data after 1958 is an extension of ice core data before 1958. Figure 4-14 presents 400,000 years of atmospheric carbon dioxide concentration data by combining Mauna Loa data with Vostok ice core data. The nearly vertical data points on the right hand side of the figure are the Mauna Loa data.

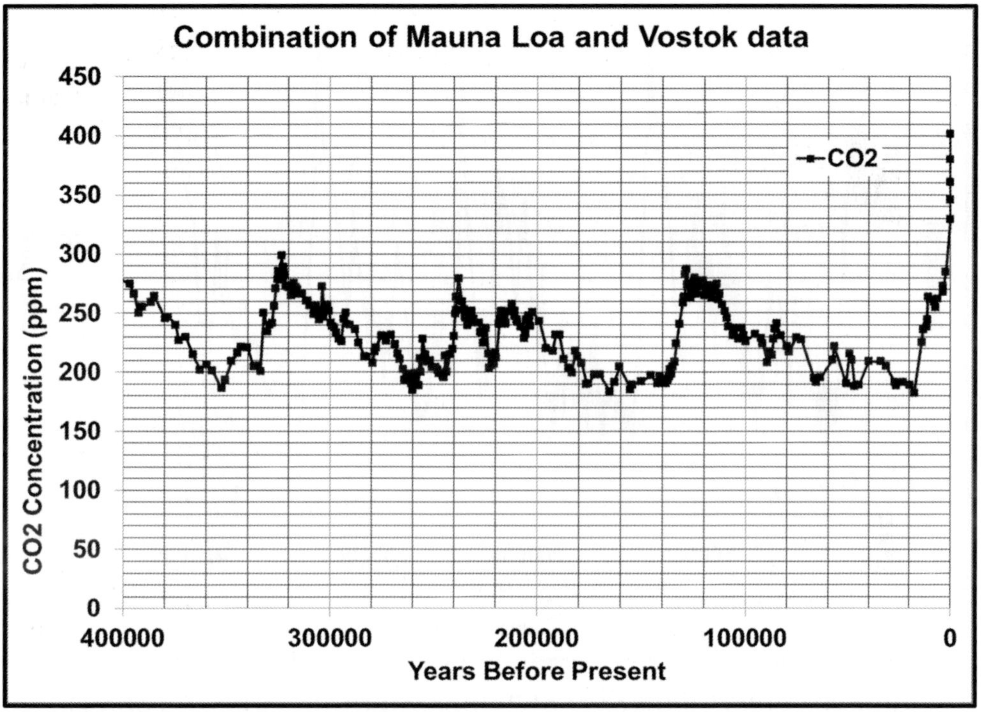

Figure 4-14. CO_2 Concentration for the Past 400,000 Years [NOAA Keeling, 2015]

The concern that temperature will increase as a consequence of increased atmospheric carbon dioxide concentration is based on the assumption that an increase in carbon dioxide concentration will drive an increase in temperature. Figure 4-15 presents approximately 450,000 years of Vostok ice core data. The present is on the left hand side of the figure. The temperature change (DT) is displayed on the left hand axis and the carbon dioxide concentration in parts per million is displayed on the right hand axis. A cursory look at approximately 450,000 years of Vostok data shows an apparent correlation between the curves.

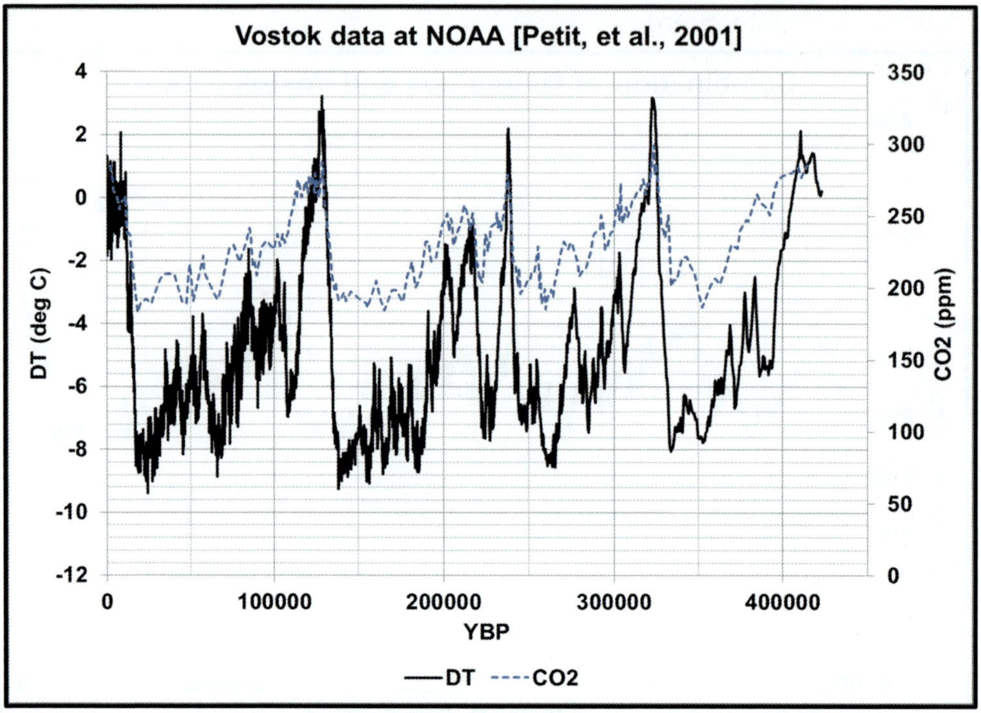

Figure 4-15. 450,000 Years of Temperature Change (DT) and CO_2 Concentration [Source: Petit, J.R., et al., 2001]

A closer look at the data leads to a question raised by climate change skeptics: does an increase in carbon dioxide concentration drive the increase in temperature or does the increase in temperature drive the increase in carbon dioxide concentration? Another possibility is that an unknown factor is driving both the change in temperature and carbon dioxide concentration.

4.4.3 The Long View of Climate Change

The change in composition of the atmosphere since the earth was formed shows that life has been changing the atmosphere since long before humanity evolved. The motion of the earth through space has also contributed to climate change. Mulutin Milankovitch, a Serbian geophysicist and climatologist (1879 – 1958), provided the modern explanation of long-term climate changes.

Milankovitch recognized three cycles were associated with the motion of the earth. The earth moves like a top spinning around the sun. The longest cycle lasts approximately 100,000 years and is called the eccentricity cycle. Eccentricity refers to the earth's position relative to the sun as a consequence of its elliptical orbit. The distance between the sun and the earth changes over time. The second cycle is approximately 41,000 years long and is called the obliquity cycle. It is due to the variation of the tilt of the earth from approximately 21.5° to 24.5° relative to an axis that is perpendicular to the orbital plane of the earth. A precession or wobble occurs in conjunction with the rotation of the tilted earth. The duration of the third cycle, known as the precession cycle, is approximately 23,000 years. Table 4-2 summarizes the duration of the cycles [Reynolds, et al., 2008; Press and Seiver, 2001].

Table 4-2
Duration of Milankovitch Cycles

Cycle	Approximate Duration of Cycle (yrs)
Eccentricity (distance)	**100,000**
Obliquity (tilt)	**41,000**
Precession (wobble)	**23,000**

Annual seasons are related to the variation of the sunshine impacting different parts of the earth each year. The axial tilt has the greatest effect on seasons, followed by the variation of distance from the earth to the sun associated with the eccentricity of the orbit. Other mechanisms that effect the seasons are the gravitational attraction of the planets, especially Jupiter, and the albedo (reflectivity) of the earth. These mechanisms are not as significant as axial tilt and eccentricity.

In addition to driving annual seasons, Milankovitch cycles drive ice age cycles. Table 4-3 shows the timeline for five major ice ages during the past 2.4 billion years. The principal cause of ice ages is the eccentricity cycle. Axial tilt and precession are secondary effects.

Table 4-3
Major Ice Ages
[Eldredge and Biek, 2010]

#	Ice Age	MYA
5	Current Ice Age (aka Quaternary Glaciation)	~ 3 to now
4	Karoo Ice Age	~ 350 to 260
3	Andean-Saharan glaciation	~ 460 to 430
2	Sturtian/Marinoan glaciation	~ 850 to 630
1	Huronian glaciation	~ 2400 to 2100

Ice ages contain glacial and interglacial periods. Figure 4-16 shows these periods for the past 450,000 years. The present is on the left hand side of the figure. Glaciation occurs during the cooler glacial periods, and warming occurs during the interglacial periods.

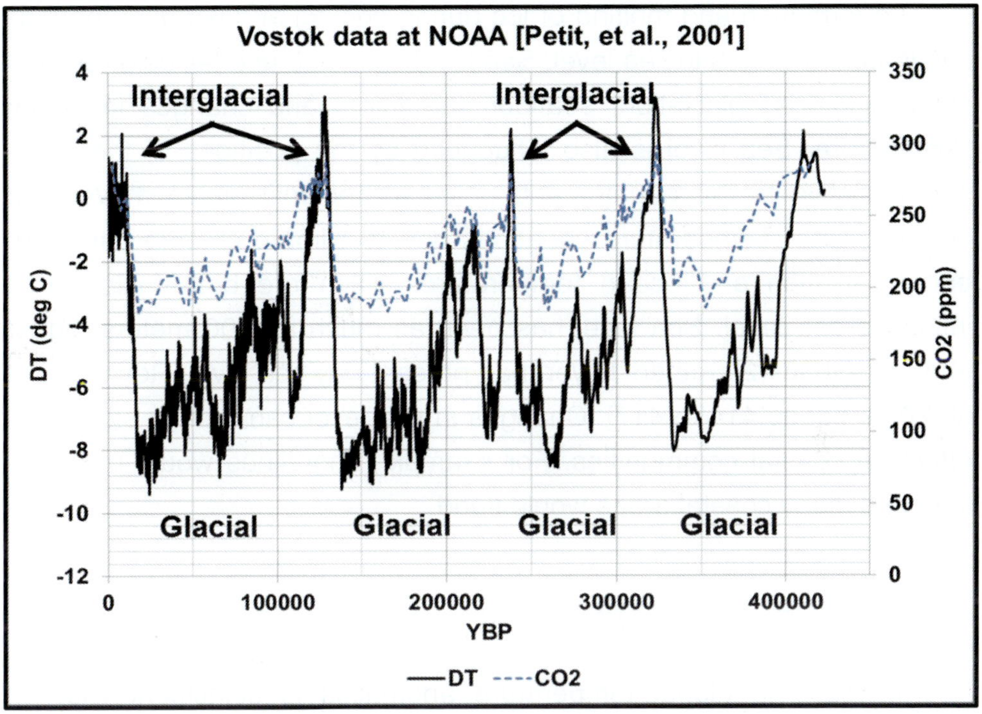

Figure 4-16. 450,000 years of Glacial and Interglacial Periods [Source: Petit, J.R., et al., 2001]

We are currently in the middle of an ice age called the Current Ice Age or Quaternary Glaciation. The variation of average global temperature of the ice ages has been approximately 14°C. For comparison, the temperature increase we are seeing today is a little greater than 1°C, which is less than the temperature variation seen during the five major ice ages.

The change in temperature also varies within an ice age. The temperature change during the past 450,000 years has varied by approximately 9°C. The temperature change over the last 12,000 years has been approximately 6°C. The modern temperature change has been about 1°C, or maybe 2°C if we go back to the coolest days of the Little Ice Age (ca 1500-1600 A.D.).

The formation and melting of ice during cooling and warming periods can lead to changes in sea level. Sea levels can rise and fall as a result of such factors as glaciation, seafloor spreading, and changes in ocean temperature. Sea level is also influenced by the positions of the continents relative to the equator and poles as tectonic plates move over geologic time. Temperature variations due to climate change can influence glaciation and ocean temperature, which are both factors that influence sea level. Sea level has been as much as 200 meters higher and more than 120 meters lower than present day sea level [Reynolds, et al, 2008, pages 414 to 417]. Changes in sea level can have a significant effect on coastal communities. For example, shorelines would move inland and many coastal areas would flood.

4.4.4 Global Population and Climate Change

The increase in global temperature anomalies is coincident with an increase of CO_2 emission into the atmosphere, and an increase in global population (Figure 4-17). The emission of CO_2 into the atmosphere is presented in Figure 4-17 relative to 2009 emission. Both the increase of CO_2 emission into the atmosphere and the increase in global temperature anomalies can be correlated with global population.

The use of fuels with energy densities larger than biomass energy density made it possible for societies to support larger populations. Scientists dating from Thomas Malthus in the 1800's to modern scientists such as Albert A. Bartlett and Paul B. Weisz have argued that growth in the world's population is a major cause of societal problems [Bartlett, 2004 and 2011; Weisz, 2004]. The growth in human population has also led to the growth in fossil fuel consumption.

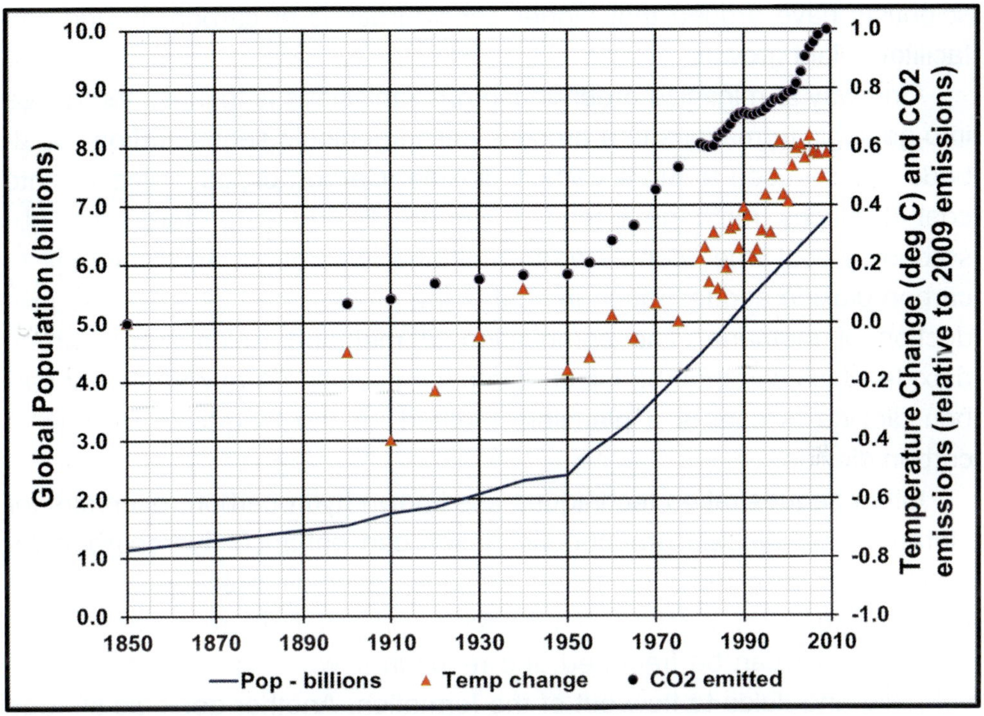

Figure 4-17. Comparison of Global Population, Temperature Change, and CO_2 Emissions [US EIA Population, 2012; NOAA Temperature, 2015; US EIA Emissions, 2012]

4.4.5 Mitigating Greenhouse Gas Emissions

Wigley, et al. [1996] projected ambient CO_2 concentration through the 21st century. They argued that society would have to reduce the concentration of greenhouse gases in the atmosphere to achieve their recommended concentrations. A U.S. Department of Energy study [US DoE Sequestration, 1999] reported that 550 parts per million would be a stable concentration of atmospheric carbon dioxide. The pre-industrial concentration was 288 parts per million and the December 2015 value measured at Mauna Loa exceeded 400 parts per million. The stable value of 550 ppm would be higher than the current value and the pre-industrial value. The optimum concentration is a source of contention. For example, some scientists have argued that higher concentrations of carbon dioxide can facilitate plant growth.

One way to minimize greenhouse gas emissions is to inject the gases into geologic formations in a process called geologic sequestration. Coalbeds can be used to store carbon dioxide. Carbon dioxide injection into coal displaces methane because carbon dioxide is preferentially absorbed by coal when compared to methane absorption. An advantage of carbon dioxide sequestration in coal is the process can increase the production of methane. The combustion of methane yields more carbon dioxide which can then be injected into the coalbed. Sequestration of carbon dioxide in coal is a plausible method for the long-term storage of carbon dioxide.

Other sequestration techniques use known hydrocarbon reservoirs for storage. The injected greenhouse gases can enhance hydrocarbon recovery, but care must be taken to keep reservoir pressure less than fracture pressure. If injection pressure exceeds fracture pressure, the reservoir rock can be fractured and result in flow paths out of the reservoir, allowing fluids to flow out of the formation. Another geologic storage location is a salt dome. A cavern can be formed in the salt dome by injecting water into the dome. The cavern becomes an impermeable storage tank. Geologic storage must have enough volume to take all of the carbon dioxide that will be emitted into the atmosphere.

There are alternatives to geologic storage of greenhouse gases. Some have suggested putting greenhouse gases into ocean locations where the mixture of water and gas at low temperature and high pressure can form a hydrate. Hydrates are gas molecules bounded by water molecules in ball-like structures. The problem with the idea of ocean hydrate storage is that the hydrate structures can break and release the bound gas molecule if the oceans warm up. Consequently, greenhouse gas in the hydrates would be released into the atmosphere and contribute to the greenhouse effect.

Carbon dioxide (CO_2) in the atmosphere can react with surface water (H_2O) in lakes and oceans to form carbonic acid (H_2CO_3). Increasing levels of atmospheric carbon dioxide can increase the amount of carbon dioxide in the water, which tends to increase the acidity of lakes and oceans. Some scientists are concerned that the increase in acidity will harm lake and sea life and adversely affect human food supply.

Point to Ponder: Is geologic sequestration a desirable policy?

It would seem that geologic sequestration would be a desirable policy for a society that is dependent on fossil fuels and is trying to minimize carbon dioxide emissions. On the other hand, geologic sequestration can help a society justify continued use of fossil fuels and reduce its motivation to develop an energy infrastructure that supports sustainable energy sources rather than fossil fuels. From this perspective, geologic sequestration is an undesirable policy.

Perhaps more important than the amount of carbon stored as a result of a sequestration policy is the cost it adds to fossil fuel electricity generation. This additional expense can help make renewable energy sources more cost competitive, but it comes at the expense of increasing the price of fossil fuels.

4.4.6 Anthropogenic Climate Change

Proponents of anthropogenic climate change consider a number of indicators along with greenhouse gas emissions. Two other key factors are changes in global sea level, and changes in Arctic sea ice. These are related since increasing global temperature is expected to cause melting of Arctic sea ice and lead to a rise in global sea level. Taken together, the evidence that human activity is causing the climate to change more than it would naturally change has motivated international attempts to regulate greenhouse gas emissions.

The Kyoto Protocol is an international treaty that was negotiated in Kyoto, Japan in 1997 to establish limits on the amount of greenhouse gases a country can release into the atmosphere. The Kyoto Protocol was not accepted by the United States. Some countries believed the greenhouse gas emission limits were so low that emission limits would adversely impact national economies without solving the problem of global warming. The Kyoto Protocol placed more controls on greenhouse gas emissions from developed countries than it did on emissions from developing nations like India and China, which have rapidly expanding fossil-fuel based economies and approximately a third of the world's population.

Another attempt to establish an international framework to regulate greenhouse gas emissions was held in Copenhagen in December 2009. The Copenhagen conference resulted in a non-binding statement of intent to control emissions.

The United States and China signed an agreement in November 2014 to reduce greenhouse gas emissions. The United States is expected to reduce emissions by up to 28% below its 2005 levels by the year 2025. In exchange, China will set the peak of its emissions by 2030 or earlier. They will not have to reduce their emissions before 2030, although China agreed to increase their share of energy consumption from non-fossil fuels to about 20% of their total consumption by 2030.

4.4.7 COP21 Agreement – Paris, 12 December 2015

A meeting of almost 200 nations was held in Paris from November 30 to December 11, 2015. The purpose of the meeting was to seek an international agreement to reduce the impact of human activity on the climate. It was convened as COP21, the 21st United Nations Framework Convention on Climate Change (UNFCCC) Conference of Parties.

The United States Department of Energy reported [Boden, et al., 2015] that global CO2 emissions from fossil fuel consumption and some industrial processes was greatest in China (28%), followed by the United States (16%), 28 countries in the European Union (10%), India (6%), Russian Federation (6%), and Japan (4%). The remaining 30% was emitted by the rest of the world.

The COP21 agreement was adopted by 195 countries on December 12, 2015 [COP21 Agreement, 2015] and is expected to enter into force in 2020. The agreement is "a global action plan to put the world on track to avoid dangerous climate change by limiting global warming to well below 2°C."

Key elements of the COP21 agreement are summarized below [COP21 Agreement, 2015]:

- Governments agreed to reduce emissions with a long-term goal of keeping the increase in global average temperature to well below 2°C above pre-industrial levels.
- Governments agreed to limit the increase to 1.5°C, since this is expected to significantly reduce risks and the impact of human activity on climate.
- Governments agreed that global emissions need to peak as soon as possible, recognizing that this will take longer for developing countries. Governments are expected to undertake rapid reductions in emissions as soon as possible in accordance with the best available science.
- Governments agreed to be transparent by providing reports to each other and the public.
- Governments agreed to be adaptable. They will strengthen societies' ability to deal with the impact of human activity on climate; and provide continued and enhanced international support for adaptation to developing countries.

Developed countries agreed to help developing countries take action to reduce emissions and build resilience to the impact of human activity on climate. Other countries are encouraged to provide such support voluntarily. Developed countries intend to mobilize US$100 billion per year until 2025 when a new collective goal will be set.

Point to Ponder: Does the COP21 agreement require global cooperation?

On December 9, 2015, United States Secretary of State John Kerry Kerry pointed out the importance of global cooperation while attending the COP21 meeting:

"The fact is that even if every American citizen biked to work, carpooled to school, used only solar panels to power their homes, if we each planted a dozen trees, if we somehow eliminated all of our domestic greenhouse gas emissions, guess what – that still wouldn't be enough to offset the carbon pollution coming from the rest of the world.

"If all the industrial nations went down to zero emissions – remember what I just said, all the industrial emissions went down to zero emissions – it wouldn't be enough, not when more than 65 percent of the world's carbon pollution comes from the developing world."

4.5 Case Study: Is the Climate Change debate settled?

Advocates of anthropogenic climate change want society to immediately transition to a low-carbon economy. They point out that sea level could be higher than the 1990 level by 7 inches to 23 inches by 2100 if greenhouse gas emissions continue to be added to the environment. For comparison, average sea level worldwide has been increasing at approximately 7 inches per century [US EPA Sea Level, 2015]. We can put sea level rise in perspective by noting that storm surge from a hurricane can be 20 feet higher than sea level [NOAA NHC Surge, 2015]. In the case of storm surge, flooding will usually recede within days. Sea level rise, on the other hand, will require adaptation to a long-term change in sea level.

The cost to convert every nation from a carbon-based economy to a low-carbon economy is enormous. On the other hand, ACC advocates argue that the cost of not converting will also be enormous. Is an immediate transition necessary, or do we have some time?

The mainstream view of climate is provided by the United Nation's Intergovernmental Panel on Climate Change (UN IPCC). A second, independent group called the Nongovernmental International Panel on Climate Change (NIPCC) has formed to independently evaluate the impact of increasing atmospheric carbon dioxide on the earth's biosphere. Both groups include scientists and engineers with good credentials in science and technology.

A comparison of the findings of both groups show that the groups agree that climate is changing but they disagree about the cause. The UN IPCC argued that climate change is due to human activity [IPCC, 2007]. The NIPCC said that the UN IPCC overstated their case and that human activity is not causing net harm to the global environment or human well-being [NIPCC CCR-II, 2013]. The NIPCC argued that the changing climate is a natural variation of climate.

The UN IPCC said global climate models are trustworthy whereas the NIPCC said that they are unreliable because they do not adequately model all of the mechanisms that affect climate behavior. For example, one mechanism that is being studied is the exchange of gas molecules between the ocean and the atmosphere. Attempts to validate the models using historical data have shown the limitations of the models.

We can conclude that two groups of individuals have come to different conclusions. We have seen that temperature change in the past couple of centuries is less than temperature changes that have occurred naturally during periods of glaciation. We have also seen the unusual increase in greenhouse gas concentrations since the mid-1800s. It is also worth noting that governments are continuing to provide substantial funding for the study of climate change, which suggests there are still unanswered questions. Advocates of ACC argue that research is needed to understand the details of ACC.

Another view has been expressed by Steven Koonin, formerly Undersecretary for Science in the United States Department of Energy during

President Barack Obama's first term. Koonin had prior experience as a physics professor and provost at Caltech. In an opinion piece in the Wall Street Journal [Koonin, 2014], Koonin said "We are very far from the knowledge needed to make good climate policy." He observed that "the idea that climate science is settled" is misguided. He argued that the idea has distorted public and policy debate, and inhibited scientific and policy discussions.

Koonin supported his opinion by pointing out that global climate models provide widely varying forecasts of how climate behaves. For example, Koonin said that global climate models "roughly describe the shrinking extent of Arctic sea ice observed over the past two decades, but they fail to describe the comparable growth of Antarctic sea ice." Koonin also wrote that "natural influences and variability are powerful enough to counteract the present warming influence exerted by human activity." Prior to the COP21 conference in Paris, Koonin [2015] said that it would be wise to consider allocating resources that would allow us to adapt to climate change rather than immediately disrupting carbon-based economies in an attempt to control the climate. Koonin pointed out that adaptation is "indifferent to whether the climate change is natural or human-induced." Examples of adaptation include raising the height of sea walls or shifting to drought-resistant crops.

Point to Ponder: Is it true that 97% of scientists believe that anthropogenic climate change (ACC) is occurring?
The evidence supporting the claim that "97% of scientists believe that ACC is occurring" is not compelling. A study conducted by Cook, et al. [2013] examined 11,944 climate abstracts from 1991–2011. The climate abstract had to match the topics 'global climate change' or 'global warming'. The authors found that 66.4% of abstracts expressed no position on anthropogenic global warming (AGW), 32.6% endorsed AGW, 0.7% rejected AGW and 0.3% were uncertain about the cause of global warming. The authors stated that "Among abstracts ex-

pressing a position on AGW, 97.1% endorsed the consensus position that humans are causing global warming. In a second phase of this study, we invited authors to rate their own papers. Compared to abstract ratings, a smaller percentage of self-rated papers expressed no position on AGW (35.5%). Among self-rated papers expressing a position on AGW, 97.2% endorsed the consensus."

Another interpretation is possible. You could conclude from Cook, et al. [2013] that 67.4% of articles with an abstract that matched the topics 'global climate change' or 'global warming' did not explicitly support AGW.

It is possible that AGW, now in the form of ACC, is the widely accepted working hypothesis, or paradigm, of scientists who conduct climate research. We can put this into perspective by keeping in mind that virtually every scientist believed in Isaac Newton's concept of absolute time before one scientist, Albert Einstein, published his special theory of relativity in 1905. Einstein's concept of relative time replaced Newton's absolute time and overthrew the Newtonian paradigm. The validity of a scientific hypothesis is not decided by a vote or a survey.

4.6 Energy Security

French economist J-M Chevalier said in **The New Energy Crisis** that "The challenge of the [21st] century is to provide enough food, water and energy without further damaging the environment" [Chevalier, 2009, page 2]. One of the key concerns about energy is security of supply. For example, the EU Supergrid is a European Union project designed to provide clean energy throughout the EU by the middle of the 21st century [EU Supergrid, 2015]. This project will require developing an electricity distribution infrastructure that is suitable for carrying renewable energy to consumers. The renewable energy will include geothermal energy from Iceland, offshore wind energy from northern Europe, and solar energy from southern Europe and cooperating nations in Africa and the Middle East. The goal is to achieve an energy portfolio that is environmentally acceptable, economically competitive, and secure. Although security is listed last, geopolitical events have demonstrated the importance of achieving a secure supply of energy.

The EU Supergrid is discussed in more detail in Chapter 11 after we have considered renewable energy, and the topic of energy security is discussed further in Chapter 13.

4.7 Activities

True-False

Specify if each of the following statements is True (T) or False (F).

1. Water vapor is a greenhouse gas.
2. Vostok ice cores are used to measure sea level changes along the Beaufort Sea.
3. The storage of greenhouse gases in subsurface formations is called geological sequestration.
4. Carbon dioxide can be used to increase oil recovery in some oil reservoirs.
5. The chemical energy in fossil fuels is released by combustion.
6. The Deepwater Horizon sank in the North Sea.
7. The ice ages are explained by Milanković cycles.
8. The Brent Spar was abandoned in the Atlantic Ocean.
9. On a per capita basis, world oil production peaked in the 1970s.
10. M. King Hubbert predicted that world oil production would peak between 1965 and 1970.

Questions

1. What gas is both a greenhouse gas and a gas exhaled by people?
2. What does the Keeling curve measure?

3. Suppose Country A imports 55% of its oil. Of this amount, 24% is imported from Region A. What is the percent of oil imported into Country A from Region A?
4. What is anthropogenic climate change?
5. A. A project is expected to recover 500 million STB of oil. The project will require installing an infrastructure (e.g. platforms, pipelines, etc.) that costs $1.8 billion, and another $2 billion in expenses (e.g. royalties, taxes, operating costs). Break even occurs when revenue = expenses. Assume the value of money does not change with time. What price of oil (in $/STB) is needed to achieve break even?
B. Suppose an unexpected environmental disaster occurs that adds another $20 billion to project cost. Assume the value of money does not change with time. What price of oil (in $/STB) is needed to achieve break even?
6. A. Suppose 100,000 gallons of oil is spilled at sea. How many barrels of oil were spilled?
B. The density of oil is 0.9 g/cc or 900 kg/m^3. Determine the mass of oil spilled. Express your answer in kg and lbm.
C. Estimate the area covered by the spill if the thickness of the spill is 1mm. Express your answer in m^2, km^2, mi^2, and acres.
7. Suppose $2 trillion was spent on the military over a 10-yr period to protect the delivery of 10 billion barrels of oil/yr to the global market. How much did the military budget add to the cost of a barrel of oil?
8. How many quads of energy were provided by fossil fuels in 2010?
9. Vostok ice cores are used to measure _________.
10. Will the COP21 Paris Agreement signed in December 2015 keep global average temperature below 2^0C above pre-industrial levels? Explain your answer.

CHAPTER 5

NUCLEAR ENERGY

After reading this chapter, you should be able to...

- Describe the discovery of the nucleus
- Define radioactivity
- Describe the history of nuclear power
  - *P2P: Can we put the nuclear genie back in the bottle?*
- Explain how nuclear fission reactors work
- Identify the most common fuel for nuclear fission, and explain how it is produced
- Explain how nuclear fusion works
- Differentiate between nuclear fission and nuclear fusion
  - *P2P: Why is nuclear fusion so desirable?*
  - *P2P: When will nuclear fusion be commercially available?*
- Discuss the global dependence on nuclear power
- Describe the social and environmental impact of nuclear energy
  - *P2P: Can nuclear waste be stored safely?*
  - *P2P: Is nuclear power socially acceptable?*

Fossil fuels were the dominant energy source in the 20th century. Concerns about the limited supply of fossil fuels, national security, and environmental impact are motivating the search for other sources of energy. One energy source that is receiving renewed interest is nuclear fission energy. Nuclear fission is the process in which a large, unstable nucleus splits into two smaller fragments. Energy from nuclear fission generates heat, which is typically used to generate electric power. Energy consumed by the United States included over 8% nuclear fission energy in 2010 (Appendix B) and world energy consumption was approximately 5% nuclear fission energy in 2010 (Appendix C). Nuclear fission energy consumption is not expected to change significantly by 2020.

Nuclear energy can be provided by the fission process and the fusion process. The nuclear fission process occurs when large nuclei split into smaller fragments. Energy can also be provided by combining, or fusing, two small nuclei into a single larger nucleus in the nuclear fusion process. Nuclear fusion reactions are the source of energy supplied by the sun.

Both nuclear fission and nuclear fusion reactions release large amounts of energy. Significant quantities of energy need to be controlled. Some of the energy is waste heat and needs to be dissipated. Some of the energy is useful energy and needs to be transformed into a more useful form. Decay products from the fission process can be highly radioactive for long periods of time and need to be disposed of in an environmentally acceptable manner. On the other hand, the byproducts of the fusion process are relatively safe. The purpose of this chapter is to discuss the history of nuclear energy, the role of nuclear energy in the current energy mix, and the potential value of nuclear energy in a future energy mix. We begin by presenting the history of nuclear energy.

5.1 History of Nuclear Energy

Nuclear energy became an important contributor to the global energy mix in the latter half of the 20th century. We obtain nuclear energy from fission reactions and fusion reactions. In both reactions, significant amounts of energy can be released. Nuclear energy in the past and present energy mix has been the result of nuclear fission reactions. Nuclear fusion for commercial application is a technology that is still under development. We consider here the historical development of nuclear energy.

5.1.1 Discovery of the Nucleus

German physicist Wilhelm Roentgen observed a new kind of radiation called x-rays in 1895. Roentgen's x-rays could pass through a human body and create a photograph of the person's internal anatomy. Frenchman Henri Becquerel discovered radioactivity in 1896 while looking for Roentgen's x-rays in the fluorescence of a uranium salt. French physicist and physician Marie Curie (originally from Warsaw, Poland) and her husband Pierre Curie were the first to report the discovery of a new radioactive element in 1898. They named the element polonium after Marie's homeland.

Ernest Rutherford identified the "rays" emitted by radioactive elements and called them alpha, beta, and gamma rays. Today we know that the alpha ray is the helium nucleus, the beta ray is an electron, and the gamma ray is an energetic photon. By 1913, Rutherford and his colleagues at the Cavendish Laboratory in Cambridge discovered the atomic nucleus by using alpha particles to bombard thin metallic foils. Scattering of alpha particles by atoms in the foil showed that most of the atomic mass is located in the center of the atom. The large mass inside the atom is the nucleus. The constituents of the nucleus were identified as the proton and a new, electrically neutral particle called the neutron. James Chadwick discovered the neutron in 1932 while working in Rutherford's laboratory.

The proton and neutron are classified as nucleons, or constituents of the nucleus. The number of protons in the nucleus is the atomic number of the nucleus. The mass number of the nucleus is the sum of the number of protons and the number of neutrons. The symbol X^A denotes the element X and the mass number A. Another notation is ${}_Z^AX_N$. In this notation, the uranium nucleus U^{238} with mass number 238 can be written as ${}_{92}^{238}U_{146}$ which indicates that the nucleus has 92 protons, and 146 neutrons.

In an electrically neutral atom, the number of negatively charged electrons is equal to the number of positively charged protons. Isotopes are nuclei with the same atomic number, but different numbers of neutrons. Carbon-12 is the isotope of carbon with a mass number of 12 (six protons plus six neutrons). The trio of isotopes that is important in nuclear fusion is the set of isotopes of hydrogen: hydrogen with a proton nucleus, deuterium with a deuteron nucleus (one proton and one neutron), and tritium with a triton nucleus (one proton and two neutrons). Each of the electrically neutral hydrogen isotopes has one electron.

The number of protons in the nucleus of an element is used to classify the element in terms of its positive electric charge. Isotopes are obtained when electrically neutral neutrons are added to or subtracted from the nucleus of an element. Changes in the number of neutrons occur by several mechanisms: emission of helium nuclei (alpha particles), electrons (beta particles), or highly energetic photons (gamma rays). The emission of alpha, beta, or gamma rays by an unstable nucleus is considered radioactive decay. Elements produced by radioactive decay are called decay products.

The nucleus is held together by binding energy. Some binding energy is released when the nucleus splits into two fragments. This splitting, or fission, of the nucleus can occur when the mass number is sufficiently large. The separation of a large nucleus into two comparable fragments is an example of spontaneous fission. The original large nucleus is called the parent nucleus, and the fission fragments are called daughter nuclei. By contrast, the fusion process releases energy when two light nuclei with relatively small mass combine to form a larger nucleus.

5.1.2 Radioactivity

Several measures of radioactivity exist. The radiation dose unit that measures the amount of radiation energy absorbed per mass of absorbing material is called the Gray (Gy) in the SI system of units. One gray equals one Joule of radiation energy absorbed by one kilogram of absorbing material.

The measure of radiation used to monitor the biological effects of radiation for different types of radiation is the Sievert (Sv) in the SI system of units. Dose equivalent (in Sv) is the product of the radiation dose (in Gy) times a radiation weighting factor, formerly known as quality factor. The radiation weighting factor is a dimensionless factor that depends on the type and energy of the radiation causing the dose. Table 5-1 illustrates several radiation weighting factors. The alpha particle referred to in Table 5-1 is the helium nucleus. The units of energy used in the table are keV (thousand electron volts) and MeV (million electron volts).

Table 5-1
Typical Radiation Weighting Factors
[after ENS, 2016]

Radiation Type and Energy	Radiation Weighting Factor
Photons (all energies)	1
Neutrons	
< 10 keV	5
10 keV to 100 keV	10
> 100 keV to 2 MeV	20
> 2 MeV to 20 MeV	10
> 20 MeV	5
Protons (> 2 MeV)	5
Alpha particles, fission fragments, heavy nuclei	20

Murray [2001, pg 214] wrote that a single, sudden dose of four sieverts can be fatal, while the typical annual exposure to natural and manmade radiation is 3.6 millisieverts (a millisievert is one thousandth of a sievert). Manmade radiation includes medical and dental radiation, which are typically on the order of a few millisieverts.

5.1.3 History of Nuclear Power

Physicist Leo Szilard conceived of a neutron chain reaction in 1934. Szilard knew that neutrons could interact with radioactive materials to create more neutrons. If the density of the radioactive material and the number of neutrons were large enough, a chain reaction could occur. A chain reaction occurs when neutrons released by a reaction cause more reactions, which release more neutrons that cause even more reactions. Szilard thought of two applications of the neutron chain reaction: a peaceful harnessing of the reaction for the production of consumable energy; and an uncontrolled release of energy (an explosion) for military purposes. Recognizing the potential significance of his concepts, Szilard filed patents (e.g. Figure 5-1) in an attempt to hinder widespread development of the military capabilities of the neutron chain reaction. This was the first attempt in history to control the proliferation of nuclear technology.

PATENT SPECIFICATION **630,726**

Application Date: June 28, 1934. No. 19157/34.
" " July 4, 1934. No. 19721/34.

One Complete Specification left (under Section 16 of the Patents and Designs Acts, 1907 to 1946): April 9, 1935.

Specification Accepted: March 30, 1936 (but withheld from publication under Section 30 of the Patent and Designs Acts 1907 to 1932)

Date of Publication: Sept. 28, 1949.

Index at acceptance:—**Class 39(iv),** P(1:2:3x).

PROVISIONAL SPECIFICATION
No. 19157 A.D. 1934.

Improvements in or relating to the Transmutation of Chemical Elements

Figure 5-1. British Chain Reaction Patent filed by Leo Szilard in 1934

Italian physicist Enrico Fermi and his colleagues in Rome were the first to bombard radioactive material using low-energy ("slow") neutrons in 1935. The spatial extent of the nucleus is often expressed in terms of a unit called the Fermi, in honor of Enrico Fermi. One Fermi is roughly the diameter of a nucleus and is the range of the nuclear force that holds the nucleus together. The correct interpretation of Fermi's results as a nuclear fission process was provided in 1938 by Lise Meitner and Otto Frisch in Sweden, and Otto Hahn and Fritz Strassmann in Berlin. Hahn and Strassmann observed that neutrons colliding with uranium could cause the uranium to split into smaller elements in a process called nuclear fission. Figure 5-2 illustrates the fission reaction $U^{235} + n \rightarrow Ba^{141} + Kr^{92} + 3n$ with target nucleus U^{235} (uranium), and fission products Ba^{141} (barium) and Kr^{92} (krypton).

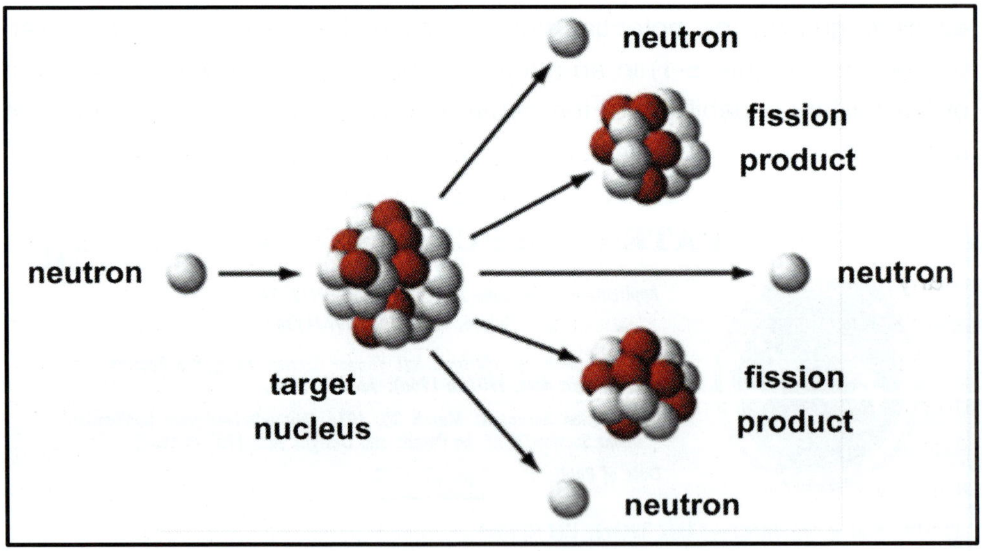

Figure 5-2. Illustration of a Fission Reaction
$U^{235} + n \rightarrow Ba^{141} + Kr^{92} + 3n$

The fission process produces smaller nuclei from the break-up of a larger nucleus, and can release energy and neutrons. If a neutron interacts with the nucleus of fissionable material, it can cause the nucleus to split and produce more neutrons, which can then react with more fissionable material in a chain reaction. Fermi succeeded in operating the first sustained chain reaction on December 2, 1942 in a squash court at the University of Chicago.

The scientific discovery of radioactivity and nuclear fission did not occur in a peaceful society, but in a world threatened by the militaristic ambitions of Adolf Hitler and Nazi Germany. In 1939, Hitler's forces plunged the world into war. Many prominent German scientists fled to the United States and joined an Allied effort to develop the first nuclear weapons. Their effort, known as the Manhattan Project, culminated in the successful development of a bomb utilizing nuclear fission. The resulting bomb was widely known as an atomic bomb, although the term "nuclear bomb" is more correctly used today. The first nuclear bomb was detonated in the desert near Alamogordo, New Mexico in 1945.

By this time, Hitler's Germany was in ruins and there was no need to use the new weapon in Europe. Japan, however, was continuing the fight against the Allied forces in the Pacific and did not seem willing to surrender without first being defeated on its homeland. Such a defeat, requiring an amphibious assault against the Japanese islands, would have cost many lives, both of combatants and Japanese non-combatants. United States President Harry Truman decided to use the new weapon.

The first nuclear bomb to be used against an enemy in war was dropped by a United States airplane on the Japanese city of Hiroshima on August 6, 1945. Approximately 130,000 people were killed and 90% of the city was destroyed. When the Japanese government refused to surrender unconditionally, a second nuclear bomb was dropped on the Japanese city of Nagasaki on August 9, 1945. The combined shocks of two nuclear attacks and a Soviet invasion of Manchuria on August 9, 1945 prompted the Japanese Emperor to accept the Allies' surrender terms on August 14, 1945 and to formally surrender aboard the USS Missouri (Figure 5-3) on September 2, 1945.

Figure 5-3. The USS Missouri berthed at Pearl Harbor, Hawaii (Fanchi, 2004)

People throughout the world realized that nuclear weapons had significantly altered the potential consequences of an unlimited nuclear war. By the early 1950's, the Soviet Union had acquired the technology for building nuclear weapons and the nuclear arms race was on. As of 1972, the United States maintained a slight edge over the Soviet Union in strategic nuclear yield and almost four times the number of deliverable warheads. Yield is a measure of the explosive energy of a nuclear weapon. It is often expressed in megatons, where one megaton of explosive is equivalent to one million tons of TNT (trinitrotoluene).

During the decade from 1972 to 1982, the size of the United States nuclear arsenal did not change significantly, but the quality of the weapons, particularly the delivery systems such as missiles, improved dramatically. By 1982, the Soviet Union had significantly closed the gap in the number of deliverable warheads and had surged ahead of the United States in nuclear yield. The Soviet Union and the United States had achieved a rough nuclear parity. The objective of this parity was a concept called "deterrence."

Deterrence is the concept that neither side will risk engaging in a nuclear war because both sides would suffer unacceptably large losses in life and property. This is the concept that underlies the doctrine of Mutual Assured Destruction: the societies of all participants in a nuclear war would be destroyed. A consequence of this policy is that the societies involved in such a pact must remain vulnerable to one another; otherwise the less vulnerable society has a strategic advantage that could threaten the more vulnerable society and lead to instability and war. The "star wars" concept advanced by United States President Ronald Reagan in the 1980's is a missile defense system that provided an alternative strategy to Mutual Assured Destruction, but threatened the global balance of nuclear power.

The Soviet Union was unable to compete economically with the United States and dissolved into separate states in the late 1980's. Some former states of the Soviet Union retained nuclear weapons, notably Russia and Ukraine. Ukraine gave up its nuclear weapons between 1994 and 1996. Other nations around the world have developed nuclear technology for peaceful purposes, and possibly for military purposes as well.

Point to Ponder: Can we put the nuclear genie back in the bottle?

Some historians argue that nuclear weapons were not needed to end World War II. Perhaps this is true, but it is worth noting that there was a race to militarize nuclear energy, and it is likely that nations other than the United States would have developed nuclear weapons. Germany and Soviet Union had already made progress developing nuclear weapons. Therefore the answer is no, the proverbial genie cannot be put back in the bottle.

Nevertheless, the world is arguably a more dangerous place because of the development of nuclear weapons. Today, nations with nuclear weapons arsenals must be concerned about the security of every weapon in their arsenals. Even one nuclear weapon can be a weapon of mass destruction. In addition, many nations are seeking to acquire nuclear technology, which increases the risk of nuclear weapons proliferation. Now that the world knows about nuclear energy, society must learn to govern its use. One forecast discussed in Chapter 14 is based on the premise that nuclear technology can be safely controlled and used as a long-term source of energy.

5.2 NUCLEAR REACTORS

The primary purpose of a nuclear reactor is to generate electric power from the energy released by a nuclear reaction. Nuclear reactors provide electrical power for communities, ships such as submarines and aircraft carriers, and serve as facilities for training and research. Our focus here is on the use of nuclear reactors to generate electricity. We then discuss global dependence on nuclear energy.

5.2.1 Nuclear Fission Reactors

The neutrons produced in fission reactions typically have energies ranging from 0.1 million electron volts (1.6×10^{-14} Joules) to 1 million electron volts (1.6×10^{-13} Joules). Neutrons with energies this high are called fast neutrons. Fast neutrons can lose kinetic energy in collisions with other materials in the reactor. Less energetic neutrons with kinetic energies on the order of the thermal energy of the reactor are called slow neutrons or thermal neutrons. Some nuclei tend to undergo a fission reaction after they capture a slow neutron. These nuclei are called fissile materials and are valuable fuels for nuclear fission reactors. Fission is induced in fissile materials by slow neutrons. Fission reactors that depend on the nuclear capture of fast neutrons exist, but are beyond the scope of this book.

Materials like light water (light water is ordinary water – H_2O), heavy water (deuterium oxide – D_2O), graphite, and beryllium control the number of neutrons in the reactor. They are called moderating materials because they moderate, or control, the nuclear reaction. Moderating materials slow down fast neutrons by acquiring kinetic energy in collisions when a fast neutron collides with the more massive, slower moderating material. Some of the kinetic energy of the fast neutron is transferred to the moderating material and the speed of the fast neutron decreases.

Coolant materials transport the heat of fission away from the reactor core. Coolants include light water, carbon dioxide, helium, and liquid sodium. In some cases, moderating materials can function as coolants.

Charged beta particles (electrons) are emitted during fission reactions. Some of the emitted electrons have enough kinetic energy to travel faster than light in a dielectric medium like water. The passage of high energy electrons through the reactor pool releases energetic photons known as Cerenkov radiation. The Cerenkov photons give the reactor pool a blue glow during operation of the fission reactor [MST Cerenkov, 2016].

A pressurized water reactor heats coolant water to a high temperature and then sends it to a heat exchanger to produce steam. A boiling water reactor provides steam directly from water in contact with the fissile materials. The steam from each of the reactors turns a turbine to generate electricity. A pressurized water reactor is sketched in Figure 5-4. A nuclear reactor is housed in a containment building (Figure 5-5). The containment building, which is typically a dome, provides protection from internal leaks as well as external dangers, such as an airplane crash.

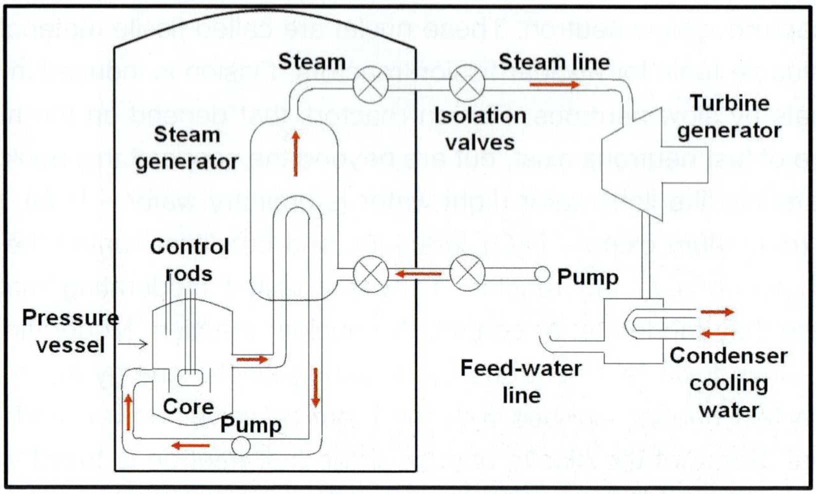

Figure 5-4. Schematic of a Pressurized Water Reactor [after Cassedy and Grossman, 1998, page 177; and United States Department of Energy Report Energy Technologies and the Environment, Report Number DOE/EP0026, 1981]

Nuclear fission reactor design and operation depends on the ratio of neutrons in succeeding generations of decay products. The ratio of the number of neutrons in one generation to the number of neutrons in the preceding generation is called the neutron multiplication factor. For example, if 100 neutrons produced in one generation of nuclear reactions interact with fissile material and produce 110 neutrons in the next generation, the neutron multiplication factor is 1.1 (that is, 110/100). The neutron multiplication factor is a positive quantity that depends on several factors [Lilley, 2001, Chapter 10].

If the multiplication factor for a reactor is less than one, the number of fission reactions will diminish from one generation to the next and the chain reaction will die out. The reactor level is considered subcritical. A critical mass of fissile material is present in the reaction chamber when there are enough neutrons being produced by nuclear reactions to balance the loss of neutrons. When the reaction is at the critical level, the

Figure 5-5. Containment Building at San Onofre Nuclear Generating Station, California (Fanchi, 2002)

multiplication factor for a reactor is equal to one and the chain reaction can be sustained. If the multiplication factor is greater than one, the number of neutrons produced by nuclear reactions exceeds the loss of neutrons and the number of fission reactions will increase from one generation to the next. The reactor level is considered supercritical. The chain reaction in a supercritical reactor will accelerate and, if it proceeds too fast, can have explosive consequences. The nuclear fission bomb is a supercritical reactor.

Products of fission reactions include nuclei that emit biologically harmful radiation. The nuclei are contained in spent fuel rods. Some of the nuclei, such as radioactive isotopes of cesium and iodine, can be radioactive for thousands of years. This material is considered radioactive waste and poses a significant environmental problem. The disposal of radioactive waste is discussed further in Section 5.4.

5.2.2 Availability of Nuclear Fuel

The most abundant fuel for nuclear fission is uranium. Uranium exists in the crust of the earth as the mineral uraninite. Uraninite is commonly called pitchblende and is a uranium oxide (U_3O_8). It is found in veins in granites and other igneous rocks. It is possible to find uranium in sedimentary rocks. In this case, scientists believe that uraninite was precipitated in sedimentary rocks after being transported from igneous rocks by the flow of water containing dissolved uraninite.

Uranium is obtained by mining for the mineral uraninite. Mining methods include underground mining, open pit mining, and in situ leaching. Leaching is a process of selectively extracting a metal by a chemical reaction that creates a water-soluble molecule that can be transported to a recovery site. The isotope of uranium that undergoes spontaneous fission (uranium-235) is approximately 0.7% of naturally occurring uranium ore. Uranium must be separated from mined ore and then enriched for use in nuclear fission reactors.

Mining uranium oxide ore is the first stage in the development of nuclear fission reactor fuel. The stages in nuclear fuel development are mining, milling, conversion and enrichment. Once uranium oxide ore has been mined, it goes through a milling process. The ore is crushed and bathed in acid so that uranium can be separated and dried to form a powder called yellowcake. Uranium yellowcake is converted to uranium fluoride (UF_6) gas by combining yellowcake with fluorine gas. When the conversion process is complete, uranium fluoride gas is put in spinning centrifuges to purify the uranium-235 isotope in an enrichment process. Enriched uranium fluoride gas is converted to uranium dioxide (UO_2) powder. The powder is used to form fuel pellets, which are inserted into tubes to form fuel rods. Several fuel rods are bundled together to make a fuel assembly for use in a nuclear fission reactor.

The degree of enrichment, or purification, of uranium-235 is often the main obstacle faced by countries seeking nuclear fission technology for use in electricity generation or weapons. As a rule of thumb, 5% enrichment is enough for commercial nuclear fission reactors; 20% enrichment is used in research reactors; and 90% enrichment is used for nuclear weapons.

The amount of uranium that can be recovered from the earth is called uranium reserves. Table 5-2 presents estimates of uranium reserves in the United States for the most common mining methods. Grade is the average percent of U_3O_8 per ton of ore. Grade is approximately 0.14% for underground mining and 0.08% for open pit mining and in situ leaching. Estimates of uranium reserves have been made and have many of the same uncertainties associated with estimates of fossil fuel reserves. Factors that are not well known include the distribution and extent of uranium deposits and the price people are willing to pay to recover the resource.

Table 5-2
United States Uranium Reserves by Mining Method and Cost, December 2008 Estimate [US EIA Uranium, 2016]

Mining Method	US$50 per pound	US$100 per pound
	Uranium Oxide (U_3O_8) million pounds	Uranium Oxide (U_3O_8) million pounds
Underground	215	608
Open Pit	93	325
In Situ Leaching	230	289
Low grade ore, etc.	1	5
TOTAL	539	1,227

Uranium is considered a nonrenewable resource because it exists as a finite volume within the earth. A comparison of the two columns of uranium oxide reserves in Table 5-2 shows that the amount of uranium oxide reserves increases as price per pound increases. The supply of uranium increases if you are willing to pay for it.

The resource of recoverable uranium is the amount of uranium that can be economically recovered from an orebody. The geological distribution of uranium is illustrated by listing the ten countries with the largest known recoverable uranium resources in the world. The list of countries in Table 5-3 is based on information from the Organization for Economic Cooperation and Development (OECD) and a recovery cost of approximately US$60 per pound. Resources listed in Table 5-3 do not include uranium in sea water.

Table 5-3
Distribution of Known Recoverable Uranium Resources
Reserves to approx. $60/lb as of 1/1/2013
[WNA, 2016]

Country	Million Tons U	Percent of World
Australia	1.877	29
Kazakhstan	0.747	12
Russian Fed	0.556	9
Canada	0.543	8
Niger	0.445	7
Namibia	0.421	6
South Africa	0.372	6
Brazil	0.304	5
USA	0.228	4
China	0.219	4
World total	6.493	

Global uranium usage in 2015 was approximately 73 tons/yr. Comparing this usage with the world total in Table 5-4 implies that the known recoverable uranium resource could last approximately 90 years if global uranium usage remains constant.

Fissile materials other than uranium, such as plutonium-239 and thorium-232, can be used in the fission process. Fissile thorium is thought to exist in nature in greater abundance than fissile uranium. Specialized reactors called breeder reactors are designed to operate with fuels other than uranium. A breeder reactor is a nuclear fission reactor that produces more fissile material than it consumes.

Two more sources of uranium that could be used for generating electrical power are sea water and nuclear weapons. Reduction in nuclear weapons stockpiles and dismantling nuclear warheads would provide some enriched uranium. A larger source of uranium is seawater. Table 5-4 shows typical concentrations of uranium in parts per million (ppm) from the World Nuclear Association website. The concentration of uranium in sea water is small, but the volume of seawater is huge. Technology for extracting uranium from seawater exists on the laboratory scale, but is currently too expensive for large-scale commercial application.

Table 5-4
Concentration of Uranium [WNA, 2016]

Very high-grade ore (Canada) - 20% U	200,000 ppm U
High-grade ore - 2% U,	20,000 ppm U
Low-grade ore - 0.1% U,	1,000 ppm U
Very low-grade ore (Namibia) - 0.01% U	100 ppm U
Granite	3-5 ppm U
Sedimentary rock	2-3 ppm U
Earth's continental crust (av)	2.8 ppm U
Seawater	0.003 ppm U

5.2.3 Nuclear Fusion Reactors

One of the appealing features of nuclear fusion is the relative abundance of hydrogen and its isotopes compared to fissile materials. The idea behind nuclear fusion is quite simple: fuse two molecules together and release large amounts of energy in the process. Examples of fusion reactions include collisions between protons, deuterons (deuterium nuclei), and tritons (tritium nuclei). Protons are readily available as hydrogen nuclei. Deuterium is also readily available. Ordinary water contains approximately 0.015 mole % deuterium [Murray, 2001, page 77], thus one atom of deuterium is present in ordinary water for every 6700 atoms of hydrogen.

The fusion reaction can occur only when the atoms of the reactants are heated to a temperature high enough to strip away all of the atomic electrons and allow the bare nuclei to fuse. The state of matter containing bare nuclei and free electrons at high temperatures is called plasma. Plasma is a gas of ionized particles. The temperatures needed to create plasma and allow nuclear fusion are too high to be contained by conventional building materials. Two methods of confining plasma for nuclear fusion are being considered: magnetic confinement, and inertial confinement.

Magnetic confinement is a confinement method that relies on magnetic fields to confine the plasma. The magnetic confinement reactor is called a tokamak reactor. Tokamak reactors are magnetic bottles that contain the plasma that is to be used in the fusion reaction. The magnetic bottle may have the shape of a cylinder that is bent into a circle, or a cylinder that is twisted and then bent into a circle. The former design is a toroidal (donut shaped) design, and the latter design is a helical stellarator. The plasma trajectories are illustrated in Figure 5-6.

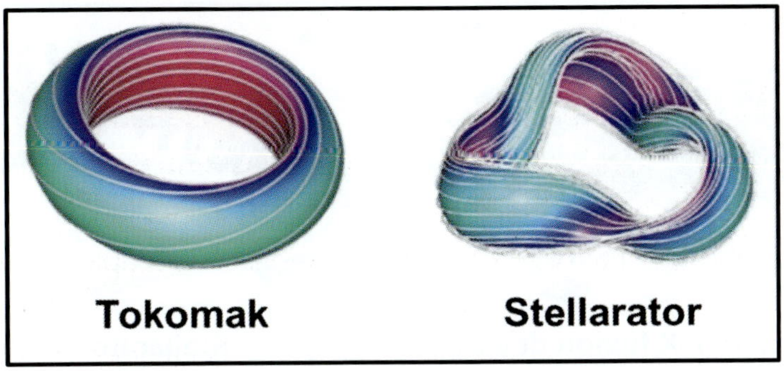

Figure 5-6. Comparison of Plasma Trajectories [after ORNL, 2016]

The Joint European Torus (JET) in Oxfordshire, United Kingdom is a magnetic confinement fusion reactor based on the tokamak concept. JET produced 16 MW of fusion power from a total input power of 24 MW. We estimate the return on energy by the Q factor defined as $Q = \frac{Output\ Power}{Input\ Power} = \frac{16MW}{24MW} = 0.67$. Nuclear fusion power was sustained for one to two seconds [Feder, 2009].

Two magnetic fields confine the plasma in a tokamak: one is provided by cylindrical magnets that create a toroidal magnetic field; and the other is a poloidal magnetic field that is created by the plasma current. Combining these two fields creates a helical field that confines the plasma. Existing tokamak reactors inject deuterium and tritium into the vacuum core of the reactor at very high energies. Inside the reactor, the deuterium and tritium isotopes lose their electrons in the high energy environment and become plasmas. The plasmas are confined by strong magnetic fields until fusion occurs.

The Wendelstein 7-X stellarator shown in Figure 5-7 is being tested for its suitability as a power plant at the Max Planck Institute of Plasma Physics in Germany [IPP, 2016]. Earlier stellarator designs have shown that the design can confine plasma with a quality comparable to a tokomak. The goal of the stellarator is to operate for longer periods of time than the tokomak.

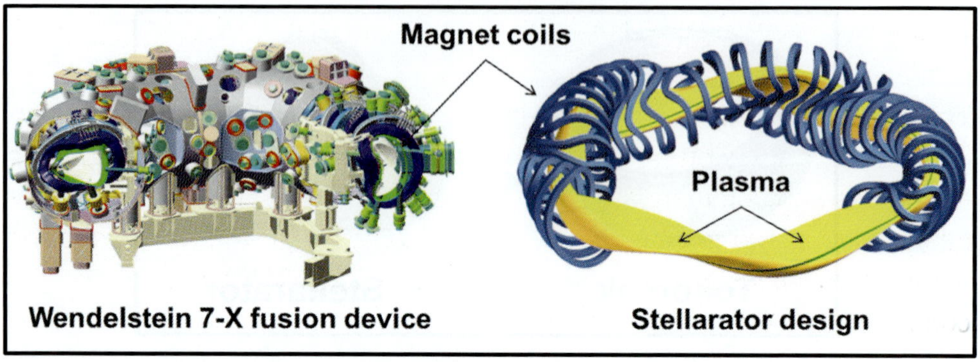

Figure 5-7. Wendelstein 7-X Fusion Device (images courtesy of Max Planck IPP, 2016)

The second confinement method is inertial confinement. Inertial confinement uses pulsed energy sources such as lasers to concentrate energy onto a small pellet of fusible material, such as a frozen mixture of deuterium and tritium. The pulse compresses and heats the pellet to ignition temperatures.

Point to Ponder: Why is nuclear fusion so desirable?
Nuclear fusion is considered an environmentally clean source of energy. Unlike nuclear fission, nuclear fusion does not generate waste products that are lethal for thousands of years. The material needed for nuclear fusion is much more plentiful than the known supply of fissionable materials, such as uranium.

The fusion reactor concept can be combined with the fission process to create a nuclear reactor system that is a fusion-fission hybrid. A fusion reactor would serve as a neutron source for a fission reactor. The hybrid concept could produce power, burn nuclear waste, and produce fissile material. The combination of fusion and fission technologies would, however, be more complicated and expensive than working with either a fusion system or a fission system.

5.2.4 ITER: A Prototype Nuclear Fusion Reactor

The goal of nuclear fusion research and development is to provide commercial nuclear fusion power sometime during the 21st century. Attempts to harness and commercialize fusion energy have so far been unsuccessful because of the technical difficulties involved in igniting and controlling a fusion reaction. Nuclear fusion must be controlled before it can be commercialized. A prototype nuclear fusion reactor based on the tokomak design is the International Thermonuclear Experimental Reactor (ITER) shown in Figure 5-8. ITER is presently under construction in Cadarache, France [ITER, 2016].

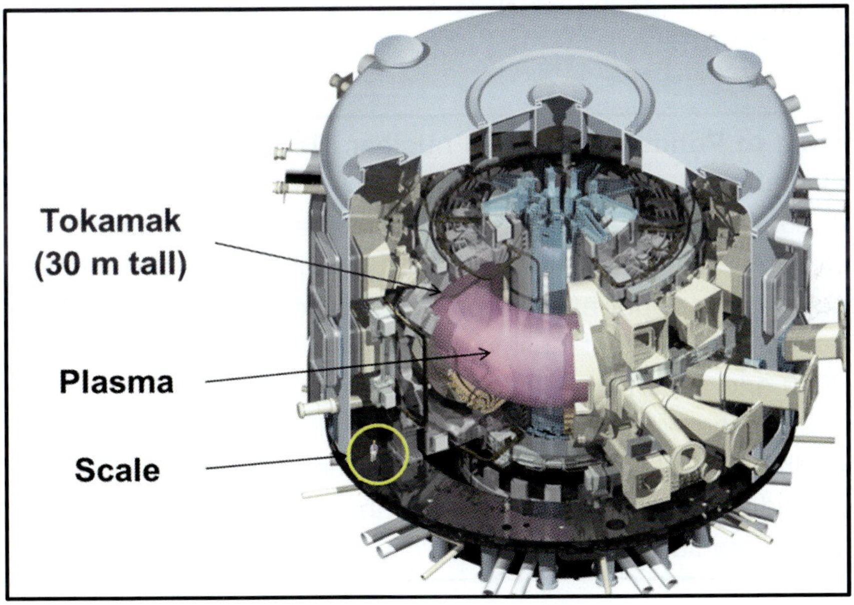

Figure 5-8. ITER Facility
(image courtesy of ITER organization, 2016)

The ITER project being built in Southern France is sponsored by China, the European Union, India, Japan, Korea, Russia, and the United States. The acronym ITER is Latin for "the way" and was selected in part to signify the way to new energy. ITER is a commercial-scale tokamak reactor that will use deuterium and tritium as nuclear fuel. The ITER nuclear fusion reactor is designed to produce 500 MW fusion power from 50 MW input power with a Q factor given by $Q = \frac{Output\ Power}{Input\ Power} = \frac{500MW}{50MW} = 10$. ITER is the precursor of a demonstration nuclear fusion power plant and will be the first fusion reactor to provide net energy, that is, $Q = \frac{Output\ Power}{Input\ Power} > 1$.

People have been working to develop commercial fusion power for decades. No one can say with certainty when nuclear fusion power technology will be ready for peaceful commercial use. Other nuclear fusion concepts are being studied. For example, Lockheed Martin is developing a compact fusion reactor which is a nuclear fusion reactor on a smaller scale than ITER.

Point to Ponder: When will nuclear fusion be commercially available?

A commercial nuclear fusion reactor has not yet been constructed. Nuclear fusion reactors are still in the development stage and are not expected to provide commercial power for a generation. The first plasma stream at ITER is expected around 2020 and deuterium-tritium fusion operation is expected in the 2020s. The waste product from deuterium-tritium fusion is the environmentally benign, inert gas helium. Fusion energy is expected to contribute significantly to the energy mix by the end of the 21st century.

5.3 Global Dependence on Nuclear Power

The first commercial nuclear power plant was built on the Ohio River at Shippingport, Pennsylvania, a city about 25 miles from Pittsburgh. It began operation in 1957 and generated 60 megawatts of electric power output [Murray, 2001, page 202]. Today, nuclear power plants generate a significant percentage of electricity in some countries.

Table 5-5 lists the top ten producers of electric power from nuclear energy and their percentage of the world's total electric power generation from nuclear energy for the year 2012. Japan was in the top ten producers of electric power from nuclear energy until a nuclear disaster occurred at the Fukushima Daiichi nuclear facility on March 11, 2011. The disaster is discussed in more detail below.

Table 5-5
Top Ten Producers of Electric Power from Nuclear Energy in 2012
[US EIA Nuclear, 2016]

Country	Electric Power from Nuclear Energy (billion kWh)	Percentage of World Electric Power from Nuclear Energy (World = 2,345 billion kWh)
United States	769.3	32.8
France	407.4	17.4
Russia	166.3	7.1
South Korea	143.6	6.1
Germany	94.1	4.0
China	92.7	4.0
Canada	89.1	3.8
Ukraine	84.9	3.6
United Kingdom	64.0	2.7
Sweden	61.5	2.6

The total electric power generated from nuclear energy in the world was 2,345 billion kilowatt-hours in 2012. Total electric power generated in the world that same year was 21,532 billion kilowatt-hours. Consequently, electric power generated from nuclear energy provided approximately 11% of the electric power generated in the world in 2012.

Table 5-6 shows the percentage of electric power generated from nuclear energy compared to total electric power generation for the top ten producers of electric power from nuclear energy for the year 2012. Nuclear energy was used to generate three quarters of the electric power generated in France. Ukraine uses nuclear power to generate almost half of its electricity.

Table 5-6
Dependence of Nations on Nuclear Energy in 2012
[US EIA Nuclear, 2016]

Country	Electric Power from Nuclear Energy (billion kWh)	Total Electricity Generation (billion kWh)	Nuclear Share (% National Total)
United States	769.3	4047.8	19.0
France	407.4	533.3	76.4
Russia	166.3	1012.5	16.4
Korea, South	143.6	499.7	28.7
Germany	94.1	585.2	16.1
China	92.7	4768.3	1.9
Canada	89.1	616.2	14.5
Ukraine	84.9	187.1	45.4
United Kingdom	64.0	335.7	19.1
United States	769.3	4047.8	19.0

Some nations that use nuclear energy for peaceful purposes also have nuclear weapons, while other nations do not. For example, the United States, Russia, and France have nuclear weapons while South Korea and Germany do not. The decision to enrich uranium so that it can be used in a nuclear weapon depends on factors such as the history of the nation and its politics. The United States, for example, developed nuclear weapons in an effort to quickly end World War II. Germany, on the other hand, was defeated in World War II and its military was subject to restrictions imposed by Allied victors led by the United States, Russia, Great Britain, and France.

5.4 NUCLEAR ENERGY AND THE ENVIRONMENT

The selection of an environmentally benign primary fuel is not a trivial problem. Two energy sources of special concern for the 21st century energy mix are nuclear fission and nuclear fusion. We consider their environmental impact in more detail here.

5.4.1 Nuclear Fission

Compared to fossil fuel driven power plants, nuclear fission plants require a relatively small mass of resource to fuel the nuclear plant for an extended period of time. Nuclear fission plants rely on a nonrenewable resource: uranium-235. The earth's inventory of uranium-235 will eventually be exhausted. Breeder reactors use the chain reaction that occurs in the reactor control rods to produce more fissionable material (typically plutonium-239). Thorium fuels can be used to breed fissile uranium-233 for use in nuclear fission reactors.

One of the main concerns of nuclear fission technology is to find a socially and environmentally acceptable means of disposing of fuel rods containing highly radioactive waste. The issue of waste is where most of the public debate about nuclear energy is focused. The waste generated by nuclear fission plants emits biologically lethal radiation and can contaminate the site where it is stored for thousands of years. On the other hand, environmentally compatible disposal options are being developed. One disposal option is to store spent nuclear fuel in geologically stable environments.

A nuclear fission power plant can contaminate air, water, the ground, and the biosphere. Air can be contaminated by the release of radioactive vapors and gases through water vapor from the cooling towers, gas and steam from the air ejectors, ventilation exhausts, and gases removed from systems having radioactive fluids and gases. The radiation released into the air can return to the earth as radiated rain, which is analogous to acid rain generated by fossil fuel combustion. Liquid water may be contaminated when radioactive materials leak into coolant water. The contaminated water can damage the environment if it is released into nearby bodies of water such as streams or the ocean. Water and soil contamination can occur when radioactive waste leaks from storage containers and seeps into underground aquifers. The biosphere (people, plants, and animals) is affected by exposure to radioactive materials in the environment. The effect of exposure is cumulative and can cause the immune system of an organism to degrade.

Nuclear fission plants may be vulnerable to disruption by natural disasters. Most are built to specifications that protect them from potential disasters in their geographic region. Computer models are run periodically to ensure that the specifications are suitable. These models often lead to plant modifications. Despite attempts to design facilities for natural contingencies, natural disasters can still occur.

An example of a nuclear reactor facility that suffered a disruption from a natural disaster was the Fukushima I Nuclear Power Plant in Japan. The plant is located on the east coast of the main Japanese island of Honshu, north of Tokyo. Tokyo Electric Power Company (TEPCO) operated six boiling water nuclear fission reactors at the plant. A nuclear disaster occurred at the Fukushima Daiichi nuclear facility on March 11, 2011, when a tsunami struck the facility following an earthquake offshore. The tsunami caused equipment failures and structural damage to six boiling water reactors. Hydrogen gas bubbles developed in the reactor buildings and then exploded. The explosions resulted in the release of toxic levels of radioactive material. Large volumes of contaminated water from the tsunami had to be disposed. There was widespread concern about contaminated land, water, crops, and sea life along the affected coast. Trust between the public and the authorities (the government and TEPCO) broke down. Subsequent studies found that the plant was not designed to withstand natural events of this magnitude.

5.4.2 Radioactive Waste

Nuclear fission power plants generate radioactive wastes that require long-term storage. Some of the end products of nuclear fission are highly radioactive and have a half-life measured in thousands of years. Half-life of a radioactive substance is the length of time it takes half of the original quantity to decay. A radioactive substance with a long half-life must be disposed of in a way that offers long-term security.

Low level radioactive waste (LLRW) generally emits radioactivity at levels that decay to natural background levels in approximately 500 years [NEI, 2016]. LLRW includes items that have come in contact with radioactive materials, such as items of clothing, plant hardware, and waste from reactor cooling-water cleanup systems. Licensed disposal facilities are available for storing LLRW. LLRW disposal facilities in the United States are located in a few locations, including Andrews County, Texas, Clive, Utah, and Hanford, Washington.

High level radioactive waste (HLRW) is a byproduct of recycling used nuclear fuel. HLRW is presently stored on-site in the United States until permanent disposal facilities are available.

One method for disposing high level radioactive waste is to place spent uranium fuel rods containing plutonium and other dangerously radioactive compounds in an isolated location that is geologically stable. The location should not be significantly affected by low levels of radiation and it needs to be large enough to handle the amount of nuclear waste generated by the industry. The Nuclear Energy Institute estimated that the inventory of nuclear waste increases by 2,000 to 2,300 metric tons each year [NEI, 2016]. Nuclear waste is continuing to be stored at or near reactor sites because no disposal facility is available for commercial nuclear waste.

The United States and France are considering geologic disposal of high level radioactive waste. A possible location for storing nuclear waste in France is the Champagne-Ardenne region of eastern France, near the village of Bure. The site is a geologic storage site in compacted clay rocks.

A possible geologic storage location in the United States is Yucca Mountain, Nevada. Figure 5-9 shows an aerial view of Yucca Mountain and Figure 5-10 shows a train inside the Yucca Mountain tunnel. Yucca Mountain was chosen for long-term nuclear waste storage by the Federal government, but local public resistance has delayed the waste disposal project.

Figure 5-9. Yucca Mountain

Figure 5-10. Train Inside Yucca Mountain Tunnel

A Waste Isolation Pilot Plant (WIPP) in the Chihuahuan Desert near Carlsbad, New Mexico was certified for disposal operations in 1998 by the United States Environmental Protection Agency [WIPP, 2016]. It was built by the United States Department of Energy and carved out of the Permian Salt Formation nearly a half mile below the surface. The WIPP site is used to dispose of defense-generated radioactive waste.

Point to Ponder: Can nuclear waste be stored safely?
Natural nuclear fission reactors were discovered by French physicist Francis Perrin at the Oklo uranium mine in the equatorial West African republic of Gabon in 1972 [WNA, 2016; Coogan and Cullen, 2009]. Scientists believe that the ore deposit, which contained about 3% uranium-235, began a self-sustaining nuclear fission chain reaction millions of years ago. The natural reactors created radioactive waste that included fission products and elements found today in used nuclear fuel. Geological isolation at Oklo occurred almost 2 billion years ago

at Oklo. According to the World Nuclear Association, "several spontaneous nuclear reactors operated within a rich vein of uranium ore. (At that time the concentration of U-235 in all natural uranium was about 3%.) These natural nuclear reactors continued for about 500,000 years before dying away. They produced all the radionuclides found in high-level waste, including over 5 tonnes of fission products and 1.5 tonnes of plutonium, all of which remained at the [Oklo] site and eventually decayed into non-radioactive elements." Natural nuclear fission reactors like Oklo are natural analogues that demonstrate that it is possible to dispose nuclear waste in a secure and safe manner for an indefinite period of time.

5.4.3 Containment Failure

Radioactive materials can be released from nuclear reactors if there is a failure of the containment system. Containment can be achieved by the reactor vessel and by the containment dome enclosing the reactor vessel. Two widely publicized nuclear power plant incidents were containment failures. They occurred at Three Mile Island, Pennsylvania in 1979 and at Chernobyl, Ukraine in 1986.

The Three Mile Island power plant was a pressurized water reactor that went into operation in 1978 and produced approximately a gigawatt of energy. The containment failure at Three Mile Island occurred on March 29, 1979. It began when coolant feedwater pumps stopped and temperature in the reactor vessel began to rise. An increase in pressure accompanied the increasing temperature and caused a pressure relief valve to open. The reactor shut down automatically. Steam from the reactor flowed through the open relief valve into the containment dome. The valve failed to shut at a pre-specified pressure and vaporized coolant water continued to flow out of the reactor vessel through the open valve. The water-steam mixture flowing through the coolant pumps caused the pumps to shake violently. Plant operators did not realize they were losing coolant and decided to shut off the shaking pumps. A large volume of steam formed in the reactor vessel and the overheated nuclear fuel melted the metal tubes holding the nuclear fuel pellets. The exposed pellets reacted with water and steam to form a hydrogen gas bubble. Some of the hydrogen escaped into the containment dome. The containment dome did not fail; it contained the hydrogen gas bubble and pressure fluctuations. The operators were eventually able to disperse the hydrogen bubble and regain control of the reactor.

The Chernobyl containment failure occurred on April 26, 1986 in a boiling water reactor that produced a gigawatt of power and had a graphite moderator. Operators at the plant were using one of the reactors, Unit 4, to conduct an experiment. They were testing the ability of the plant to provide electrical power as the reactor was shut down. To obtain measurements, the plant operators turned off some safety systems, such as the emergency cooling system, in violation of safety rules. The operators then withdrew the reactor control rods and shut off the generator that provided power to the cooling water pumps. Without coolant, the reactor overheated. Steam explosions exposed the reactor core and fires started. The Chernobyl reactors were not encased in massive containment structures that are common elsewhere in the world. When the explosions exposed the core, radioactive materials were released into the environment and a pool of radioactive lava burned through the reactor floor. The Chernobyl accident was attributed to design flaws and human error.

The disparity in the death toll associated with the Chernobyl incident illustrates the range of conclusions that can be drawn by different people with different perspectives. M.W. Carbon [1997, Chapter 5] reported that the known death toll at Chernobyl was less than fifty people. R.A. Ristinen and J.J. Kraushaar [1999, Section 6.9] reported an estimate that approximately 47,000 people in Europe and Asia will die prematurely from cancer because they were exposed to radioactivity from Chernobyl.

5.4.4 Nuclear Fallout

Nuclear fallout is the deposition of radioactive dust and debris that was carried into the atmosphere by the detonation of a nuclear weapon. Nuclear fallout can also be generated by the release of radioactive material into the atmosphere. Radioactive material can be released by a variety of events such as reactor containment failure or detonation of a dirty bomb made by combining conventional explosives and radioactive waste. The pattern of deposition of the fallout depends on climatic conditions such as ambient air temperature and pressure, and wind conditions.

5.4.5 Nuclear Winter

Nuclear winter is the phrase used to describe a decline of air temperature resulting from an increase of particulates in the atmosphere following the detonation of many nuclear weapons. The temperature decline would generate wintry conditions, hence the phrase "nuclear winter." A decline in air temperature may have occurred during extinction events associated with meteors striking the earth. Many scientists believe that the disappearance of the dinosaurs approximately 65 million years ago is an example of an extinction event that was caused by a massive extraterrestrial object colliding with the earth.

In the nuclear winter scenario, an increase of particulates in the atmosphere increases the earth's albedo, which quantifies the reflection of incident sunlight away from Earth. The albedo increases when the amount of incident sunlight reflected by the atmosphere increases. Proponents of the nuclear winter scenario argue that an increase in the reflection of incident solar radiation by the atmosphere can cause a reduction of air temperature. Critics of the nuclear winter scenario argue that the greenhouse effect may tend to increase the surface temperature of the earth. It is now believed that the decline of air temperature would not be as severe as originally thought.

5.4.6 Nuclear Fusion

The isotopes of hydrogen needed for nuclear fusion are abundant. The major component of fusion, deuterium, can be extracted from water, which is available around the world. Tritium, another hydrogen isotope, is readily available in lithium deposits that can be found on land and in seawater. Unlike fossil fuel driven power plants, nuclear fusion does not emit air pollution.

Nuclear fusion reactors are considered much safer than nuclear fission reactors. The amounts of deuterium and tritium used in the fusion reaction are so small that the instantaneous release of a large amount of energy during an accident is highly unlikely. The fusion reaction can be shut down in the event of a malfunction with relative ease. A small release of radioactivity in the form of neutrons produced by the fusion reaction may occur, but the danger level is much less than that of a fission reactor.

The main problem with nuclear fusion energy is that the technology is still under development; commercially and technically viable nuclear fusion reactor technology does not yet exist. The ITER project discussed previously is an example of a nuclear fusion concept that has potential for future commercialization.

Point to Ponder: Is nuclear power socially acceptable?
Nuclear power is socially acceptable in some political jurisdictions around the world. Some nations, notably France, are reliant on nuclear fission power. The acceptability of nuclear power depends on such factors as finding an environmentally acceptable solution for the storage of nuclear wastes and the cost of energy using non-nuclear energy sources. The tsunami in Japan and ensuing nuclear disaster at Fukushima has revived concerns about the safety of nuclear facilities in the face of natural disasters. If the price of non-nuclear energy to the

consumer continues to increase, nuclear energy may become more appealing. Environmental concerns about the geologic storage of radioactive waste and reactor damage or failure must be weighed against environmental concern about the emission and accumulation of greenhouse gases in the atmosphere.

5.5 ACTIVITIES

True-False

Specify if each of the following statements is True (T) or False (F).

1. Nuclear fission is the splitting of a large nucleus into smaller nuclei.
2. Most of the uranium ore that is mined can be used for nuclear fission.
3. Leo Szilard was the first person to attempt to prevent nuclear technology proliferation.
4. Water can be used as both a moderating and cooling material in nuclear fission reactors.
5. A pressurized water reactor provides steam directly from the reactor to turn a turbine.
6. A breeder reactor consumes more fissile material than it produces.
7. Nuclear fission is a nonrenewable energy source.
8. The fuel for nuclear fusion is cleaner and more plentiful than the fuel for nuclear fission.
9. Isotopes are nuclei with the same atomic number but different numbers of neutrons.
10. Energy from the sun is the result of nuclear fusion.

Questions

1. List 5 nations that produce electric energy from nuclear energy. Which of these countries use commercial nuclear fusion power plants?

2. Match the site to the events by writing the letter of the event in the column to the left of the site.

	Site		Event
	Alamogordo	A	Japanese nuclear power plant flooded by a tsunami
	Chernobyl	B	First city bombed by a nuclear weapon in World War II
	Fukushima	C	First nuclear weapon detonation
	Hiroshima	D	Possible nuclear waste storage site
	Yucca Mountain	E	Ukrainian nuclear power plant accident with many deaths

3. What is nuclear winter?
4. What is Mutual Assured Destruction?
5. How do you control the number of neutrons in a nuclear fission reactor?
6. What is the purpose of a coolant material in a nuclear fission reactor?
7. What is the purpose of a containment building?
8. What is the difference between nuclear fission and nuclear fusion?
9. List the two methods of confining plasma used in nuclear fusion reactors.
10. What is a tokamak reactor and what does it do?

CHAPTER 6

RENEWABLE ENERGY – SOLAR ENERGY

After reading this chapter, you should be able to...

- Define renewable energy
- Describe how solar energy reaches the earth
  - *P2P: Can we collect light from outer space?*
- Discuss energy conversion efficiency and the concept of intermittency
- Define and differentiate between different solar energy types (solar active, solar passive, solar thermal, solar electric)
- Describe multiple types of solar power plants
  - *P2P: Are solar thermal power plants becoming more popular in sunny, deserted areas?*
- Explain how solar electric technologies work
  - *P2P: What are some current uses of photovoltaics?*

In Chapter 1, we discussed the history of energy use and mentioned that major technological advances are often the catalyst for change in the global energy mix. These advances include such developments as the steam engine, discovery of oil's use as an illuminant, Edwin L. Drake's first successful oil well, and more recently the discovery of nuclear power. Today, the catalyst for change is different, but the need for new technology is still a limiting factor. The modern transition to a new energy mix is being driven by the desire to replace nonrenewable fossil fuels, the concern that fossil fuel combustion is having an adverse effect on the environment, and the desire to achieve a secure energy supply. These factors have motivated interest in renewable energy.

Renewable energy is energy obtained from sources at a rate that is less than or equal to the rate at which the source is replenished. Energy consumed by the United States in 2010 included about 8% renewable energy (Appendix B), and world renewable energy consumption was also approximately 8% in 2010 (Appendix C). Renewable energy consumption is expected to increase significantly by 2020. Most of the energy from renewable energy has been hydroelectric and biomass. The growth in renewable energy supply is expected to come from solar and wind energy sources. Solar energy is discussed in this chapter, and the other renewable energy sources are discussed in subsequent chapters.

The production of electrical power from renewable energy sources does not emit as much pollution as power production from fossil fuels. As for cost, most renewable energy will become less expensive over time as the infrastructure for harvesting and delivering energy matures. The most likely sources of renewable energy that are expected to contribute to the future energy mix are solar energy, wind energy, energy from water, and biofuels. Many of these are significant contributors to the global energy supply today.

The term "renewable" is somewhat of a misnomer when applied to solar energy. Solar energy is provided by the sun as the result of nuclear fusion reactions. The nuclear fusion process in the sun consumes isotopes of hydrogen to form helium and release energy. Eventually, the fuel for the nuclear fusion process will be exhausted. Since the remaining lifetime of the sun is expected to be a few billion years, many people consider solar energy an inexhaustible supply of energy. In fact, solar energy from the sun is finite, but should be available for such a long period of time that is considered renewable. Energy sources that are associated with solar energy, such as wind and biomass, are also considered renewable.

Solar radiation may be converted to other forms of energy by several conversion processes. Thermal conversion relies on the absorption of solar energy to heat a cooler surface. Biological conversion of solar energy relies on photosynthesis, which is a process that uses light to synthesize biologically useful chemical compounds. Photovoltaic conversion generates electrical power by the generation of an electrical current as a result of the interaction of light photons with certain types of conducting materials. Wind power and ocean energy conversion rely on atmospheric pressure gradients and oceanic temperature gradients to generate electrical power. In this chapter we focus on thermal conversion and photovoltaic conversion of solar energy in three forms: passive, active, and electric. Passive and active solar energy are generally used for space conditioning, such as heating and cooling, while solar electric energy is used to generate electrical power. Each of these forms is discussed below.

6.1 Source of Solar Energy

Solar energy is energy emitted by a star. Figure 6-1 shows the anatomy of a star. The source of solar energy is nuclear fusion. The fusion process occurs in the core, or center, of the star. Energy released by the fusion process propagates away from the core by radiating from one atom to another in the radiation zone of the star. As the energy moves away from the core and passes through the radiation zone, it reaches the part of the star where the energy continues its journey towards the surface of the star as heat associated with thermal gradients. This part of the star is called the convection zone. The surface of the star, called the photosphere, emits light in the visible part of the electromagnetic spectrum. The photosphere is surrounded by an atmosphere of hot gases called the chromosphere.

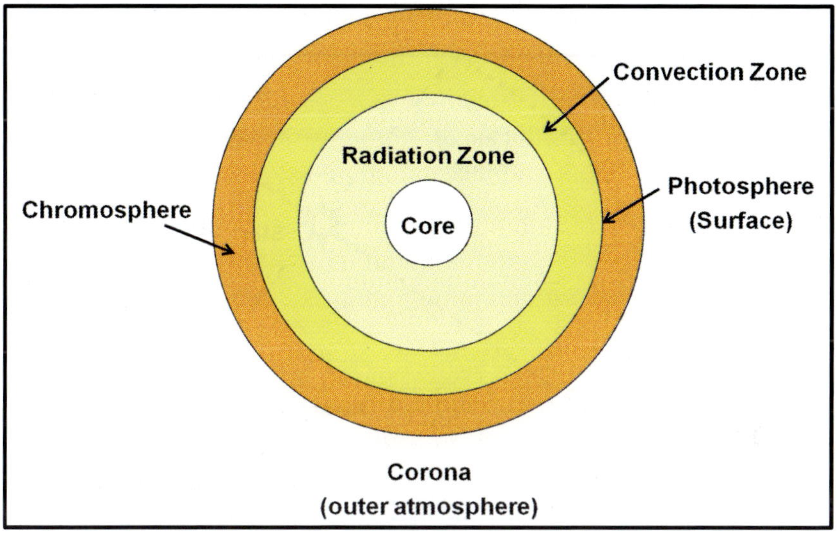

Figure 6-1. Cross-section of a Star

The amount of sunlight that reaches the earth depends on the motion of the earth around the sun (Figure 6-2). When viewed from earth, the sun appears to follow a path relative to distant, background stars as earth orbits the sum. This path is an ellipse and the ellipse lies in a geometric plane called the ecliptic plane. The orbit of the earth around the sun lies in a geometrical plane called the orbital plane. The line of intersection where the orbital plane passes through the ecliptic plane is the line of nodes. The ecliptic plane and the orbital plane are shown in Figure 6-2.

The luminosity of a star is the total energy radiated per second by the star. The amount of radiation from the sun that reaches the earth's atmosphere is called the solar constant. The solar constant varies with time because the earth follows an elliptical orbit around the sun and the axis of rotation of the earth is inclined relative to the plane of the earth's orbit. Distances between points on the surface of the earth and the sun vary throughout the year.

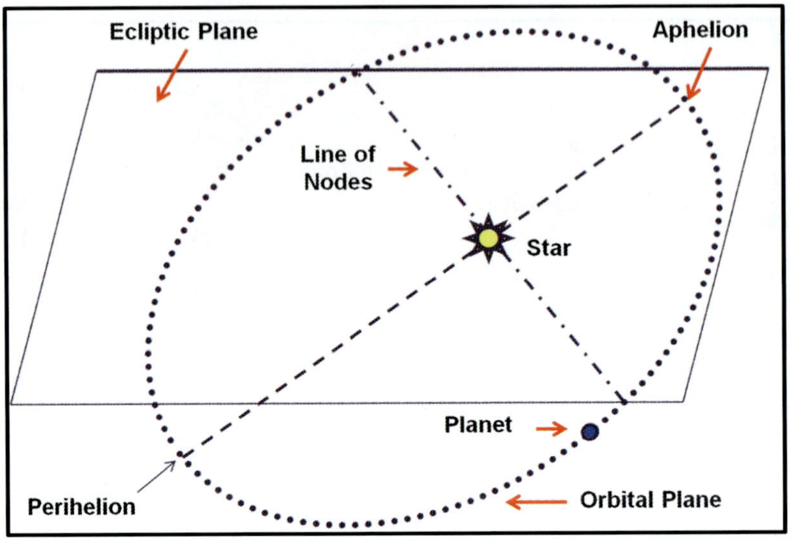

Figure 6-2. Planetary Orbit and the Ecliptic Plane

The amount of solar radiation that reaches the surface of the earth depends on the factors illustrated in Figure 6-3. The flux of solar radiation incident on a surface placed at the edge of the earth's atmosphere depends on the time of day and year, and the geographical location of the surface. Some incident solar radiation is reflected by the earth's atmosphere. The fraction of solar radiation that is reflected back into space by the earth-atmosphere system is called the albedo.

Approximately 35% of the light from the sun does not reach the surface of the earth. This is due to clouds (20%), atmospheric particles (10%), and reflection by the earth's surface (5%). The solar flux that enters the atmosphere is reduced by the albedo. Once in the atmosphere, solar radiation can be absorbed in the atmosphere or scattered away from the earth's surface by atmospheric particulates such as air, water vapor, dust particles, and aerosols. Some of the light that is scattered by the atmosphere eventually reaches the surface of the earth as diffused light. Solar radiation that reaches the earth's surface from the disk of the sun is called direct solar radiation if it has experienced negligible change in its original direction of propagation.

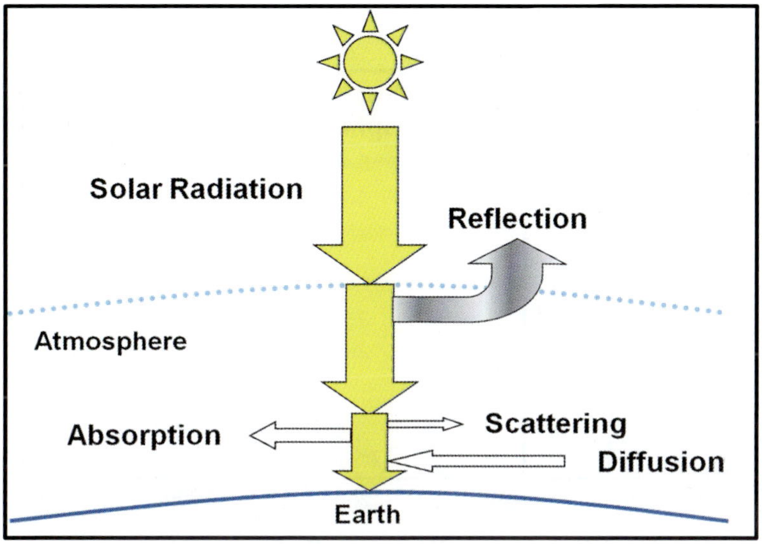

Figure 6-3. Solar Radiation and the Earth-Atmosphere System

Light is electromagnetic radiation. It can be used to transfer energy by the propagation of electromagnetic waves from one location to another. Two other energy transfer processes are conduction and convection. Conduction is the transfer of energy as the result of a temperature difference between substances that are in contact. Convection is the transfer of energy by the movement of a heated substance.

The sunlight arriving at the surface of the earth differs from the sunlight arriving at the top of the earth's atmosphere. Figure 6-4 illustrates the difference by showing the solar spectral irradiance as a function of wavelength of light. Solar irradiance is the solar energy striking an area of surface each second. Solar spectral irradiance is solar irradiance per unit wavelength of light. Wavelength in the figure is expressed in nanometers (nm), and solar spectral irradiance is expressed in $W/m^2/nm$.

Figure 6-4 shows the range of wavelengths that are visible to the human eye. The red (top) line in the visible wavelength range shows solar spectral irradiance of sunlight measured at the top of the atmosphere.

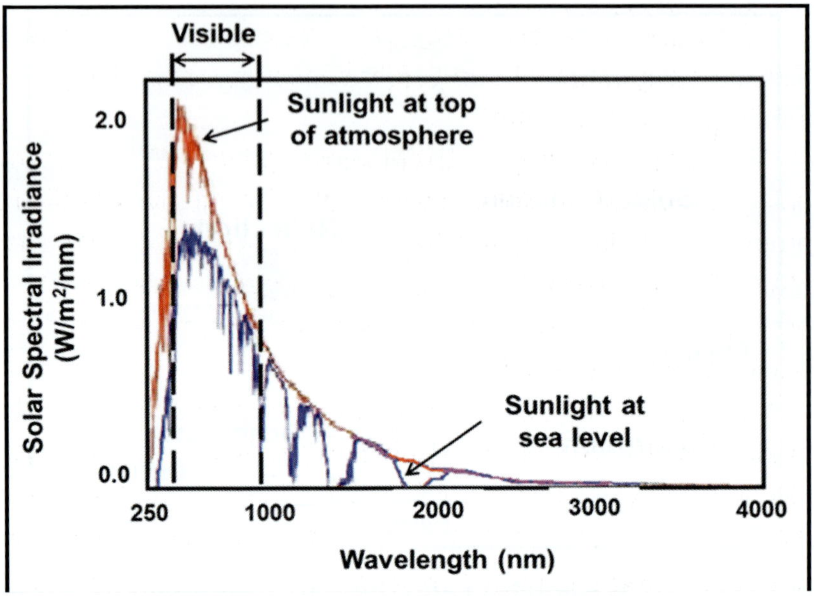

Figure 6-4. Solar Spectral Irradiance

The blue (lower) line in the visible wavelength range is the solar spectral irradiance of sunlight measured at sea level. The two lines are different because there is more solar energy per wavelength of light arriving at the top of the atmosphere than there is at sea level. The gaps in sunlight at sea level indicate that some wavelengths of light are reflected by the atmosphere or absorbed in the atmosphere. They do not make it to sea level. Some of the gaps are the result of light absorption by greenhouse gases, notably water vapor and carbon dioxide.

Point to Ponder: Can we collect light from outer space?

We have seen that the atmosphere reduces the amount of sunlight reaching the surface of the earth. It is conceivable to collect sunlight above the surface of the earth and transmit it as an intense beam through the atmosphere. The goal would be to reduce the effective albedo of the atmosphere. Obviously, this type of space-age technology will be expensive. There would also be environmental and safety concerns for anything crossing the path of the beam.

The Japan Aerospace Exploration Agency (JAXA) is developing a multi-billion dollar Space Solar Power System (SSPS) project which is designed to gather solar power from space [JAXA SSPS, 2016]. The project is a large-scale renewable energy project by one of the world's most technologically advanced countries with limited natural energy resources. Japan has been an oil importer for most of its modern history because it lacks oil reserves and the Fukushima nuclear fission disaster reduced public support for nuclear fission energy. The lack of energy resources, concerns about nuclear safety, and a desire to be energy independent has motivated Japanese leaders to develop renewable energy.

SSPS plans to launch and test a large flexible photovoltaic structure with 10 MW power capacity by 2020, followed by a 250 megawatt prototype around 2030. Solar photovoltaic technology is described below. The SSPS expects to beam collected solar energy to the ground using a collection of lasers or microwaves. Microwaves are electromagnetic waves that can pass through atmospheric hindrances such as clouds, rain, dust, and smog.

6.2 Energy Conversion Efficiency and Intermittency

Efficiency is a concept that facilitates the comparison of different energy sources. Estimates of efficiency can help us quantify the total amount of energy that is needed to provide a smaller amount of energy in a useful form. Intermittency is a measure of the availability of the resource. An intermittent energy resource is a resource that is not always available. Here we consider energy conversion efficiency and intermittency.

Energy conversion efficiency is the ratio of the useful energy output by an energy conversion system to the energy input to the system. For example, if a power plant consumes 100 J of energy to generate 50 J of energy, the energy conversion efficiency is 50%. The Q factor introduced in our discussion of nuclear reactors is $Q = \frac{Output\ Power}{Input\ Power} = \frac{Output\ Energy}{Input\ Energy} = \frac{50J}{100J} = 0.5$ for this example. The energy that does not appear as useful output energy is used to operate the system or it is lost as waste energy. Waste energy often appears as heat. The energy balance must include useful energy output as well as work and energy loss.

Another factor that must be considered when comparing different energy systems is the availability of the system. Fossil fuels and nuclear energy can provide energy on demand. Renewable energy sources such as solar and wind are considered intermittent energy sources because they are not always available. Solar energy relies on access to sunlight, which is not available at night and is reduced by atmospheric conditions such as dust, clouds and smog. Wind varies in magnitude and direction depending on weather conditions. One way to mitigate intermittency is to use energy storage systems that harvest energy when the energy source is available, and then release the stored energy when it is needed.

6.3 Passive Solar Energy

Passive solar energy technology integrates building design with environmental factors that enable the capture or exclusion of solar energy. Mechanical devices are not used in passive solar energy applications. We illustrate passive solar energy technology by considering two simple but important examples: the roof overhang and thermal insulation.

6.3.1 Roof Overhang

Sunlight that strikes the surface of an object and causes an increase in temperature of the object is an example of direct solar heat. Direct solar heating can cause an increase in temperature of the interior of buildings with windows. The windows that allow in the most sunlight are facing south in the northern hemisphere and facing north in the southern hemisphere. Figure 6-5 illustrates two seasonal cases. The figure shows that the maximum height of the sun in the sky varies from season to season because of the angle of inclination of the earth's axis of rotation relative to the ecliptic plane. The earth's axis of rotation is tilted from approximately 21.5° to 24.5° relative to a line that is perpendicular to the ecliptic plane.

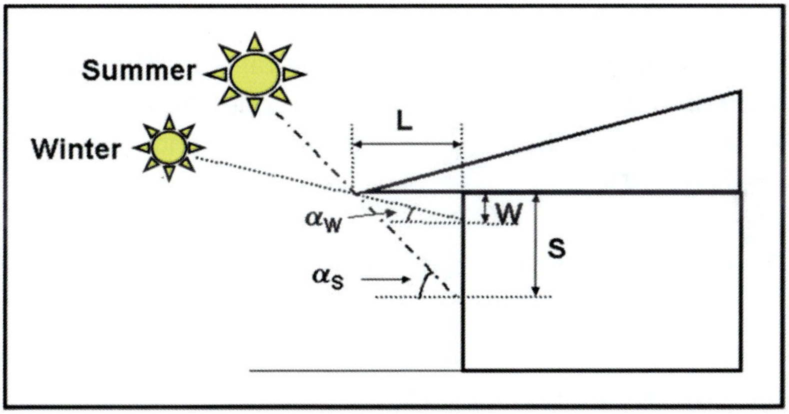

Figure 6-5. Roof Overhang

One way to control direct solar heating of a building with windows is to build a roof overhang that can control the amount of sunlight entering the windows. Figure 6-5 illustrates a roof overhang. The longest shadow from a roof overhang is obtained during the summer, while the shortest shadow from a roof overhang is obtained during the winter. The shadow length can be changed by changing the length of the roof overhang. The impact of the shadow on the window is affected by the size and position of the window on the wall.

Passive solar cooling is achieved when the roof overhang casts a shadow over the windows facing the sun. In this case, the roof overhang is designed to exclude sunlight and its associated energy from the interior of the building. Alternatively, the windows may be tinted with a material that reduces the amount of sunlight entering the building. Another way to achieve passive solar cooling is to combine shading with natural ventilation. For example, the Maui airport uses sea breezes to cool the airport terminal.

Passive solar heating is the capture and conversion of solar energy into thermal energy. The technology for passive solar heating can be as simple as using an outdoor clothesline to dry laundry or designing a building to capture sunlight during the winter. In the latter case, the building should be oriented to collect sunlight during cooler periods. Sunlight may enter the building through properly positioned windows that are not shaded by a roof overhang, or through skylights. The sunlight can heat the interior of the building and it can provide natural light. The use of sunlight for lighting purposes is called daylighting. An open floor plan in the building interior maximizes the effect of daylighting and can substantially reduce lighting costs.

6.3.2 Thermal Conductivity and Insulation

Solar energy may be excluded from the interior of a structure by building walls that have good thermal insulation. The quality of thermal insulation for a wall with the geometry shown in Figure 6-6 can be expressed in terms of thermal conductivity and thermal resistance.

The rate of heat flow through the insulated wall shown in Figure 6-6 depends on wall thickness, the cross-sectional area perpendicular to the direction of heat flow, and the temperature difference between the exposed and inside faces of the wall. The rate of heat flow through the insulated wall depends on a property of the wall called thermal conductivity. Thermal conductivity is a measure of heat flow through a material. Metals have relatively high thermal conductivities and allow heat to flow more rapidly than low thermal conductivity materials like fiberglass or asbestos.

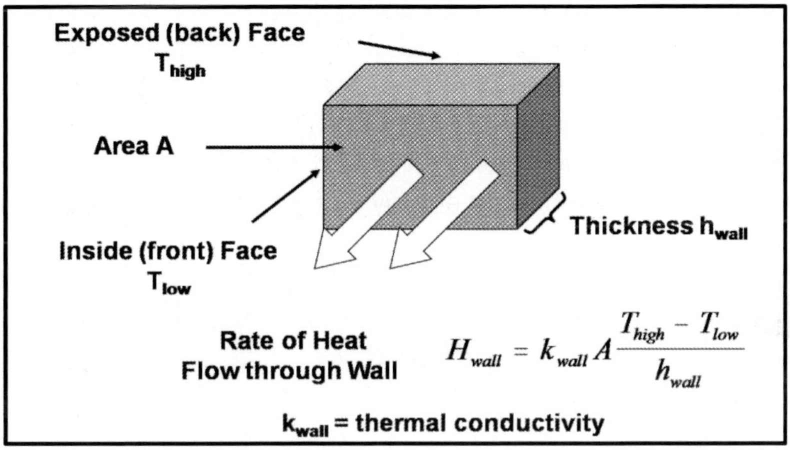

Figure 6-6. Thermal Conductivity of an Insulated Wall

We have considered thermal insulation as an example of passive solar technology. Thermal insulation is also an energy conservation technology. Thermal insulation in walls can keep heat out of a structure during the summer and keep heat in during the winter. Consequently, thermal insulation can reduce the demand for energy to cool a space during the summer and heat a space during the winter. This reduces the demand for energy and makes it possible to conserve, or delay, the use of available energy.

6.4 Active Solar Energy

Active solar energy refers to the design and construction of systems that collect and convert solar energy into other forms of energy such as heat and electrical energy. Active solar energy technologies are typically energy conversion systems that are used to collect and concentrate solar energy, and convert it to a more useful form of energy. We discuss solar heat collectors and solar power plants as illustrations of active solar energy technology.

6.4.1 Solar Heat Collectors

Solar heat collectors capture sunlight and transform radiant energy into heat energy. Figure 6-7 is a diagram of a solar heat collector. Sunlight enters the collector through a window made of a material like glass or plastic. The window is designed to take advantage of the observation that sunlight is electromagnetic radiation with a distribution of frequencies. The window in a solar heat collector is transparent to incident solar radiation and opaque to radiant heat energy in the form of infrared radiation.

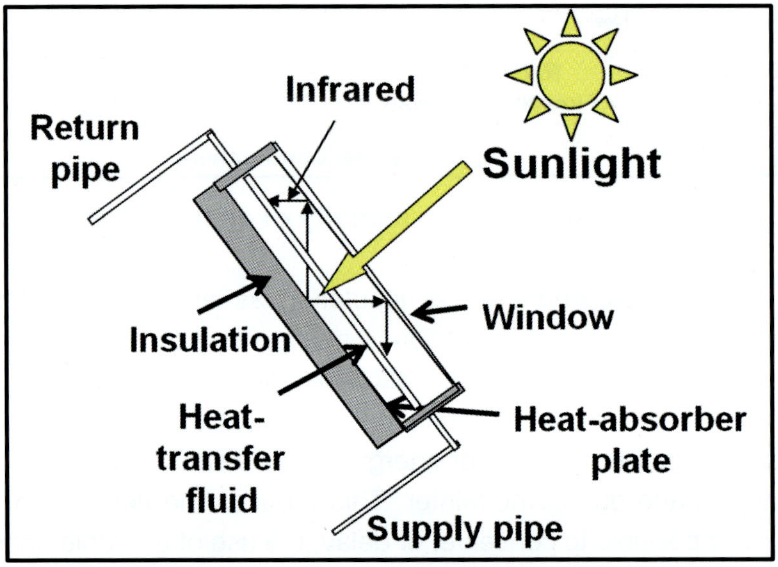

Figure 6-7. Solar Heat Collector

The heat absorber plate in the solar heat collector is a dark surface, such as a blackened copper surface, that can be heated by the absorption of solar energy. The surface of the heat absorber plate emits infrared radiation as it heats up. Sunlight enters through the window, is absorbed by the heat absorber plate, and is reradiated as infrared radiation. Greenhouses work on the same principle: the walls of a greenhouse allow sunlight to enter and then trap reradiated infrared radiation. The window of the solar heat collector is not transparent to infrared radiation so the infrared radiation is trapped in the collector.

The solar heat collector must have a means of transferring collected energy to useful energy. A heat transfer fluid such as water is circulated through the solar heat collector in Figure 6-7 and carries heat away from the solar heat collector for use elsewhere.

The solar heating system sketched in Figure 6-8 uses solar energy to heat a liquid coolant such as water or anti-freeze. The heat exchanger uses heat from the liquid coolant in the primary circulation system to heat water in the secondary circulation system. The control valve in the lower right of the figure allows water to be added to the secondary circulation system. An auxiliary heater in the upper right of the figure is included in the system to supplement the supply of heat from the solar collector. It is a reminder that solar energy collection is an intermittent source. A supplemental energy supply or a solar energy storage system must be included in the design of the heating system to assure continuous availability of heat from the solar heating system.

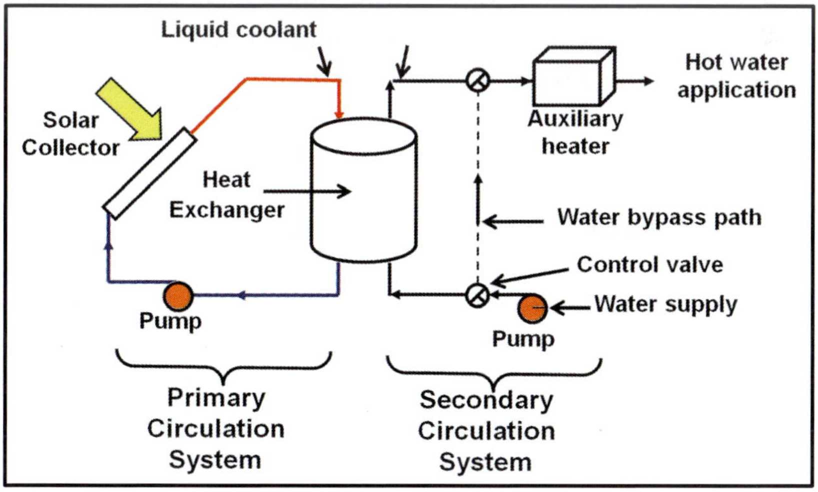

Figure 6-8. Solar Heating System [after Cassedy and Grossman, 1998, page 282]

The energy conversion efficiency of a solar heat collector depends on weather conditions. Energy conversion efficiency is the ratio of the useful energy output by the solar heat collector to the solar energy input to the collector. The energy conversion efficiency depends on the increase in temperature relative to ambient temperature, the intensity of solar radiation, and the quality of thermal insulation. The temperature of a solar heat collector does not increase indefinitely because the window and walls of the solar heat collector cannot prevent energy from escaping by convection and radiation. The loss of energy by convection and radiation causes a decrease in energy conversion efficiency.

6.5 Solar Thermal Power Plants

Solar thermal power plants are becoming a significant component in the energy portfolios of many nations. The plants are designed to provide electrical power on a scale that is comparable to power plants that rely on nuclear or fossil fuel. They use reflective materials like mirrors to concentrate solar power. The solar power tower, the Solar Electric Generating Station in Southern California, and concentrated solar power (CSP) systems using dish concentrators are examples of solar thermal power plants. They are described below.

6.5.1 Solar Power Tower

Figure 6-9 is a sketch of a solar power tower with a heliostat field. The heliostat field is a field of large, sun-tracking mirrors called heliostats arranged in rings around a central receiver tower. The heliostats concentrate sunlight on a receiver at the top of the tower. The solar energy heats a fluid inside the receiver.

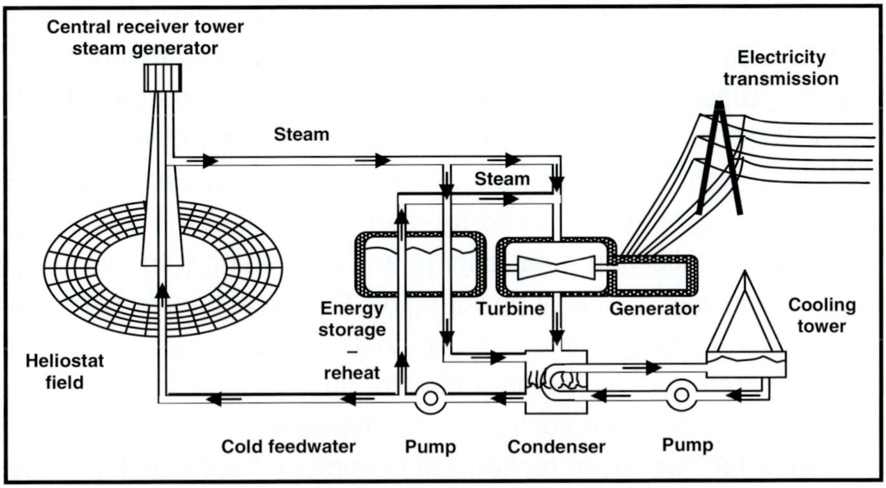

Figure 6-9. Solar Power Tower Schematic [after Kraushaar and Ristinen, 1993, page 172; and Solar Energy Research Institute (now National Renewable Energy Laboratory), Golden, Colorado]

Figure 6-10 is a sketch of the geometry of the sun-tracking mirrors relative to the central receiving station. The heliostats must be able to rotate to optimize the collection of light at the central receiving station. Computers control heliostat orientation. As a ring of heliostats gets farther away from the tower, the separation between the ring and adjacent, concentric rings must increase to avoid shading one ring of mirrors by an adjacent ring.

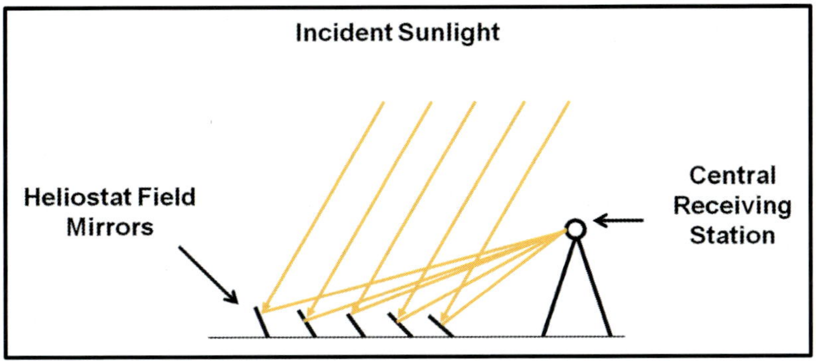

Figure 6-10. Solar Tower Sketch

The first solar power plant based on the solar power tower concept was called Solar One and built in the Mojave Desert near Barstow, California in the 1980's. The solar-thermal power plant at Barstow used 1900 heliostats to reflect sunlight to the receiver at the top of a 300-foot tall tower. The sunlight generates heat to create steam. The steam is used to drive a turbine or it can be stored for later use. Solar One demonstrated the feasibility of collecting solar energy and converting it to electrical energy. Solar One was a 10 MW power plant. The initial heat transfer fluid in Solar One was steam. The Solar One installation was later modified to use molten nitrate salt as the heat transfer fluid. Molten salt can be used to store heat for use when sunlight is not available. The modified installation, called Solar Two, used molten salt to improve heat transfer efficiency and thermal storage for the 10 MW demonstration project. The hot salt mitigated the intermittency of the solar power tower because the heat transfer fluid could be retrieved when needed. A heat exchanger can use energy from the heat transfer fluid to boil water into steam. The steam was used, in turn, to drive a turbine and generate electricity.

6.5.2 Case Study: Ivanpah Solar Power Plant

The Ivanpah Solar Thermal Power Plant is an example of a modern solar thermal power project that uses solar towers. The plant is located in the Ivanpah Dry Lake of the California Mojave Desert near the border with Nevada. Funding for the Ivanpah Project was provided by private enterprises (Brightsource, Google and NRG) and the federal government. The Ivanpah Project received $1.6 billion in Department of Energy loan guarantees in addition to private funding to build the power plant.

The Ivanpah Project in southeastern California is designed to generate a total of 370 MW of power. It consists of three separate solar power tower plants built over a 3-year period that began in late 2010 [Ivanpah, 2012]. The towers in the project are 459 feet tall and are surrounded by approximately 100,000 heliostat mirrors each [Ivanpah Fact Sheet, 2012].

The computer-controlled mirrors concentrate sunlight on a receiver atop each solar tower. The heat is used to produce superheated steam, which drives a turbine and generates electrical power. Water use is minimized by producing steam in a closed-loop system and using air in a dry-cooling process to condense steam to water for reheating. The plant generates enough power to serve 140,000 homes. The tower and heliostat system includes a natural gas-fired boiler as an auxiliary boiler.

Some hazards have been identified at the Ivanpah Concentrated Solar Power Plant. Reflected light from the heliostat field can heat the air around the solar receiver tower. Anything that enters the region, such as birds and bats, are exposed to intense radiant heat. Vapor and heat plumes around the tower can rise hundreds of feet above the tower and create a hazard for low flying aircraft and interfere with communication.

A one year study at Ivanpah from October 2013 to October 2014 found 695 dead birds from 83 species and 32 dead bats [Fairley, 2015]. Biologists adjusted for missing birds and other correction factors and estimated a mortality rate of 3,500 birds. Causes of death included collisions with site structures, burns from solar heating, and natural causes in the desert. The biologists concluded that bird mortality rate would have a low impact and bat mortality rate would have a negligible impact on local, regional and national populations of relevant species. Deterrence effects introduced at Ivanpah are expected to reduce wildlife mortality rates further. Fairley [2015] said that "deterrents include broadcasts of recorded distress calls for species observed on site, LED lighting (which draws fewer of the insects that can attract birds), and spiked surfaces to make perching near the tower undesirable."

Solar thermal power plants can supply electrical power to local utilities during peak demand periods. In Southern California, a peak demand period would be hot summer afternoons when the demand for air conditioning is high. This is a good match for a solar thermal power plant because solar irradiance is high. Peak demand periods also correspond to periods of high pollution. One benefit of a solar thermal power plant is its ability to provide electrical power without emitting fossil fuel pollutants such as nitrous oxide (a component of smog) and carbon dioxide (a greenhouse gas). Other types of solar thermal power plants could also be used to meet this demand.

6.5.3 Solar Electric Generating Systems

A Solar Electric Generating System (SEGS) consists of a large field of solar heat collectors and a conventional power plant. The SEGS plant in Southern California used rows of parabolic trough solar heat collectors. The collectors were sun-tracking reflector panels, or mirrors (Figure 6-11). The sunlight reflected by the panels was concentrated on tubes carrying heat transfer fluid. The fluid was heated and pumped through a series of heat exchangers to produce superheated steam. The steam spun a turbine in a generator to produce electricity.

Figure 6-11. Solar Mirrors near Barstow, California (© Fanchi, 2002)

For extended periods of poor weather, solar power plants must use auxiliary fuels in place of sunlight. A prototype SEGS plant used natural gas as an auxiliary fuel. Goswami, et al. [2000, Section 8.7] reported that, on average, 75% of the energy used by the plant was provided by sunlight, and the remaining 25% was provided by natural gas. They further reported that solar collection efficiencies ranged from 40% to 50%, electrical conversion efficiency was on the order of 40%, and the overall efficiency for solar to electrical conversion was approximately 15%.

The overall efficiency of a solar electric generating system is the product of optical efficiency, thermal conversion efficiency, and thermodynamic efficiency. Optical efficiency is a measure of how much sunlight is reflected into the system. Thermal conversion efficiency is a measure of how much sunlight entering the system is converted to heat in the system. The thermodynamic efficiency is a measure of how much heat in the system is converted to the generation of electricity.

6.5.4 Dish Concentrators

Concentrated solar power (CSP) using dish concentrators are discussed in this section. We focus on dish concentrators and trough concentrators such as those illustrated in Figure 6-12.

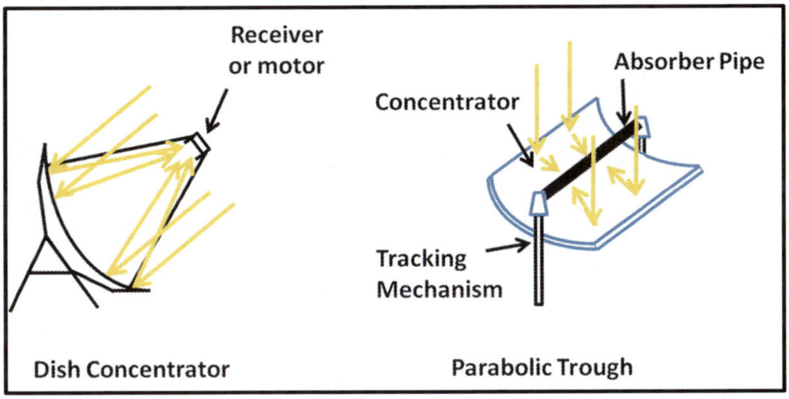

Figure 6-12. Dish Concentrators [after Pitz-Paal, 2008]

Dish concentrator power plants are fields of individual dishes, like a satellite dish people might mount on their house for a television signal, only larger. Each dish is designed to reflect incident solar thermal energy onto a receiver. This allows each dish to function as if it were an individual power tower, with the dish itself acting as a heliostat. One benefit of this form of concentrator is that it is self-contained and suitable for distributed generation application. For example, a small number of dishes can be installed in an area to provide localized power, or a single dish could be used to provide supplemental power to a home or building. These applications are not feasible with a solar power tower since the size of the heliostat field must be able to provide enough power to justify the expense of constructing the project. Similarly, parabolic trough concentrators require too much infrastructure and maintenance to be built on a small scale.

A benefit of the dish concentrator is that each dish may move independently to track the sun. As with heliostat fields associated with solar power towers, the dish could be programmed to move on its own to follow the sun and maximize output during daylight hours. Dish concentrators do not need to be integrated into the electrical grid, so they can be used for distributed generation and can provide small scale power to isolated areas.

Point to Ponder: Are solar thermal power plants becoming more popular in sunny, deserted areas?

One of the major advantages of solar thermal power is that the energy is free. There are, however, several issues to consider that impact cost and social acceptability. The free energy from the sun has to be collected and transformed into commercially useful energy. Solar thermal power plants presently require a significant amount of land and produce a fraction of the power associated with a fossil fuel fired power plant. This means that solar thermal power plants will require large areas of land to provide enough power to replace a fossil fuel power plant. Such plants may also be considered eyesores by some people. The technology of maintaining the collectors, and collecting, transforming, and transmitting solar energy is still relatively expensive. The construction and operation of facilities like Ivanpah Solar Thermal Power System are demonstrating that solar thermal power facilities are commercially viable.

6.5.5 Case Study: Spanish Solar Power

The southern coast of Spain is a sun-drenched tourist attraction. Spain has recognized that this natural resource can also be used to provide energy for itself and the members of the European Union. Consequently, Spain has placed a premium on solar power.

According to Solar PACES, an international organization that coordinates solar power projects under the auspices of the International Energy Agency, Spain first created a solar power premium payment in 2002. This premium came in the form of a feed-in tariff that guaranteed an above-market payment to renewable energy producers from utilities for electricity generated. The tariff was designed to encourage utilities to adopt solar power by granting a government subsidy for solar-generated electricity. The premium was increased by 50% from its original level and guaranteed for 25 years in 2004 in an effort to generate more investment. The tariff led to large investment in solar technology [IEA Solar PACES, 2012]. The feed-in tariff law prevented utilities from passing increased electricity costs on to consumers. This created a large deficit for the utilities which was eventually backed by the Spanish government. As a consequence, the financial burden was passed on to Spanish taxpayers. Concern about national debt motivated the Spanish government in 2012 to cease granting new subsidies for solar power [IER Spain, 2012].

Among the projects the tariff helped launch is PS10, a solar power tower plant near Seville, Spain, that has a capacity of 11 MW and is considered the first commercially viable solar power tower plant. Adjacent to PS10 is PS20, a solar power tower plant built after PS10 was completed. PS20 has a capacity of 20 MW. It was the largest solar power tower plant in the world when it was completed. PS10 and PS20 are both projects of Abengoa Solar, a Spanish company that operates many of the solar power plants in Spain [Abengoa Solar, 2016].

Abengoa is operating a series of connected solar power plants called the Solucar Complex. The Complex is located near Sanlúcar la Mayor, Seville, Spain. Three different solar technologies (tower, parabolic trough and PV) in commercial operation, pilot projects, and laboratories are gathered together at the Complex. Along with the PS10 and PS20 plants, three solar trough plants with 50 MW capacity each, a concentrated solar power (CSP) plant using dish technology with 80 MW capacity, and solar photovoltaic projects are part of the Complex. An image of the area from orbit is shown in Figure 6-13.

Figure 6-13. NASA Image of Seville, Spain, including PS10 and PS20 [NASA Earth Observatory, 2009]

Abengoa developed and is co-owner of the Solana Solar Generating Plant approximately 70 miles southwest of Phoenix, Arizona. It uses parabolic troughs to collect solar energy and molten salt to store energy for up to 6 hours. The plant's solar field covers 1,920 acres, provides gross turbine capacity of 280 MW power, net turbine capacity of 250 MW, and generates enough electricity to power 70,000 homes. The plant became operational in 2013 [NREL Solana, 2015] and has a 30-year power supply contract with Arizona Public Service.

6.6 Solar Electric Technology

Solar electric technologies are designed to convert light from the sun directly into electrical energy. Two of the most important solar electric processes are the photoelectric effect and photovoltaics. These processes are introduced here.

6.6.1 Photoelectric Effect

Heinrich Hertz discovered in 1887 that electrons could be ejected from a metal exposed to light, but the effect depended on the frequency of the light. The effect was called the photoelectric effect. Albert Einstein used the concept of a quantum of energy to explain the photoelectric effect in 1905, the same year he published his special theory of relativity.

The view of light propagation at the beginning of the 20th century was that light propagates as electromagnetic waves. Einstein used the then new quantum theory to postulate that light also propagates as a particle. The particle of light is now called the photon, and is a packet, or quantum, of energy. The energy of the photon is proportional to the frequency of light. Einstein realized that a collision between a photon and an electron could transfer enough energy from the photon to the electron to eject the electron from the metal (Figure 6-14).

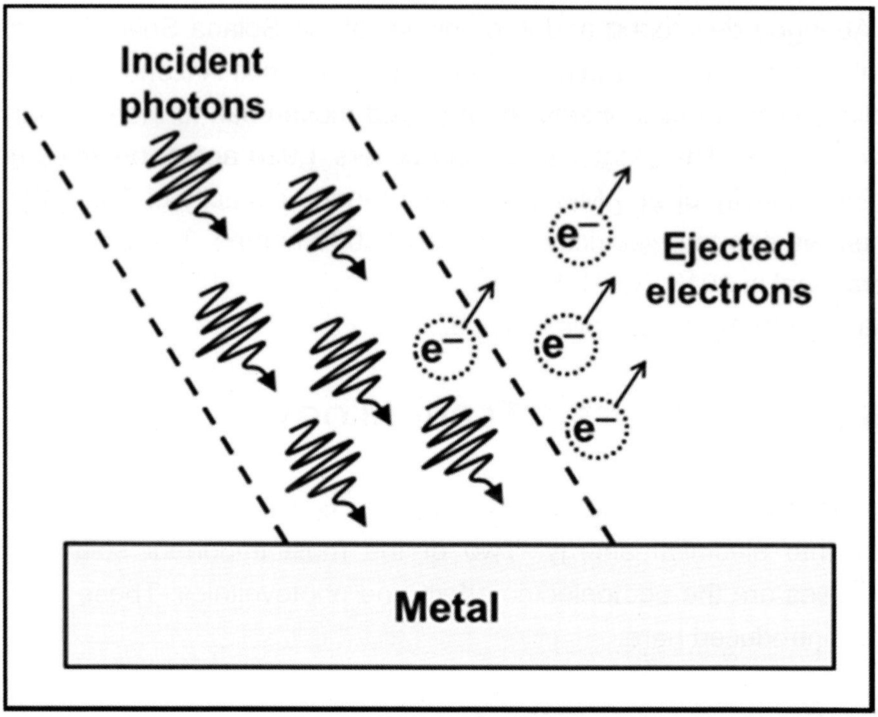

Figure 6-14. Photoelectric Effect

The photoelectric effect requires that the photon have enough energy (a high enough frequency) to overcome the work function of the metal, where the work function is the smallest energy (smallest frequency) needed to extract an electron from the metal. The work function depends on the type of material and the condition of its surface. An electron ejected from a metal because of a collision with a photon is called a photoelectron, and the photoelectric effect is the mechanism of ejecting an electron with a suitably energetic photon. If we note that an electrical current is the movement of electrons, the ejected electrons can be directed to move through a circuit as an electrical current.

As a historical note, Einstein received the Nobel Prize for his 1905 work on the photoelectric effect, not for his work on relativity.

6.6.2 Photovoltaics

Photovoltaics is an application of the photoelectric effect. We can describe photovoltaics as the use of light to generate electrical current. A photovoltaic (PV) cell, or photocell, can be made using semiconductors. A semiconductor is a material, such as a crystalline solid, that can support the movement of electrons. There is more resistance to the flow of electrons in insulators like rubber than in semiconductors, but there is more resistance to the flow of electrons in semiconductors than metallic conductors.

Silicon is often used as a semiconducting material. Some of the silicon atoms can be replaced by other atoms known as dopants in the crystalline structure to form specific types of semiconductors. For example, if we use phosphorous as a dopant in a silicon crystal, the phosphorous atom provides an extra negatively charged electron to the valence (outer) layer of electrons in the crystal and the semiconductor is known as an n-type semiconductor. By contrast, if we use boron as the dopant in a silicon crystal, the boron atom provides one less electron to the valence (outer) band of electrons than would have been provided by a silicon atom, and the semiconductor is known as a p-type semiconductor. The spot where the electron is missing in the p-type semiconductor is known as a "hole".

A common PV cell is made by placing an n-type semiconductor adjacent to a p-type semiconductor (Figure 6-15). The extra electron in the n-type semiconductor can be mobilized when sunlight shines on the semiconductor. Electrons flow from the n-type semiconductor through a circuit to the p-type semiconductor. If we place a load in the circuit, such as a lightbulb, the moving electrons are an electrical current.

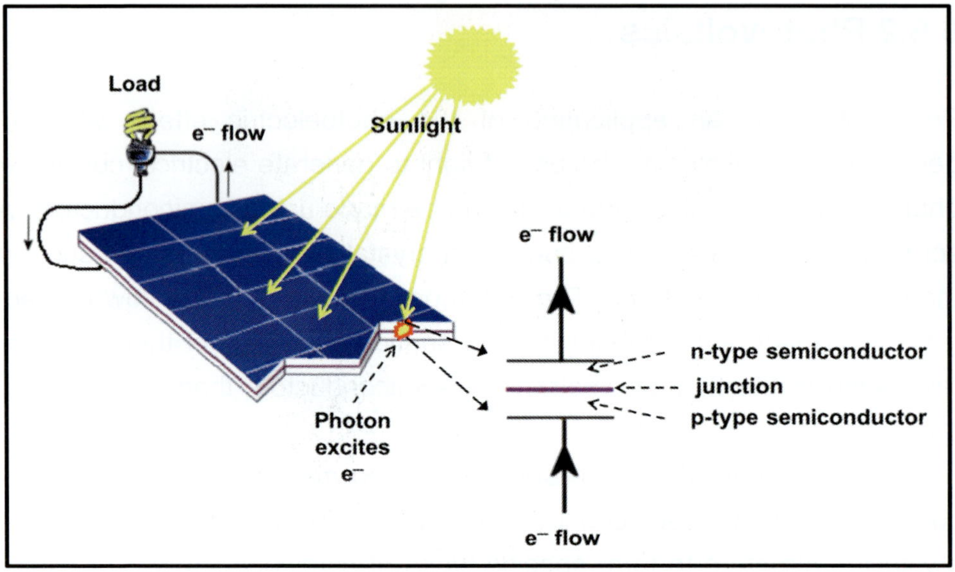

Figure 6-15. Photovoltaic Effect

PV cells are not sources of energy and they do not store energy. PV cells transform sunlight into electrical energy. When sunlight is removed, the PV cell will stop producing electricity. If a PV cell is used to produce electricity, an additional system is also needed to provide energy when light is not available. The additional system can be an energy storage system that is charged by sunlight transformed into electrical energy and then stored, or it can be a supplemental energy supply provided by another source.

Point to Ponder: What are some current uses of photovoltaics?

There are many applications of photovoltaics. They include solar powered calculators, solar powered camping equipment, solar powered satellites, the International Space Station, and solar powered cars. A solar powered traffic system for helping enforce speed limit laws is shown in Figure 6-16 and a solar powered car is shown in Figure 6-17.

Figure 6-16. Solar Powered Traffic Control, Golden, Colorado (© Fanchi, 2002)

Figure 6-17. Experimental Solar Powered Car, Fort Worth, Texas (© Fanchi, 2012)

6.6.3 Transparent Solar Cells

Conventional photovoltaic (PV) cells discussed so far are designed to capture as much light as possible and are opaque. They require a substantial footprint for harvesting sunlight. Researchers are developing alternative solar cells that allow visible light to pass through the cell while capturing ultraviolet and infrared light [Lendino, 2015]. The new cell has been called transparent solar cell (Figure 6-18) or transparent luminescent solar concentrator (TLSC). According to Lendino [2015], the TLSC "consists of organic salts that absorb specific non-visible wavelengths of ultraviolet and infrared light, which they then luminesce (glow) as another wavelength of infrared light (also non-visible). This emitted infrared light is guided to the edge of plastic, where thin strips of conventional photovoltaic solar cell convert it into electricity."

Figure 6-18. Transparent Solar Cell (Courtesy Richard Lunt, 2016; Photo by Yimu Zhao)

Figure 6-19 compares the absorption characteristics of conventional PV cells and transparent solar cells [Stauffer, 2014]. The visible spectrum is approximately between 400 nm and 700 nm wavelengths. Light absorption curves for the PV cells are sketched over the solar spectral irradiance curves shown in Figure 6-4. Transparent solar cells do not absorb as much solar energy as conventional PV cells. Researchers such as Richard Lunt at Michigan State University and Vladimir Bulović at Massachusetts Institute of Technology believe that transparent solar cells are capable of achieving 10% power conversion efficiency and allow as much as 90% of the light in the visible range to pass through the cell so the transparent solar cells appear transparent to the human eye [Ubiquitous Energy, 2016].

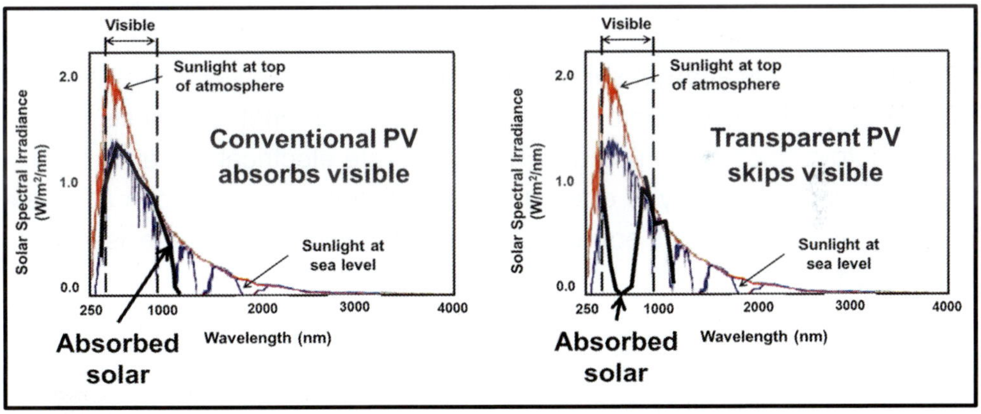

Figure 6-19. Absorption of Light by Solar PV Cells

Transparent solar cells can be used to replace other transparent materials such as glass. This creates an opportunity to use transparent solar cells in such applications as windows and cell phone screens. The integration of transparent solar cells into the economy depends on a range of factors such as transparency, power conversion efficiency, and cost.

6.7 Activities

True-False

Specify if each of the following statements is True (T) or False (F).

1. A solar heat collector is a passive solar energy device that captures sunlight and transforms radiant energy into heat energy.
2. Most of the energy that is radiated by the sun reaches the surface of the earth.
3. Passive solar technology integrates building design with environmental factors to enable the capture or exclusion of solar energy.
4. The efficiency of solar heat collectors depends on weather conditions.
5. Solar energy is inexhaustible.
6. A supplemental energy supply is often needed to accompany solar heating systems.
7. A heliostat is a large, sun-tracking mirror.
8. Biological conversion of solar energy relies on photosynthesis.
9. Photovoltaics is the use of light to generate electrical current.
10. Renewable energy is energy that is obtained from sources at a rate that is less than or equal to the rate at which the source is replenished.

Questions

1. What is a heliostat?
2. What is the difference between active and passive solar energy?
3. Which of the following systems use active solar energy?
   - Roof overhang
   - Solar heat collector
   - Shade tree
   - Solar tower
   - Photovoltaic cell
4. What is the photoelectric effect?

5. How does a solar electric generating system work?
6. What causes sunlight to be reflected into space by the earth's atmosphere?
7. If a power plant consumes 250 J of energy and generates 150 J of energy, what is the energy conversion efficiency of that plant?
8. What are the benefits of dish concentrators as a form of Concentrated Solar Power?
9. What kind of power plants are the PS10 and PS20 in Spain?
10. List the layers of a star, from innermost to outermost.

CHAPTER 7

RENEWABLE ENERGY – WIND ENERGY

After reading this chapter, you should be able to...

- Describe the history of wind power
- Define and differentiate between vertical axis and horizontal axis wind turbines
- Describe a modern wind turbine and how it works
- Describe wind farms and the logistics involved in constructing them
    - *P2P: Would you want a wind farm in your backyard?*
- Explain how Europe is pioneering the development of offshore wind power
- Discuss the potential for wind power to meet global energy needs
- Describe the environmental impact of wind power

Wind is air in motion. Differences in atmospheric pressure from one location to another cause air to move. The kinetic energy of moving air, or wind, is an indirect form of solar energy and is considered renewable. Wind turbines harness the kinetic energy of wind and convert the mechanical energy of a rotating blade into electrical energy from a generator. The objective of this chapter is to discuss the use of wind as a source of energy for generating useful power. We begin our discussion of wind energy by reviewing the history of wind power.

7.1 History of Wind Power

Wind has been used as an energy source for thousands of years. Historical applications include sails for ship propulsion and windmills for grinding grain and pumping water. Wind is still used today as a source of power for sailing vessels, parasailing, and wind surfing.

The earliest known applications of wind as an energy source come from Persia [Manwell, et al., 2002]. Around 900 A.D., wind was used to drive early vertical axis windmills. Modern wind turbines are classified as either horizontal axis turbines or vertical axis turbines. A vertical axis turbine has blades that rotate around a vertical axis and its visual appearance has been likened to an eggbeater. A horizontal axis turbine has blades that rotate around a horizontal axis (see Figure 7-1). Horizontal axis turbines are the most common turbines in use today. Early vertical axis windmills had a simple design and were particularly susceptible to damage in high winds.

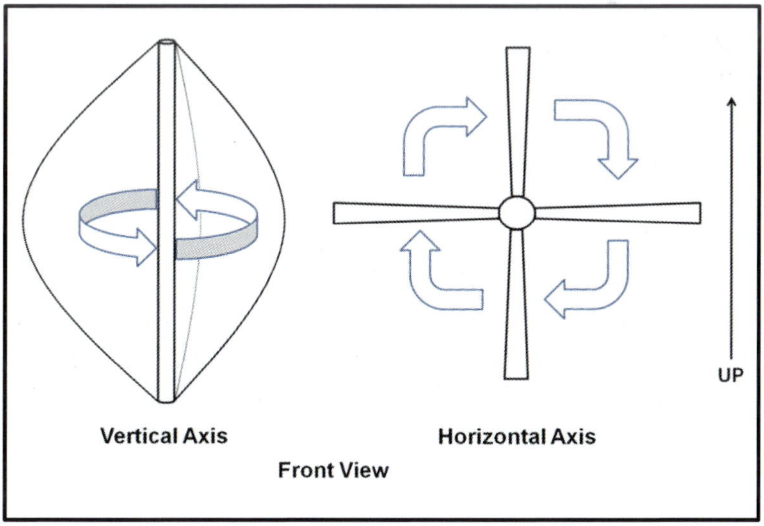

Figure 7-1. Comparison of Vertical Axis and Horizontal Axis Windmills

The first known use of horizontal axis windmills, a precursor to the technology seen predominantly today, appeared in Europe in the Middle Ages. Horizontal axis windmills could be used for a variety of simple tasks such as pumping water, grinding grain, and sawing wood. Early European windmills typically had four blades, as shown by the Dutch windmill in Figure 7-2.

Wind power was a significant contributor to European energy for centuries prior to the Industrial Revolution. As time passed, use of wind as an energy source lost favor because it was difficult to distribute and it was not always available when it was needed. Eventually coal entered the energy mix because coal had several advantages: it could be used when and where it was most needed, and helped reduce dependence on wood as a combustible fuel, which was deforesting some areas. To a lesser extent, water also overtook wind as a source of power because water could be moved through waterways such as canals and stored for more timely use in containers such as ponds.

Figure 7-2. Dutch Windmill (©Fanchi, 2003)

Blades on early European windmills were usually designed to rotate together in the wind. This design made the mill susceptible to damage because of the large number of moving parts. European windmill technology eventually matured to the point that only a small part of the mill, the rotor, turned to move the blades and adjust to wind direction without affecting the adjoining structure. Controls also were implemented to allow the blades to rotate on their own and reduce the amount of operational supervision. Scientific testing of windmills led to a more sophisticated knowledge of the forces at work in a wind energy system. Many of the advances made in Europe have been integrated into modern wind turbine designs.

As use of windmills declined in Europe, a new form of windmill came into widespread use in the Western United States (Figure 7-3). This type of multi-bladed windmill, or 'fan mill', was used primarily for pumping water in the more arid regions of the United States. The water was needed for irrigation, supplying livestock with water, and providing water for steam-driven locomotives. A simple but effective regulating system was developed for these windmills that allowed them to function for long periods unattended, and set the stage for automatic control systems that are integral parts of modern turbines.

Figure 7-3. Windmill in the Western United States (© Fanchi, 2009)

The first wind turbines for electricity generation appeared in the late 19th century in the United States. Following the development of the electric generator, it was only logical that someone would try to turn generator shafts using wind power. Early wind turbines were built to provide electricity for residential areas on a very small scale, usually one turbine for one home. The first larger-scale wind turbine was built by Marcellus Jacobs in the early 1920's. By the 1940s, the Jacobs' wind turbine had three rotor blades with airfoil shapes and resembled wind turbines installed in the 1980s and still in use today (Figure 7-4). The Jacobs wind turbine could provide approximately 2.5 kW power and could be integrated into an electrical grid to provide power for distribution to many consumers.

Figure 7-4. Wind Turbine in Tehachapi, California (© Fanchi, 2012)

A decision by the United States Rural Electrification Administration to expand the central electric grid in the 1930's may have delayed the adoption of an expanded role for wind energy in the United States energy mix. With that decision, small scale wind power lost much of its appeal since fewer areas were isolated from the electrical grid.

A large number of small-scale wind turbines with electrical production ranging from 20 kW to roughly 60 kW were developed in Denmark. Then, shortly after World War II, Johannes Juul erected the 200 kW Gedser turbine in southeastern Denmark. This turbine had several major advances built into it, including an aerodynamic design that enabled greater control of power output by varying the angle of the blade in response to changes in wind speed, and a generator that could be connected directly to the electrical grid. Also around this time, in the 1950's, German Ulrich Hütter made advances in the application of aerodynamic principles to wind turbine design. Many of Hütter's ideas led to concepts in use today. More discussion is provided by Kühn [2008] and the American Wind Energy Association [AWEA, 2009].

The use of wind energy for electrical energy began in the late 19th century when windmills were being used on a very small scale, such as for individual houses or farms. Over the course of the 20th century, wind energy received increasing attention as a potential power source, but the technology did not exist for large scale use. By the middle of the century, wind energy was still limited to small scale use despite attempts to develop the technology. Even in 1980, the total power from wind turbines for electricity generation was less than 1000 MW, which had a small impact on the energy mix. Over the next 20 years, wind technology advanced significantly. Modern turbine technology appeared between 2000 and 2005. Today, a single wind turbine can produce up to 6 MW electrical power, and researchers are seeking to increase power output up to 10 MW using new technology, such as superconducting generators [Matthews, 2009].

7.2 Wind Turbine

A schematic of a typical modern wind turbine is shown in Figure 7-5. A typical horizontal axis turbine consists of a rotor with three blades attached to a machine cabin set atop a post that is mounted on a foundation block. The machine cabin contains a generator attached to the wind turbine. The rotor blades can rotate in the vertical plane and the machine cabin can rotate in the horizontal plane.

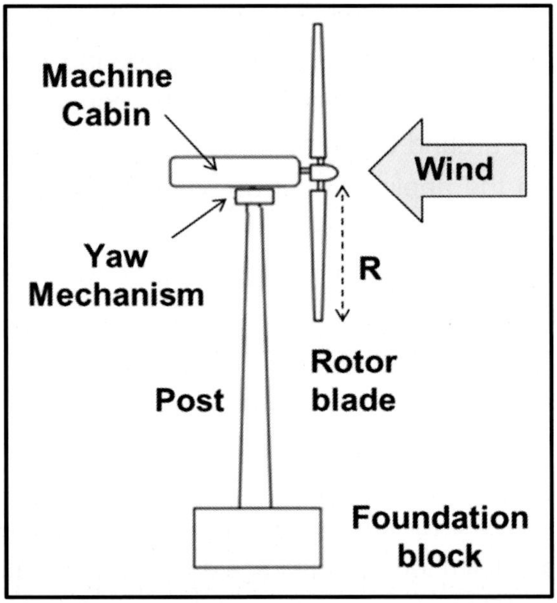

Figure 7-5. Schematic of a Wind Turbine

The machine cabin is known as the nacelle. Moving air rotates blades attached to a drive shaft and gearbox in the machine cabin (Figure 7-6). Another drive shaft is connected to a turbine in an electrical generator. The rotational energy of the rotating blades is transformed into electrical energy in the generator. Electricity is transmitted through a line in the post that connects each wind turbine to the electric grid.

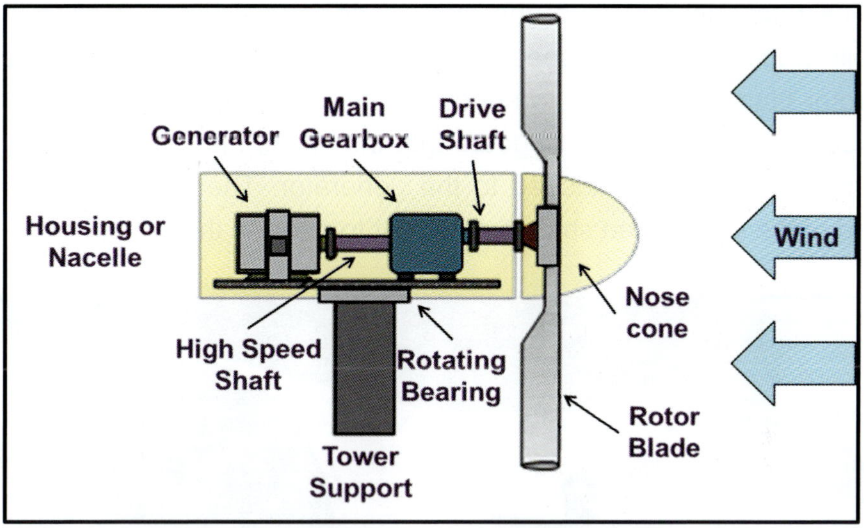

Figure 7-6. A Modern Wind Turbine Blade in Lamar, Colorado (© Fanchi, 2009)

Figure 7-7 illustrates the size of a rotor blade. Most modern horizontal axis turbines have three rotor blades instead of two. Rotor blades attached to a generator shaft, or rotor, make up a rigid body with a particular moment of inertia. Rotational properties of a rigid body, such as angular momentum and torque, depend on moment of inertia. The moment of inertia of the wind turbine depends on the number of rotor blades and their orientation. Wind turbines with two rotor blades have a higher moment of inertia when the blades are vertical than when they are horizontal. The difference in moment of inertia between the horizontal and vertical configuration of two blades introduces a mechanical imbalance that can increase wear on the system. On the other hand, the use of three equally spaced blades adds the cost of another blade but allows a symmetric placement of blades that makes it easier to balance the blades as they rotate. The improved stability of turbines with three rotor blades increases wind turbine reliability and reduces maintenance costs.

The rotational energy of the rotor blades turns the drive shaft in the nacelle (Figure 7-6). The speed of rotation of the drive shaft attached to the rotor blades is too low to effectively generate electricity, so the main gearbox contains a series of gears that increases the rotational speed of the high speed shaft attached to the generator. The resulting rotational speed of the high speed shaft is sufficient to operate the generator.

Figure 7-7. A Modern Wind Turbine Blade in Lamar, Colorado (© Fanchi, 2009)

The yaw mechanism shown in Figure 7-5 is a rotating bearing attached to the post just below the machine cabin. The purpose of the yaw orientation system is to align the rotor shaft with wind direction. Figure 7-8 illustrates the difference between yaw, pitch and roll in aeronautical terms. Yaw refers to rotation of the nose in the horizontal plane around the vertical axis. Rotation of the airplane around the B to B' line is called pitch, and rotation of the wings around the A to A' line is called roll.

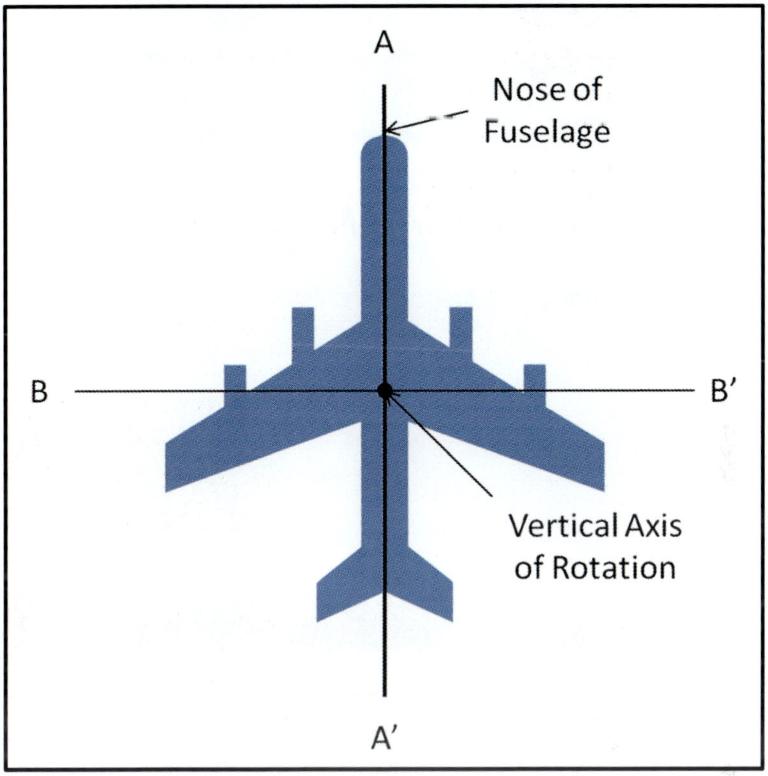

Figure 7-8. Aeronautical Orientation System

Modern wind turbines are several hundred feet tall. For example, a wind turbine that generates 1.6 MW of electrical power is approximately 113 meters (370 feet) tall from its base to the tip of the rotor blade. The storage tanks in Figure 7-9 illustrate the scale of a modern wind turbine and demonstrate that wind turbines can be erected on existing industrial properties. For example, wind turbines have been erected near fossil fuel-fired power plants, along highways, on ranchland, and in shallow waters offshore.

Figure 7-9. Modern Wind Turbine near Rotterdam, Holland (© Fanchi, 2003)

If the speed of rotation of the tip of the rotor blade is fast enough, it can be lethal to flying animals such as birds or bats that enter the fan area of the rotor blade. In addition to direct contact between wildlife and rotating rotors, air pressure disturbances between the front and back of the wind turbine can harm fragile wildlife. Environmental hazards can be minimized by selecting locations for wind turbines that avoid migration patterns and nesting areas.

7.2.1 Turbine Power Output

Wind power can be maximized when wind direction is perpendicular to the plane of rotation of the rotor blades. A change in wind direction can put stress on wind turbines.

Electrical power output from a wind turbine is affected by the reliability of the wind turbine. Wind speed is seldom constant; it can vary from still to tornado or hurricane speed. Wind turbines have three regions of power output, illustrated in Figure 7-10, that depend on wind speed. Region I occurs when the wind speed is low and little or no power is produced. Region II occurs with intermediate wind speed allowing the turbine to begin producing meaningful power output. Region III occurs when the wind speed is high and the turbine reaches maximum power. Eventually, if the wind speed is great enough, the turbine will reach a cutout point at which time it shuts down entirely. This is generally reserved for events, such as hurricanes or tornadoes, when the wind speed is dangerous to the equipment.

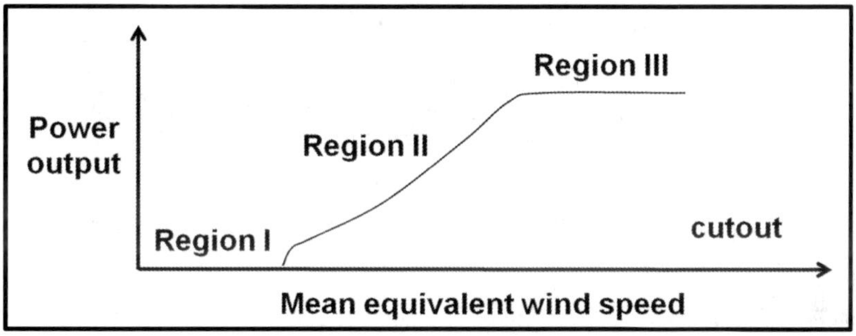

Figure 7-10. Wind Turbine Power Output Regions

The peak power of a wind turbine is its maximum power output at optimum wind speed. Since wind turbines do not usually operate at peak power, their performance is measured using capacity factor, which is the average output power of a wind turbine divided by its peak power.

One factor that affects the efficiency of a wind turbine is the efficiency of converting mechanical energy of the rotor blade into electrical energy. In 1928, Albert Betz showed that the maximum percentage of wind power that can be extracted is approximately 59.3% of the power in the wind. The rate of rotation of the rotor blade depends on wind speed and size of the blades. If the wind speed is too great, the rotor blade can turn too fast and damage the system.

7.3 WIND FARMS

A wind farm or wind park is a collection or array of wind turbines. The spacing between two wind turbines depends on whether the wind turbines are aligned parallel to the direction of wind flow (downwind spacing) or perpendicular to the direction of flow (crosswind spacing). Figure 7-11 illustrates crosswind and downwind spacing.

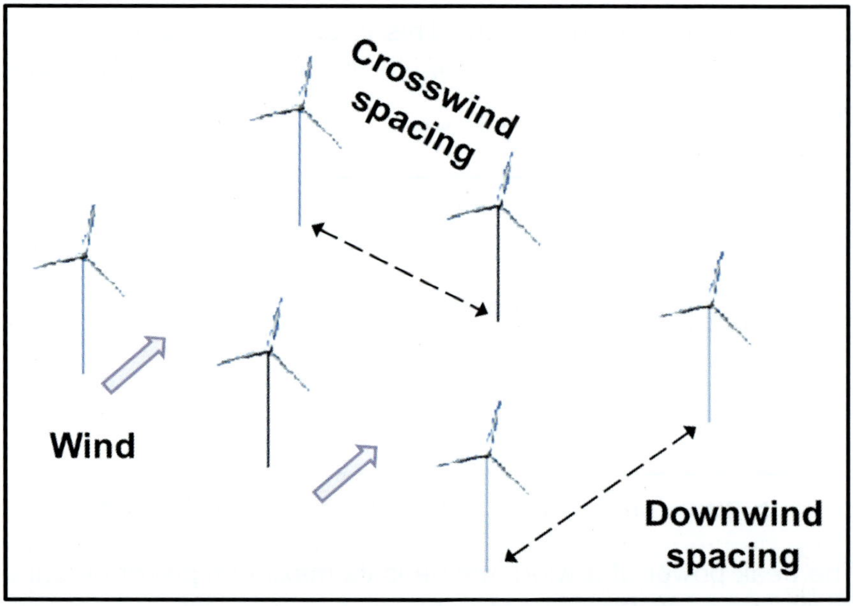

Figure 7-11. Wind Turbine Spacing

The areal extent of the wind farm depends on the radius R of the rotor blades and the effective radius R_{eff} of the wind turbine (Figure 7-12). A wind turbine must have enough space around the post to allow the fan of the rotor blade to face in any direction. The minimum spacing between the posts of two equivalent wind turbines must be $2R_{eff}$ to avoid collisions between rotor blades. Two other factors are important. Turbulence and changes in wind speed tend to cause more wear and tear on downwind turbines. The loss of wind energy from the upstream row of turbines to the downwind row reduces the power generated by downwind turbines. Spacing is a compromise between minimizing the area occupied by the wind farm and the need to reduce wind shadowing, that is, energy loss due to turbulence from upstream turbines.

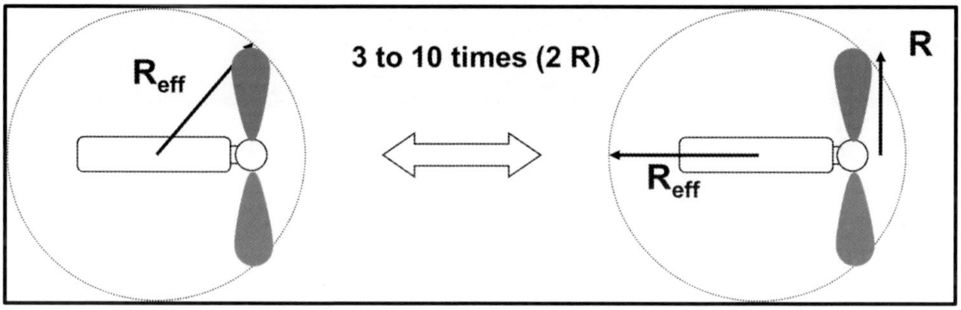

Figure 7-12. Wind Turbine Spacing

The distance between wind turbine posts is designed to minimize wind turbulence between wind turbines and enable the restoration of the wind stream to its original undisturbed state after it passes by one turbine on its way to the next turbine. Wind turbine spacing, or the distance between turbines, is an important factor in determining the surface area, or footprint, needed by a wind farm. Wind turbine spacing should be approximately three to five times rotor diameter $2R$ in the crosswind direction and eight to ten times rotor diameter in the downwind direction. Figure 7-13 shows the spacing of wind turbines in a Texas wind farm near Sweetwater, Texas. The train on the left side of the figure shows the scale of the wind turbines.

Figure 7-13. Wind Turbine Spacing on a Texas Wind Farm (© Fanchi, 2009)

7.3.1 Offshore Wind Farms

Current technology favors onshore wind farms as the most economical form of wind farm. There are drawbacks to onshore wind farms that can be mitigated by construction of offshore wind farms. For example, the sight of a 200-300 foot tall wind turbine does not appeal to everyone, and wind turbines do make some noise. Offshore wind turbines can be placed beyond visible and audible range of many populated areas.

Several countries around the world do not have good conditions for onshore wind farms but do have good conditions along nearby shoreline. Some conditions that would hinder the development of onshore wind farms include unreliable onshore wind patterns or a population density so large that access to land is too costly. Wind currents tend to be stronger and smoother offshore than on land. Onshore topographic features like mountains, valleys, and skylines can disrupt the smooth flow of air and introduce turbulence. Offshore wind farms can be built to provide power to densely populated coastal areas where the cost to access land is prohibitive.

European countries with coastlines began constructing offshore wind turbines in the early 1990's, led by Denmark and the United Kingdom. The wind resource map in Figure 7-14 shows that the Gulf Coast, Atlantic coast, Pacific coast, and Great Lakes of the United States are areas where wind conditions are better offshore than onshore. Offshore wind conditions give these areas access to wind power.

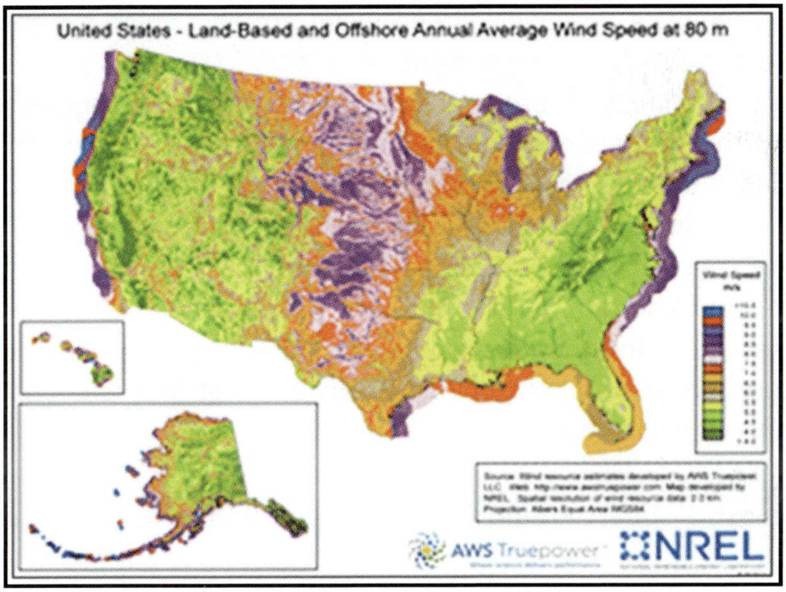

Figure 7-14. Wind Patterns in the United States [NREL Wind, 2012]

A concern that is particularly significant to the United States is the cost of transporting large, heavy components to isolated parts of the country. Transportation costs for development of offshore wind farms can be lower because marine shipping and handling equipment is better suited for the heavy components needed for wind farms.

Offshore wind farms can have many benefits compared to onshore wind farms, but there are a number of drawbacks. One issue is the difficulty of creating an infrastructure system to get the electricity from the turbine to shore without significant loss of power. This concern also exists with onshore wind farms located a significant distance from the end user. In that case, the cost of transmitting power over long distances is comparable for both onshore and offshore power generation by wind farms.

Another drawback of offshore wind farms is increased maintenance costs. Salt in seawater has a corrosive effect that increases water damage to equipment. The location of turbines offshore increases the difficulty of accessing them. Offshore turbines are subjected to more severe weather conditions, including strong wave activity, which increases the risk of serious damage to equipment.

Countries with sea coasts adjacent to shallow sea shelves include Denmark, the Netherlands and the United Kingdom. Shallow water can extend more than 12 miles (about 19 km) from shore. These countries have a geographic advantage for placement of offshore wind farms because the cost of building an offshore wind farm in shallow water is less than the cost of building an offshore wind farm in deep water. Two major difficulties of constructing an offshore wind farm in deep water are installation and control of wind turbines to ensure that they are adequately anchored to the sea floor.

Point to Ponder: Would you want a wind farm in your backyard?

The first offshore wind farm to be built in the United States was proposed for the Nantucket Sound area of Massachusetts. Cape Wind [2016] was approved by the U.S. Department of the Interior in April 2010. The farm will contain 130 wind turbines with a capacity of 3.6 MW each. The farm is located in Federal waters off the coast of Cape Cod, Massachusetts in a region called Horseshoe Shoal. The Shoal has an average wind speed of approximately 20 miles per hour at the rotor blade elevation. Wind turbines should be able to produce power about 88% of the time. The site is located in shallow water inside Nantucket Sound and is protected from Atlantic storms. Power cables from Cape Wind will be submerged offshore and buried onshore. The power will interconnect with a power substation capable of handling Cape Wind power. Each turbine will be spaced far enough apart to allow easy navigation of the surrounding waters by fishing craft and shallow water boats.

So what is the problem? The largest concern exhibited by the local population is that the aesthetic impact will drastically decrease the value of beachfront homes. A related concern is that the drop in home values would lead to a drop in tax revenue gathered from property taxes. CapeWind.org answered these concerns by pointing to a similar wind farm constructed at Tunoe Knob, Denmark in 1995. The concerns of Tunoe Knob residents were similar to those held by Massachusetts residents. Tunoe Knob home values remained steady years after construction of the wind farm, and the presence of the wind farm is no longer a major topic of discussion.

The Cape Wind project is expected to create jobs in the area and have a positive impact on the local economy. Opponents of the project argue that it could seriously damage the natural beauty of Nantucket Sound, a popular tourist destination, which in turn could reduce tourism, tourism-related jobs, and harm the local economy.

Projects such as Cape Wind can provide clean, renewable energy, create energy-related jobs, and increase energy independence. They can also have social and environmental consequences that need to be considered.

7.3.2 Advances in Offshore Wind Technology

Offshore wind turbines provide access to steady sea winds and do not require onshore surface area in countries with high population densities. Figure 7-15 shows the progression of wind turbines from onshore to deeper offshore waters. Shallow water offshore wind turbines are designed for up to 30 m of water depth. Some offshore wind turbines are designed to float and maneuver in accordance with wind and wave patterns, while others are designed to be moored to the sea floor. Seabed mounted wind turbines are being developed for transitional depths ranging from 30 m to 60 m. Floating offshore wind turbines can be used in waters that are too deep for conventional turbines, and are being developed for water depths greater than 60 m. A floating turbine can be designed to align with wind direction.

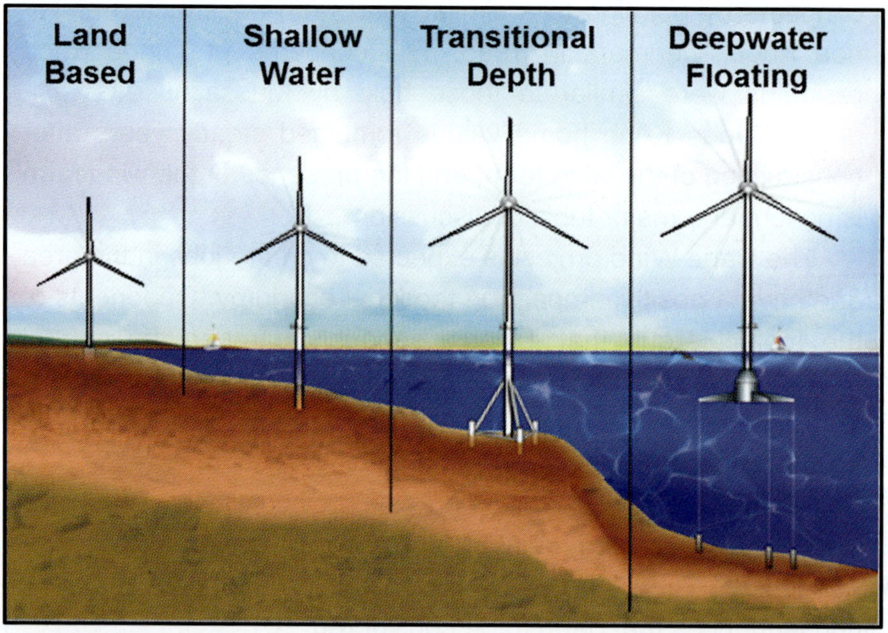

Figure 7-15. Offshore Wind Turbines [after DoE Wind, 2012]

Wind turbines must be sturdy enough to withstand bad weather conditions, and there must be a way to transmit power to the electrical grid. Many European nations have long shorelines and are pioneering the development of offshore wind turbine technology. Their efforts are discussed below.

7.4 Case Study: European Wind Power

Over the past decade, advances in the commercial and technological viability of wind have increased support for wind as an alternative to fossil fuels. Technological advances in wind turbine technology, electrical transmission technology and power grid organization have improved the efficiency and cost effectiveness of wind. With these improvements, nations lacking fossil fuel resources have begun viewing wind as a possible primary energy source for the future. While the United States is taking large strides in this direction, Europe is leading the development of wind power, especially offshore wind power.

Many members of the European Union have a high UN Human Development Index score. We showed in Chapter 1 that quality of life is correlated to energy use. European countries have relatively small conventional fossil energy reserves. Some European countries have access to unconventional fossil energy resources but there is public concern about the environmental impact of hydraulic fracturing and the emission of greenhouse gases. Overall, Europe has chosen to develop renewable sources of energy. In an attempt to secure a local energy resource with a fixed fuel cost, these countries have turned to wind and solar energy as abundant, renewable and free natural resources.

Table 7-1 [EWEA Stats, 2014, page 8] shows the change in the EU power mix between 2000 and 2014. Both fossil and nuclear fission power declined as a % of total power while renewable power has increased as a % of total power. Nuclear fission power in MW did not significantly change during this period.

Table 7-2 [EWEA Stats, 2014, page 8] shows the change in the EU power mix by source between 2000 and 2014. The table shows that coal and fuel oil consumption declined while gas consumption increased. Nuclear fission was essentially unchanged. The increase in renewable power was largely due to significant increases in wind power and solar PV power.

Table 7-1
European Union Power Mix

	2000		2014	
	Power (MW)	% of Total	Power (MW)	% of Total
Fossil	288899	53.0	410405	45.1
Nuclear	122966	22.6	122328	13.4
Renewable	133224	24.4	377360	41.5
Total	545089	100.0	910093	100.0

Table 7-2
EU Power Mix by Source

Source	2000		2014	
	Power (MW)	% of Total	Power (MW)	% of Total
Fossil				
Gas	92942	17.1	200442	22.0
Coal	132584	24.3	164390	18.1
Fuel Oil	61706	11.3	43765	4.8
Peat	1667	0.3	1808	0.2
Nuclear				
Fission	122966	22.6	122328	13.4
Fusion	0	0	0	0
Renewable				
Solar PV	125	0.02	88000	9.7
Hydroelectric	112919	20.7	140686	15.5
Wind	12887	2.4	128751	14.1
Biomass	4578	0.8	12282	1.3
Waste	2123	0.4	4254	0.5
Concentrated Solar Power	0	0	2314	0.3
Geothermal	592	0.1	810	0.1
Ocean	0	0	263	0.03
Total	545089	100.0	910093	100.0

Many European countries have limited land area to develop onshore wind farms, but have extensive coastlines. The North Sea has very strong winds and a shallow continental shelf that makes it a desirable location for installing offshore wind farms.

Global installed offshore wind capacity in 2014 is shown in Table 7-3. Five of the countries in the Top 10 are from the EU (Germany, Spain, United Kingdom, France, and Italy) and represent 25% of cumulative installed offshore wind capacity in the world. The infrastructure for bringing offshore power to the EU onshore market is evolving and is discussed in Chapter 11.

Table 7-3
Cumulative Installed Offshore Wind Capacity in 2014 [GWEC Offshore, 2014]

Country	Power (MW)	% of World Total
China	114,609	31.0
United States	65,879	17.8
Germany	39,165	10.6
Spain	22,987	6.2
India	22,465	6.1
United Kingdom	12,440	3.4
Canada	9,694	2.6
France	9,285	2.5
Italy	8,663	2.3
Brazil	5,939	1.6
Rest of the world	58,473	15.8
Total Top 10	311,126	84.2
World Total	369,599	100.0

7.5 Can Wind Meet All of Our Energy Needs?

Wind appears to have many of the advantages that would make it an appealing solution to our energy problems. We can build enough wind farms to provide the energy we need based on area and power capacity. The ability to build wind farms offshore further reduces concerns about access to land area.

Figure 7-16 shows the exponential growth of wind production in the United States. To understand the actual role of wind in the United States, we compare the production of energy from wind to coal production and total production in Figure 7-17. Although the contribution of wind energy to energy portfolios is growing, the relative contribution of energy production from wind is much less than the contribution of energy production from coal.

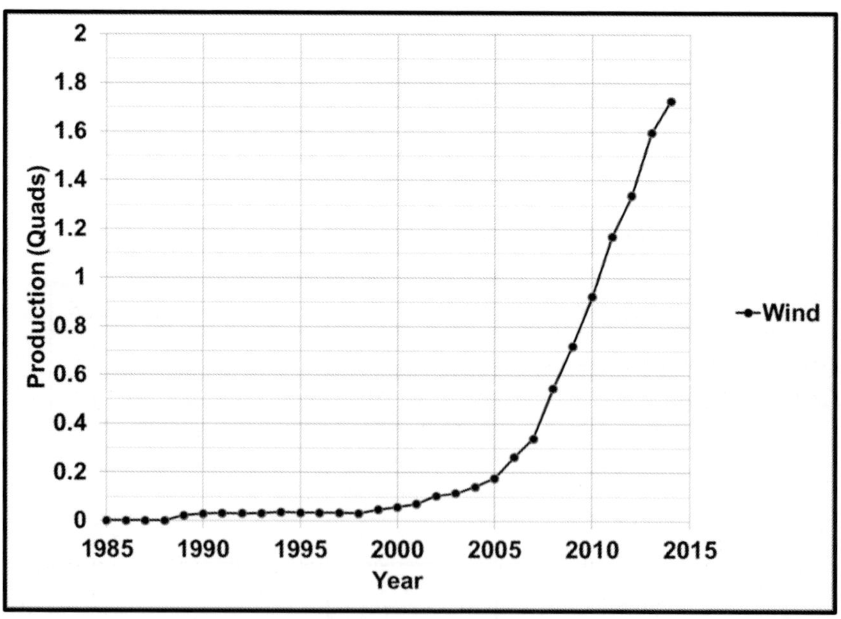

Figure 7-16. Energy Production from Wind in the United States [US EIA Energy Production, 2014]

Figure 7-17 shows that the contribution from coal began to decline after 2011. Much of this reduction can be attributed to increasing government regulation as part of an effort by the Obama administration to reduce greenhouse gas emissions. The increase in total production beginning in 2010 is due to shale oil and gas production in the United States. A decline in shale oil and gas production (not shown in the figure) began at the end of 2014 when Saudi Arabia made a decision to compete for market share by driving out competitors with high production costs.

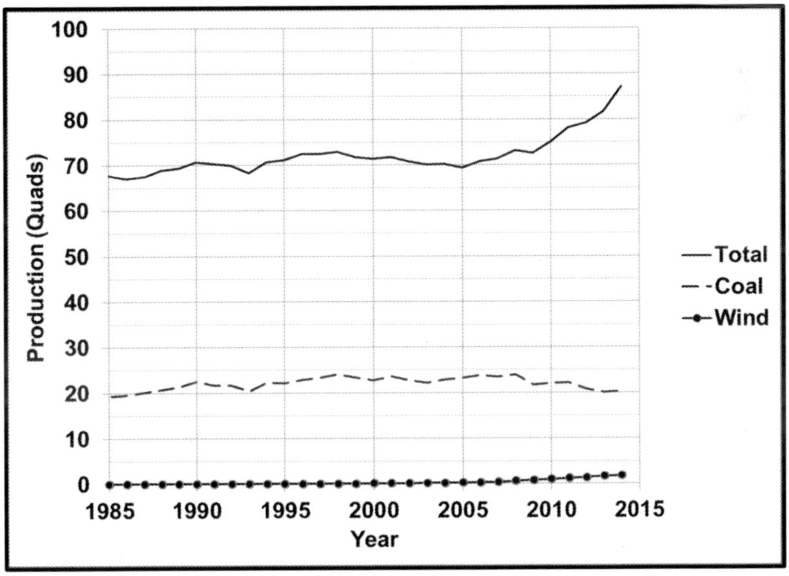

Figure 7-17. Comparison of Energy Production from Wind and Coal in the United States [US EIA Energy Production, 2014]

Integration of wind energy into the United States energy mix is not as extensive as it is in the European Union. One of the concerns about both solar energy and wind energy is the intermittency problem. Neither energy source can be relied on 24 hours a day, seven days a week. We need to provide energy when the wind does not blow or the sun does not shine. Another concern is that energy demand fluctuates. We need to distribute the energy where it is needed when it is needed. We need to provide energy in a form that best fits the need. Some possible solutions to these issues are discussed below.

7.5.1 Energy Storage

The variability of energy demand means that we can expect excess power generation capacity when demand is low. The excess power can be used to store energy. Some possible storage examples include batteries, molten salts, and storage of energy as compressed air in porous media. For example, we could use wind energy to charge batteries, develop large scale energy storage capacity, or use wind energy to produce hydrogen for use in fuel cells. One advantage of hydrogen production is that we could use hydrogen in the transportation sector. Hydrogen is considered in Chapter 10.

Stored energy can be accessed when needed to resolve both the intermittency concern and the fluctuating demand concern. Peak loads can be met, and energy will be available when conditions do not allow wind and solar power generation. Improvements in stored energy are needed to improve the efficiency and reliability of solar and wind energy systems.

7.5.2 Geographic Distribution

The intermittency problem is typically associated with localized power generation. A wind farm in the North Sea or a concentrated solar power (CSP) facility in Spain may cover a huge area, but there will be times when the wind does not blow or the sun does not shine. One way to resolve the issue is to distribute power generation geographically so that a significant portion of the power generation capacity is always able to function. The power generating facilities need to be connected by the same power distribution system that covers the entire region. For example, the North Sea wind farm would need to be connected to the same power transmission grid as the CSP facility in Spain. This is the idea behind the EU Supergrid discussed in Chapter 11.

7.5.3 Solar Wind Downdraft Tower

The solar wind downdraft tower is a hybrid solar/wind renewable energy technology designed to make wind and solar energy more reliable by resolving the issue of intermittency [SWET, 2016]. The tower is a tall hollow cylinder located in a hot dry environment such as the southwestern United States (Figure 7-18). Water is pumped from the base of the tower to an injection system near the top of the tower where it is sprayed as mist across the opening of the tower. The mist evaporates and is absorbed by hot dry air. The air inside the tower becomes denser than air outside of the tower as its humidity increases. The denser air falls through the tower cylinder and creates a downdraft heading toward the base of the tower. The downdraft is diverted into wind tunnels installed around the Tower base. Turbines inside the wind tunnels spin generators to produce electricity. The injected water is captured and recycled to minimize the need for water in arid climates.

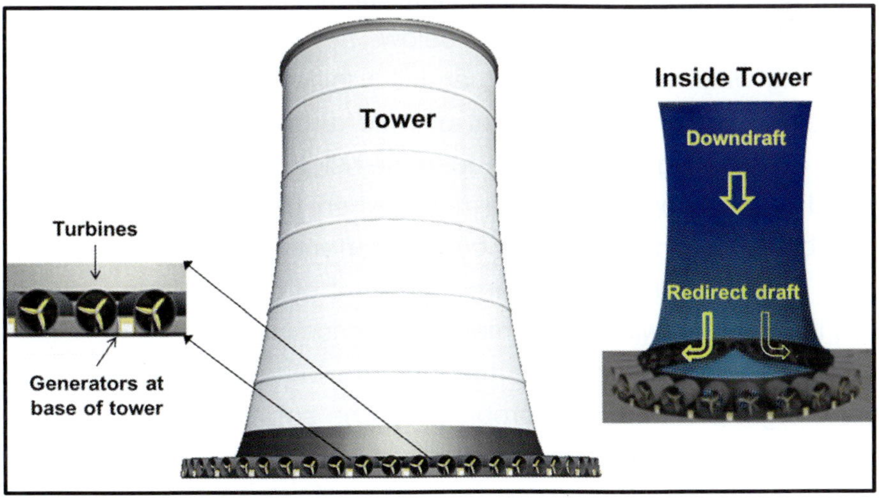

Figure 7-18. Solar Wind Downdraft Tower [SWET, 2016]

Operation of the solar wind downdraft tower generates electrical power with minimal environmental impact. The tower requires very little fuel, has negligible carbon footprint, and produces minimal waste. A facility capable of providing over 4 million MWh of energy per year is expected to occupy a 300 acre site.

7.6 Environmental Impact

Wind energy is a renewable energy that is considered a clean energy because it has a minimal impact on the environment compared to other forms of energy. Wind turbines provide electrical energy without emitting greenhouse gases. On the other hand, we have already observed that the harvesting of wind energy by wind turbines can have environmental consequences.

When properly designed and located, wind turbines have a relatively small environmental impact compared to other energy options. Some people may object to large fields filled with wind turbines, but fields of wind turbines can be built on property to serve more than one purpose. For example, wind farms can be built on West Texas ranch land or along the roadways in industrial areas of Rotterdam, Holland. In both cases the turbines rotate well above activities below.

Rotating wind turbine blades can kill wildlife and interfere with migration patterns. Judicious placement of wind turbines away from migration patterns and nesting areas can reduce the risk to wildlife.

Wind turbines produce some noise when they operate. In the past, wind turbines with metal blades could interfere with television and radio signals. Today, turbine blades are made out of composite materials that reduce interference with electromagnetic transmissions.

Modern, high capacity wind turbines are tall enough to interfere with air traffic flight patterns. This problem can be solved by regulating the locations where wind turbines are installed, and the type of wind turbines that are acceptable.

Additional concerns have arisen as more wind turbines are placed in populated areas. For example, if wind farms are built around coastal cities, such as Corpus Christi, Texas, the wind farms can begin to effect zoning and how the city grows. Suitable locations for wind turbines might also be in the path of an expanding community.

Wind farms can have a significant visual impact that may be distasteful to some people. Aside from the aesthetics of wind farms, wind turbines can "flicker." Flicker is the effect a wind turbine has on the area of land in its shadow. Turbine blades that rotate in daylight cast a shadow that causes a flickering light effect. This flicker can be irritating or distracting to people if their homes or businesses are in the shadow of the turbine.

7.7 Activities

True-False

Specify if each of the following statements is True (T) or False (F).

1. Modern wind turbines provide energy whenever the wind is blowing.
2. Solar energy and wind energy are intermittent sources of energy.
3. A sailboat can use wind energy and water currents to move.
4. Wind turbines must be spaced to minimize turbulence.
5. Two drawbacks of wind energy are the visual impact and footprint of wind farms.
6. A horizontal axis turbine has blades that rotate around a horizontal axis.
7. A wind farm is a collection of wind turbines.
8. Wind turbines have no adverse effect on the environment.
9. Use of wind as an energy source is a recent development (within the last 200 years).
10. The United States was the first country to construct an offshore wind farm.

Questions

1. A renewable energy project will generate 400 MW of wind power using 100 wind towers and will cost approximately US$ 600 million.
   a. On average, how much wind power will be provided by each wind tower?
   b. What is the average cost per MW?
2. What is the minimum distance between two wind turbine posts on a wind farm?
3. Does a wind turbine have any negative environmental consequences?
4. When and where were horizontal axis windmills first utilized?
5. Where were fan mills developed and what were their purpose?
6. What does the yaw mechanism on a wind turbine do?
7. Why do most modern wind turbines use three blades?
8. List three benefits of offshore wind farms relative to onshore wind farms.
9. A 1.5 MW wind turbine provides enough power for 350 homes. What is the average power used by each home, in kW?
10. What is "flicker"?

CHAPTER 8

RENEWABLE ENERGY – ENERGY FROM WATER

After reading this chapter, you should be able to…

- Discuss the history of hydroelectric power in the U.S.
- Explain how hydroelectric power is produced
- Describe the Three Gorges Dam (China)
  - *P2P: What is the environmental impact of a hydroelectric dam?*
- Differentiate between wave and tidal energy
  - *P2P: How much energy can be obtained from coastal waves?*
- Describe ocean thermal energy conversion
- Discuss different types of geothermal energy and their uses
  - *P2P: Can we use water to provide our energy needs?*

Water is a renewable energy source that is already making a significant contribution to worldwide energy and may have a larger impact on the future energy mix. Total renewable consumption in the United States increased from about 8.1 quads in 2010 (Appendix B) to 9.4 quads in 2015 [US EIA Renewable, 2016]. About 31% of the renewable energy consumed by the United States in 2010 was hydroelectric. The contribution of hydroelectric power to renewable energy consumed in the United

States in 2015 declined to approximately 24% primarily because of the increase in wind and solar energy consumption. Hydroelectric power, or hydropower, is an example of energy harvested from moving water. Water movement and temperature gradients in water can be used to provide energy.

The water cycle is a global cycle of moving water (Figure 8-1). Water evaporates from lakes and oceans and rises into the atmosphere where it coalesces into clouds. Clouds can move over all parts of the earth until atmospheric pressure and temperature changes lead to water precipitation in the form of rain or snow. Some of the precipitated water seeps into the earth as groundwater, and some flows along rivers and streams back to lakes and oceans, where the water cycle begins again. The hydrosphere includes groundwater and water found in oceans, glaciers, surface waters such as rivers and lakes, and atmospheric moisture.

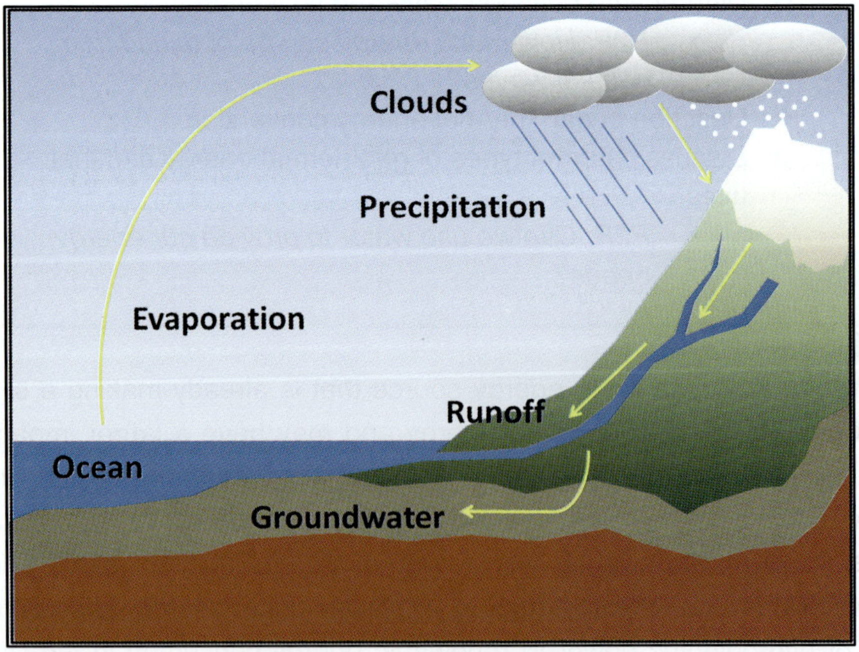

Figure 8-1. The Water Cycle

The water cycle is considered renewable since the water cycle is driven by energy from the sun in a seemingly endless series of cycles. Water in a variety of systems can be used to generate power. For example, hydropower is usually obtained by damming streams and rivers. Energy can be harvested from waves and tides, and from temperature differences in columns of water. We consider power production using energy from water in this chapter.

8.1 HYDROELECTRIC POWER

People have known for some time that falling water could be used to generate electric power. The Greeks were among the first to use hydropower. They ground wheat into flour more than 2,000 years ago by turning water wheels with flowing water. French hydraulic and military engineer Bernard Forest de Bélidor presented hydraulic principles in *Architecture Hydraulique*, his four-volume work published between 1730 and 1770. Bélidor described vertical- and horizontal-axis machines that could be rotated by flowing water. We summarize the history of hydroelectric power and describe the generation of hydroelectric power below.

8.1.1 History of Hydroelectric Power in the United States

Hydroelectric power provided approximately 40% of the electricity generated in the United States by 1940. Today hydroelectric power provides less than 10% of the electricity generated in the United States. Table 8-1 highlights key events in the history of hydroelectric power in the United States.

Table 8-1
History of Hydroelectric Power in the United States
[DoE Hydropower History, 2009]

Year	Comment
1775	U.S. Army Corps of Engineers founded.
1880	Grand Rapids Electric Light and Power Company in Michigan provided electricity from a dynamo belted to a water turbine to power lamps for theatre and storefront lighting.
1881	Niagara Falls city street lamps powered by hydropower.
1882	World's first hydroelectric power plant began operation on the Fox River in Appleton, Wisconsin.
1886	About 45 water-powered electric plants in U.S. and Canada.
1887	San Bernardino, California opens first hydroelectric plant in the west.
1889	Two hundred electric plants in the U.S. use waterpower for some or all electricity generation.
1901	Federal Water Power Act enacted.
1902	Bureau of Reclamation established.
1920	Federal Power Commission established.
1933	Tennessee Valley Authority established.
1935	Federal Power Commission covers hydroelectric utilities engaged in interstate commerce.
1937	The first Federal dam, the Bonneville Dam, begins operation on the Columbia River. Bonneville Power Administration established.

The United States Corps of Engineers was founded in 1775 with the naming of a Chief Engineer in the Continental Army. Its mission has been to provide engineering services for military and public projects that are in the national interest, which often includes canals, dams, and levees. The United States Federal Water Power Act, now known as the Federal Power Act, was enacted in 1901 to coordinate hydroelectric projects.

The Bureau of Reclamation was established in 1902. The Bureau was formerly known as the Reclamation Service and oversees water resource management in the United States. It is part of the United States Department of Interior.

The Federal Power Commission was established by the Federal Power Act in 1920 with the authority to issue licenses for hydro development on public lands. The Tennessee Valley Authority was established in 1933 during the Great Depression. The authority of the Federal Power Commission was extended in 1935 to include hydroelectric projects built by utilities engaged in interstate commerce. Congress reorganized the FPC in 1977 as the Federal Energy Regulatory Commission (FERC) [FERC, 2016].

8.1.2 Generation of Hydroelectric Power

Many of the earliest commercial electric power plants relied on flowing water as their primary energy source. A schematic of a hydroelectric power plant is presented in Figure 8-2.

Water flows from an upper elevation to a lower elevation through a pipeline called a penstock. The change in elevation of the water is a change in the potential energy of the water. Dams with turbines and generators convert the change in potential energy into mechanical energy of flowing water. The water current turns a turbine that is connected to a generator. The turbine is called the prime mover because it transforms an initial source of energy, mechanical energy of flowing water, into another form of energy, namely kinetic energy of rotation of the turbine. The rotating turbine spins a shaft that is connected to either a magnet adjacent to a coil of wire or a coil of wire adjacent to a magnet. The relative motion of the wire and the magnet produces an electric current in which the flow of electric charge periodically reverses direction. The current is known as alternating current. Direct current, for comparison, is the flow of electric charge in only one direction. The abbreviations AC and DC are often used to denote alternating current and direct current, respectively. The hydroelectric system converts mechanical energy into electrical energy.

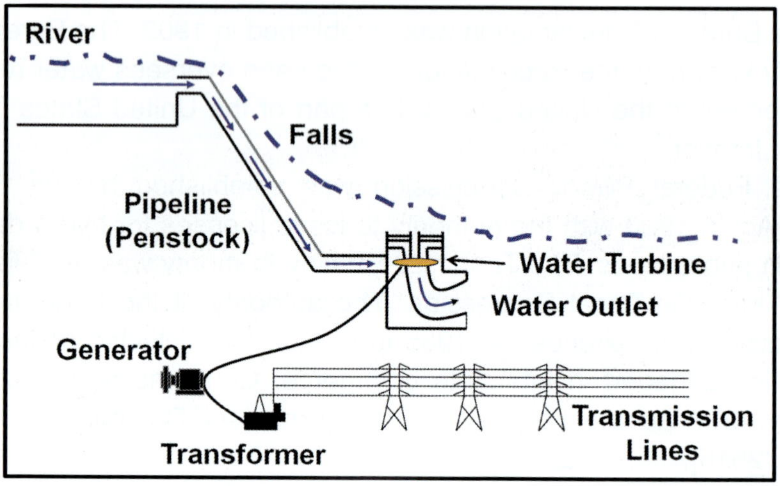

Figure 8-2. Principles of Hydroelectric Power Generation

A dam is built by diverting or blocking water flow past the planned location of the dam. When the dam is completed, the water is impounded, or blocked, so that water begins to pool and form a reservoir upstream of the dam. The elevation, or head, is the height the water falls from upstream of the dam to downstream. The head is replaced by the effective head for realistic systems. The effective head is less than the actual head, or elevation, because water flowing through a conduit such as a pipe will lose energy to friction and turbulence. The rate that water falls through the effective head depends on the volume of the penstock shown in Figure 8-2. If the penstock volume is too small, the output power will be less than optimum because the flow rate could have been larger. On the other hand, the penstock volume cannot be arbitrarily large because the flow rate through the penstock depends on the rate that water fills the reservoir behind the dam.

Figure 8-3 shows the reservoir behind the Hoover Dam at the Arizona-Nevada border in the United States. The light colored rock in the figure bounding the reservoir shows that the depth of the reservoir was below normal. The drainage region into the reservoir had experienced an extended drought at the time the photo was taken.

Figure 8-3. Reservoir behind Hoover Dam (© Fanchi, 2002)

The volume of water in the reservoir and corresponding height depends on the water flow rate into the reservoir. During drought conditions, the elevation can decline because there is less water in the reservoir. During rainy seasons, the elevation can increase as more water drains into the streams and rivers that fill the reservoir behind the dam. Hydroelectric power facilities must be designed to balance the flow of water through the electric power generator with the water that fills the reservoir as a result of natural sources such as rainfall, snowfall, and drainage.

Three hydroelectric power plant sizes are shown in Table 8-2. The plant sizes are classified by electrical power generating capacity. Hoover Dam has an installed capacity of over 1000 MW, which makes it a large hydroelectric power plant.

Table 8-2
Hydroelectric Power Plant Sizes
[DoE Hydropower, 2016]

Size	Electrical Power Generating Capacity (MW)
Micro	< 0.1
Small	0.1 to 30
Large	> 30

Figure 8-4 shows hydroelectric energy consumption in the United States [DoE Hydropower Data, 2016]. The consumption of hydroelectric power in the United States began to decline at the end of the 20th century as public awareness of the environmental impact of dams has increased.

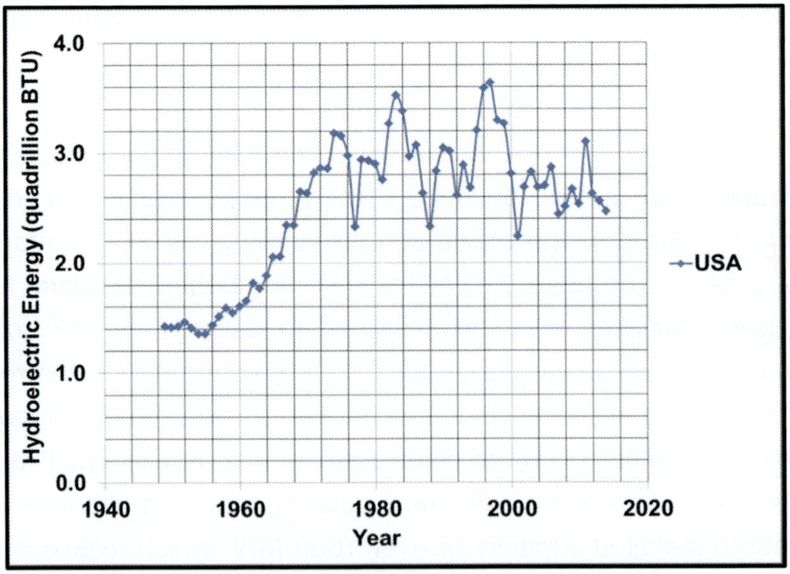

Figure 8-4. Hydroelectric Energy Consumption in the United States

Adverse environmental effects include flooding upstream of the dam, and changing river flow downstream of the dam. These effects occur wherever a dam is built. The Case Study below illustrates some adverse environmental effects of hydroelectric power for a recent hydroelectric power project in China.

8.2 Case Study: Three Gorges Dam

The Three Gorges Dam was the world's largest hydroelectric dam when it was completed [International Rivers, 2009; Three Gorges Geography, 2014, Three Gorges Power, 2016]. Built on the Yangtze River in China, dam construction began in 1994 and was completed in 2008. At completion, the Three Gorges Dam is 185 meters (607 feet) high and 2,309 meters (7,575 feet) wide. Its reservoir extends 660 kilometers (410 miles) behind the dam. The power plant portion of the dam went online in October 2008 with 26 generators. The plant provided a total capacity of 18,300 MW. Additional generators are being added. The dam with 32 main generators cost US$59 billion.

Supporters of the Three Gorges Dam project consider it to be a symbol of China's technological and economic progress. The dam helps decrease China's carbon output and creates a renewable source of energy for China's expanding economy. The dam reduces the potential for downstream floods on the Yangtze, a problem that has historically plagued inhabitants along the river's shore. Control over the river's flow makes river navigation easier.

Since its inception, the Three Gorges Dam has been highly controversial. The Yangtze River is among the world's longest rivers, and the Yangtze River valley has been heavily populated for thousands of years. When a river of this magnitude is dammed, the water must be stored somewhere. The reservoir formed behind the Three Gorges Dam forced relocation of approximately 1.3 million people. Artifacts and ruins from thousands of years of Chinese history were submerged by the reservoir.

The Chinese government promised to provide land for relocated farmers and jobs for relocated urban workers, but arable land is difficult to come by in heavily populated China. Many of the jobs that were going to be provided by state-run enterprises did not materialize. Some cities built to house resettled populations have grown and thrived, but others have become destitute with resettled populations falling into poverty. It has also been alleged that funds allocated for resettling populations have been embezzled by local officials.

The expected environmental impact of the Three Gorges Dam is already being seen, but some unexpected issues are emerging. The dam has changed the flow of the river, with upstream water becoming stagnant while the flow speed has increased in some downstream waterways. The river used to carry large loads of silt that would mitigate its speed, but that silt is being collected at the dam and the downstream river speed has increased causing damage to levees. The benefit of controlling unpredictable flooding has been undermined by this damage.

The change in water speed and composition is also causing changes in the chemical levels and acidity (pH) of the river. These changes are harming fish populations. Migratory fish are affected by the dam, and many indigenous fish species have become endangered. Trash and debris are collecting upstream of the reservoir. Downstream, seawater is beginning to force its way up the river because the river's flow is not as strong as it was previously. Encroaching seawater is threatening farm lands and drinking water supplies near the mouth of the river.

The dam and associated reservoir may be impacting seismic activity in the area. The change in the distribution of weight on the surface of the earth changes subsurface stresses. Larger dam systems seem to have the greatest negative effect. The Three Gorges Dam is constructed to be resistant to earthquake activity in the area, but office buildings, homes and schools in the dam's vicinity are not.

The reservoir created by the dam is less stable than anticipated. Much of the land surrounding the Yangtze River is made of silt and dirt deposits from millions of years of river floods and recessions. This land is not as stable as the rock canyons surrounding dams such as the Hoover Dam in the United States. Five months after the Three Gorges Dam was impounded, 150 geological events were recorded according to the Chinese business magazine *Caijing*. These events included landslides caused by erosion. There is concern that the number of events will increase and cause damage to the population around the reservoir basin. Additional relocation may be necessary.

Point to Ponder: What is the environmental impact of a hydroelectric dam?

While there are many environmental hazards associated with hydroelectric dams, many of which have only recently been observed and studied, there are still many benefits. With greenhouse gas emissions a primary environmental concern in today's society, hydroelectric power is a viable alternative to fossil fuels. According to Wengenmayr [2008, page 25], "Reservoirs can – depending on their geological and climatic situation – emit carbon dioxide and, as a result of the decomposition of plant material, also methane. But even when the production of the materials for constructing a typical hydroelectric plant is taken into account, on the average only the equivalent of a few grams of CO_2 per kilowatt hour are released, [Stefan] Hirschberg [from the Paul Scherrer Institut in Villigen, Switzerland] explains. An average coal-fired power plant, in contrast, emits a kilogram (approx. 2 lb) of CO_2 per kilowatt of power produced, and a Chinese power plant emits even as much as 1.5 kg (approx. 3 lb) per KW."

Wengenmayr went on to say that "[t]he outdated Chinese power plants also lack filtering systems. In densely-populated areas, they shorten the lifespan of the population measurably. According to Hirschberg's studies, 25,000 years of life expectancy are lost there per gigawatt of power produced per year.

> Furthermore, the coal mines degrade the overall ecological and social conditions in China. They emit enormous amounts of methane, and only between 1994 and 1999, more than 11,000 miners lost their lives in accidents."

The Three Gorges Dam project has become an example for developing countries seeking to increase their energy independence by building local hydroelectric power plants. Developing nations without the economic flexibility or natural resources of the United States view hydroelectric power as a virtually carbon-free, renewable and domestic energy option and China is viewed as the world leader in this industry.

Chinese government officials acknowledge evidence of the environmental and geological impact of the dam. Nevertheless, the dam is a technological achievement in size and scope. It also provides an enormous source of renewable, virtually carbon-free energy that decreases China's reliance on coal and other environmentally harmful energy sources.

8.3 Waves and Tides

Ocean waters are solar-powered sources of energy. The mechanical energy associated with waves and tides, and thermal energy associated with temperature gradients in the ocean can be used to drive electric generators. We consider each of these energy sources in this section.

8.3.1 Waves and Tides

In our discussion of hydropower, we saw that water can generate power when it moves from a high potential energy state to a low potential energy state, like a waterfall. If the moving water is an ocean wave, the elevation varies sinusoidally with time. The sinusoidal wave has crests and troughs. The wave amplitude, or height of the wave relative to a calm sea, depends on weather conditions: it will be small during calm weather and can be very large during inclement weather such as hurricanes. The build-up of an ocean wave off the coast of Oahu is shown in Figure 8-5. The change in the potential energy of wave motion can be transformed into energy for performing useful work.

Figure 8-5. Motion of an Ocean Wave, Oahu, Hawaii (© Fanchi, 2003)

The energy density of waves breaking on a coastline in favorable locations can average 65 megawatts/mile (40 megawatts/kilometer) of coastline [DoE Ocean, 2002]. The motion of the wave can be converted to mechanical energy and used to drive an electricity generator. The wave energy can be captured by floats or pitching devices like the Pelamis wave energy converter shown in Figure 8-6 [Pelamis, 2016].

Figure 8-6. Pelamis Wave Energy Converters [Pelamis, 2016]

A wave energy converter (WEC) system such as that used by Pelamis Wave Power looks like a long snake on the surface of the water. The system is a series of tubes that are connected by universal joints to allow movement in all directions. As the waves pass the tubes, the movement at the joints is converted into electricity. The Pelamis system is loosely tethered to the sea floor to hold it in place as well as to connect the machine to subsea power lines that bring generated electricity to shore. The design of the system allows large waves to pass by the wave energy converter so that the system can function even in very rough surf conditions.

Pelamis systems can be installed 2 to 10 km from the coast and in seawater as deep as 50 m or more. A single machine is rated at 750 kW and has a capacity factor ranging from 25 to 40 percent. The capacity factor is the ratio of average output power to peak power.

Pelamis Wave Energy Converters were used in the world's first commercial wave energy project. The project was called the Agucadoura Wave Farm and began operating in 2008. The Agucadoura Wave Farm was located five kilometers off the Agucadoura coast in Portugal.

Another approach to capturing wave energy is to install an oscillating water column (OWC) wave energy converter. The rise and fall of water in a cylindrical shaft drives air in and out of the top of the shaft. The motion of the air provides power to an air-driven turbine. The output power depends on the efficiency of converting wave motion to mechanical energy and then converting mechanical energy to electrical energy. An OWC system is presented in Figure 8-7.

A third approach to capturing wave energy is the wave surge or focusing technique. A tapered channel installed on a shoreline concentrates the water waves and channels them into an elevated reservoir. Water flow out of the reservoir is combined with hydropower technology to generate electricity.

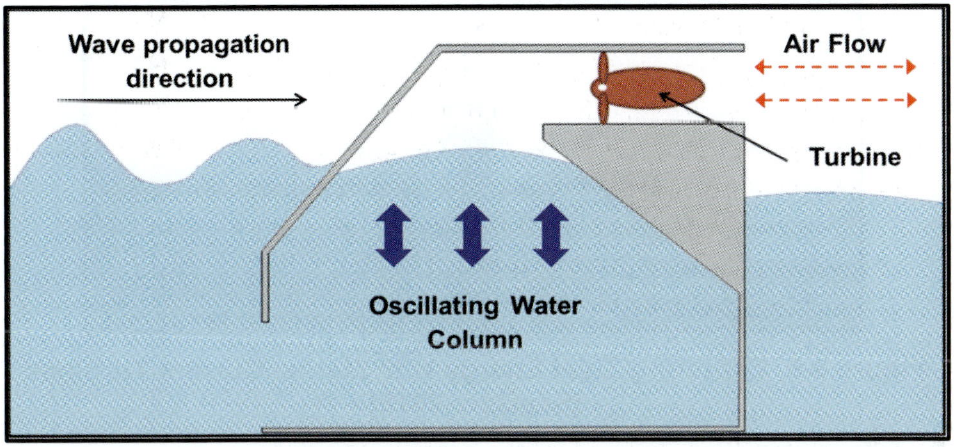

Figure 8-7. Oscillating Water Column (OWC) Wave Energy System [after Graw, 2008, page 78]

The ebb and flow of tides produces tidal energy that can be captured to produce electricity. One way to capture tidal energy is to install underwater turbines with generators in coastal waters. The motion of the tides can drive underwater turbines in a manner that is analogous to wind driving wind turbines.

The world's first commercial-scale tidal turbine was commissioned in Strangford Lough, Northern Ireland in July 2008. The 1.2 MW SeaGen project was developed by Marine Current Turbines (MCT). The project consists of two 600 kW turbines and produced approximately 56 GWh of power between July 2008 and September 2012. This was considered an important milestone because it indicated the completion of the demonstration phase of the project. Figure 8-8 shows the SeaGen with the rotor blades lifted out of the water. The rotor blades are lowered into the water to capture tidal energy.

Figure 8-8. Capturing Tidal Energy with Marine Current Turbines [SeaGen, 2016]

Open center turbines shown in Figure 8-9 are designed for deployment on the seabed. An open center turbine can have a diameter of approximately 16 m and provide 2 MW power. The turbines can be located at a depth that is deep enough to have minimal impact on navigation and not be seen or heard on the surface by nearby communities.

Figure 8-9. Capturing Tidal Energy with Open Centre Turbines [OpenHydro, 2016]

The Sustainable Marine Energy (SME) tidal energy platform is shown in Figure 8-10. The platform shown in Figure 8-10A is designed to be positioned at a depth in the water column that optimizes power production from the current. The platform is anchored to the seabed. The platform shown in Figure 8-10B has two sets of rotor blades.

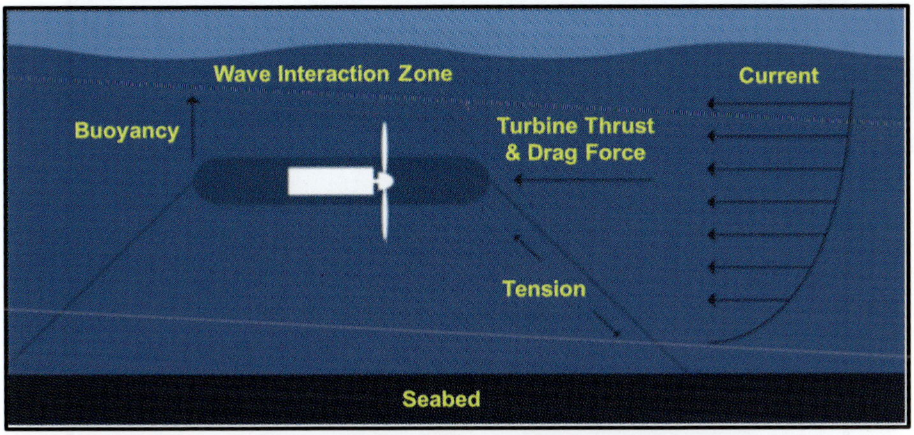

Figure 8-10A. Schematic of tidal energy platform [SME PLAT-O, 2016]

Figure 8-10B. PLAT-O tidal energy platform [SME PLAT-O, 2016]

Figure 8-11 illustrates a technique of capturing tidal energy using a tidal energy station. A dam with a sluice is erected across the opening to a tidal basin to capture tidal energy. A sluice is a channel with a gate that can be used to control the flow of water. The sluice is opened to let the tide flow into the basin and then closed when sea level drops. The water can flow through the gate four times per day because the tide rises and falls twice a day. Hydropower technology can be used to generate electricity from the elevated water in the basin.

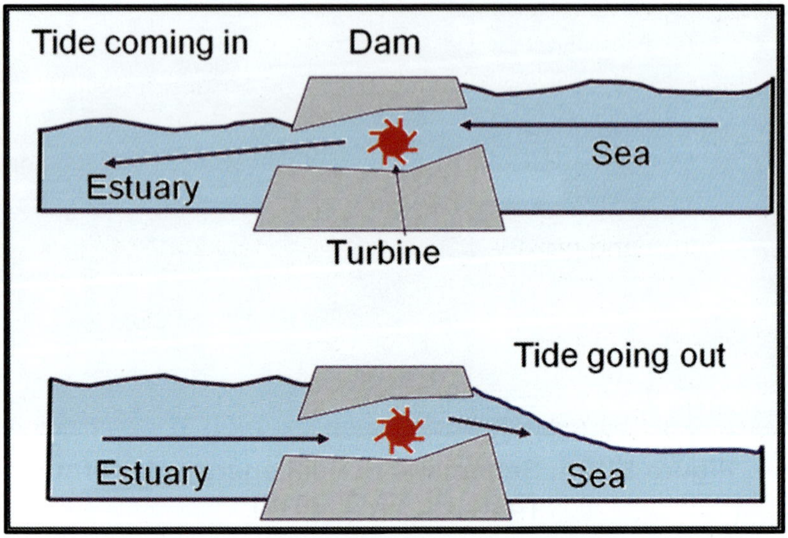

Figure 8-11. Capturing Tidal Energy
[after Shepherd and Shepherd, 1998, page 227]

La Rance tidal power station on the Rance River near St. Malo, France was built in the 1960s and opened in 1966. The power station is part of a 330 m dam built in front of a basin. The power station is capable of generating 240 megawatts of power [DoE Ocean, 2002; La Rance, 2016]. The difference between high and low tides, the tidal range, averages eight meters and can rise as high as 15 meters on the Rance River. Tidal energy stations can have an environmental impact on the ecology of the tidal basin because of reduced tidal flow and silt buildup.

Point to Ponder: How much energy can be obtained from coastal waves?

A typical fossil fuel fired power plant produces 1000 MW of power. If we assume that the average wave power along a coast is 30 MW per kilometer and assume the efficiency of power production is 25%, the equivalent of 133 kilometers (82 miles) of coastline is needed to produce 1000 MW power [Fanchi, 2004, Exercise12-6] for a single line of wave energy converters aligned parallel to the coast. The length of coastline can be reduced significantly by placing lines of wave energy converters perpendicular to the coast. For example, a sea-snake array can link several Pelamis-type generators in 1 km^2 of ocean area and provide 30 MW power [Hopkins, 2007].

8.3.2 Ocean Thermal

Solar energy warms the surface of the world's lakes, seas and oceans. The temperature of water near the surface can be much warmer than the temperature of water in deep oceans and seas, especially in deep water near the equator. The difference in temperature at different depths is referred to as a temperature gradient. Temperature gradients exist near ocean bottom geothermal vents and underwater volcanoes because there is hot water near the heat source and cooler water away from the heat source. If temperature gradients are large enough, they can be used to generate power using ocean thermal energy conversion (OTEC) power plants.

Temperature gradients in deep oceans between 20° North latitude and 20° South latitude are most suitable for OTEC systems [Plocek, et al., 2009]. This area around the globe includes the Caribbean Sea, Gulf of Mexico, Atlantic Ocean, Pacific Ocean, Indian Ocean, and Arabian Sea. Differences in temperature between warm surface water and cooler deep water can be over 20°C (40°F).

Three types of OTEC systems are recognized [DoE Ocean, 2002]: closed-cycle plants, open-cycle plants, and hybrid plants. A closed-cycle plant circulates a heat transfer fluid with a low boiling point like ammonia through a closed system. The fluid is heated with warm seawater until it is flashed to the vapor phase. The vapor is routed through a turbine and then condensed back to the liquid phase with cooler seawater.

An open-cycle plant uses pressure changes to flash warm seawater to vapor and then use the vapor to turn a turbine. An example of a pressure change that can transform liquid water to vaporized water is a vacuum chamber. Liquid water at atmospheric pressure can be placed in the vacuum chamber and flashed to the vapor phase.

A hybrid plant is a combination of an open-cycle plant with a heat transfer fluid. Warm seawater is flashed to steam in a hybrid plant and then used to vaporize the heat transfer fluid. The vaporized heat transfer fluid circulates in a closed system as in the closed-cycle plant.

An ocean thermal energy conversion system can be built on land near a coast, installed near the shore on a continental shelf, or mounted on floating structures for use offshore. In an open-cycle plant or a hybrid plant, the flashed seawater can be condensed as fresh, desalinated water while the salt and other contaminants are captured. In some parts of the world, a desirable byproduct of an OTEC system is the production of desalinated water. Salt from the desalination process must be disposed of in a manner that minimizes environmental impact.

8.4 Geothermal

Geothermal energy is energy provided by the earth. The earth's interior is divided into a solid inner core, molten outer core, mantle, and crust. Drilling in the earth's crust has shown that the temperature of the crust tends to increase linearly with depth; the interior of the earth is much hotter than the crust. The source of heat energy is radioactive decay, and the crust of the earth acts as a thermal insulator to prevent heat from escaping into space.

Heat energy harvested from geological sources is called geothermal energy. Figure 8-12 shows how thermal energy from lava creates steam on the southern coast of Hawaii. The dark surface is a solidified lava field that is cool enough to walk on but contains tubes that channel molten rock to the sea.

Figure 8-12. Lava meeting the Pacific, Hawaii (© Fanchi, 2002)

Geothermal energy is obtained by using a fluid (such as hot water or steam) to carry the internal energy of the earth to the surface of the earth. Geothermal energy can be obtained from temperature gradients between the shallow ground and surface, subsurface hot water, hot rock several kilometers below the earth's surface, and magma. Magma is molten rock in the mantle and crust that is heated by the large heat reservoir in the interior of the earth. In some parts of the crust, magma is close enough to the surface of the earth to heat rock or water in the pore spaces of rock. Magma, hot water, and steam are carriers of energy.

Some of the largest geothermal production facilities in the world are at the Geysers in California, and in Iceland. Figure 8-13 shows the Blue Lagoon and Svartsengi geothermal power plant in Keflavik, Iceland. Surplus warm water from the plant fills up the Blue Lagoon, which is a bathing resort. The power plant is on the far side of the lagoon. Water vapor can be seen in the cooler air. Geothermal plants at the Geysers and Svartsengi are built adjacent to geothermal energy sources. The technology for converting geothermal energy into useful heat and electricity can be categorized as geothermal heat pumps, direct-use applications, geothermal heating systems, and geothermal power plants. Each of these technologies is discussed below.

Figure 8-13. Svartsengi Blue Lagoon, Iceland (© A. Fanchi, 2006)

8.4.1 Geothermal Heat Pump

A geothermal heat pump uses energy near the surface of the earth to heat and cool buildings. The temperature of the upper three meters (10 feet) of the earth's crust remains in the relatively constant range of 10° Centigrade to 16° Centigrade. A geothermal heat pump for a building consists of ductwork in the building connected through a heat exchanger to pipes buried in the shallow ground nearby. The building can be heated during the winter by pumping water through the geothermal heat pump. The water is warmed when it passes through the pipes in the ground. The resulting heat is carried to the heat exchanger where it is used to warm air in the ductwork. During the summer, the direction of heat flow is reversed. The heat exchanger uses heat from hot air in the building to warm water that carries the heat through the pipe system into the cooler shallow ground. In the winter, heat is added to the building from the earth, and in the summer heat is removed from the building.

8.4.2 Direct-Use Applications

A direct-use application of geothermal energy uses heat from a geothermal source directly in an application. Hot water from a geothermal reservoir may be piped directly into a facility and used as a heating source. A direct-use application for a city in a cold climate with access to a geothermal reservoir is to pipe the hot water from the geothermal reservoir under roads and sidewalks to melt snow.

Minerals that are present in the geothermal water will be transported with the hot water into the pipe system of the direct-use application. Some of the minerals will precipitate out of the water when the temperature of the water decreases. The precipitate can form a scale in the pipes and reduce the flow capacity of the pipes. Filtering the hot water or adding a scale retardant can reduce the effect of scale. In either case, the operating costs will increase.

8.4.3 Geothermal Heating System

An example of a geothermal application with a heat exchanger is the system sketched in Figure 8-14. The geothermal reservoir is an aquifer with hot water or steam. The production well is used to withdraw hot water from the geothermal reservoir and the injection well is used to recycle the water. Recycling helps maintain reservoir pressure. If the geothermal reservoir is relatively small, the recycled, cooler water can lower the temperature of the aquifer. The electric pump in the figure is needed to help withdraw water if the reservoir pressure is not high enough to push the water to the surface. Heat from the geothermal reservoir passes through a heat exchanger and is routed to a distribution network.

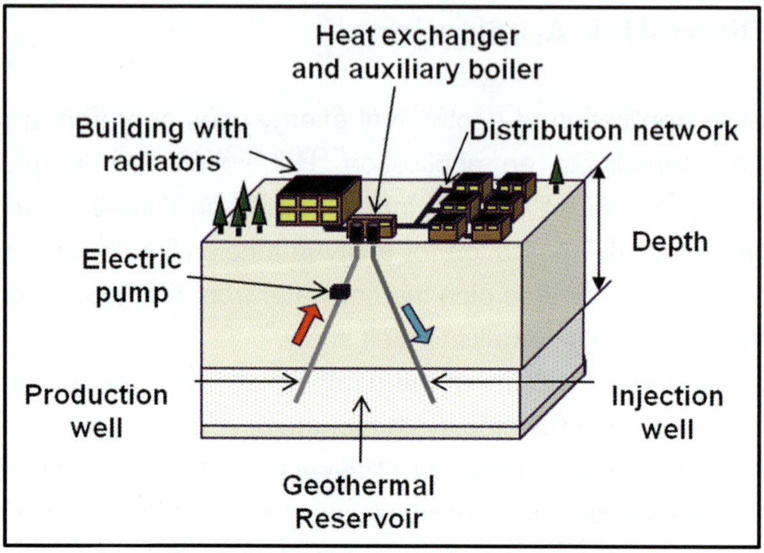

Figure 8-14. Geothermal Heating System [after Shepherd and Shepherd, 1998, page 149]

8.4.4 Geothermal Power Plants

Geothermal power plants use steam or hot water from geothermal reservoirs to turn turbines and generate electricity. Dry steam power plants use steam directly from a geothermal reservoir to turn turbines. Flash steam power plants allow high-pressure hot water from a geothermal reservoir to flash to steam in lower-pressure tanks. The resulting steam is used to turn turbines. A third type of geothermal power plant called a binary-cycle plant uses heat from moderately hot geothermal water to flash a second fluid to the vapor phase. The second fluid must have a lower boiling point than water so that it will be vaporized at the lower temperature associated with the moderately hot geothermal water. There must be enough heat in the geothermal water to supply the latent heat of vaporization needed by the secondary fluid to make the phase change from liquid to vapor. The vaporized secondary fluid is then used to turn turbines.

8.4.5 Managing Geothermal Reservoirs

Like oil and gas reservoirs, the hot water or steam in a geothermal reservoir can be depleted by production. The phase of the water in a geothermal reservoir depends on the pressure and temperature of the reservoir. Single-phase steam will be found in low pressure, high temperature reservoirs. In high-pressure reservoirs, the water may exist in the liquid phase, or in both the liquid and gas phases, depending on the temperature of the reservoir. When water is produced from the geothermal reservoir, both the pressure and temperature in the reservoir can decline. In this sense, geothermal energy is a nonrenewable, finite resource unless the produced hot water or steam is replaced. A new supply of water can be used to replace the produced fluid or the produced fluid can be recycled after heat transfer at the surface. If the rate of heat transfer from the heat reservoir to the geothermal reservoir is slower than the rate of heat extracted from the geothermal reservoir, the temperature in the geothermal reservoir will decline during production. To optimize the performance of the geothermal reservoir, it must be understood and managed in much the same way that petroleum reservoirs are managed.

8.4.6 Hot, Dry Rock

Another source of geothermal energy is hot, dry rock several kilometers deep inside the earth. Figure 8-15 illustrates a hot, dry rock facility that is designed to recycle an energy carrying fluid. Hot, dry rocks are heated by magma directly below them and have elevated temperatures, but they do not have a means of transporting the heat to the surface. In this case, it is technically possible to inject water into the rock, let it heat up and then produce the hot water. This system is known as an Enhanced Geothermal System.

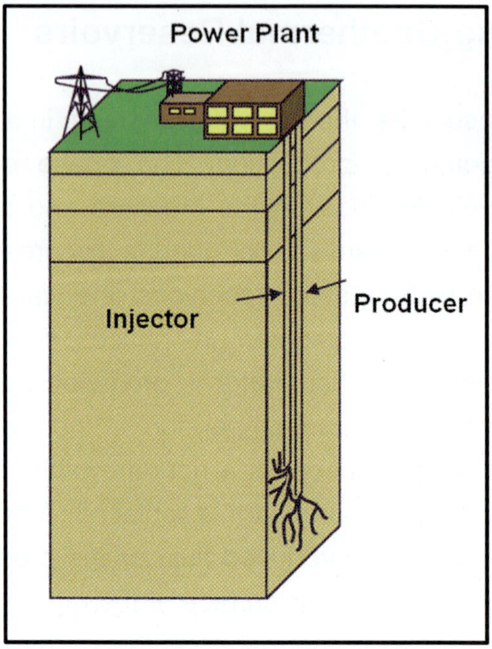

Figure 8-15. Geothermal Energy from Hot, Dry Rock

Water is injected into fissures in hot, dry rock through the injector and then produced through the producer. The fissures may need to be created by a technique such as hydraulic fracturing. The power plant at the surface uses the produced heat energy to drive turbines in a generator. After the hot, produced fluid transfers its heat to the power plant, the cooler fluid can be injected again into the hot, dry rock.

Point to Ponder: Can we use water to provide our energy needs?

Water is available worldwide and has been used to provide energy for centuries. It does have limitations, however. Hydropower is already widely used and can impact the environment. Large scale demonstration projects have shown that wave and tidal energy can provide power on a commercial scale. Geothermal energy is already used as a commercial source of energy in many locations. Improvements in technology should improve the comparative economics of wave, tidal, ocean thermal, and geothermal energy conversion projects.

8.5 Activities

True-False

Specify if each of the following statements is True (T) or False (F).

1. OTEC technology relies on the motion of water waves.
2. Hydropower is power generated by the change in potential energy of water moving through a gravitational field.
3. An OTEC system must be mounted on floating structures.
4. Dams built between the sea and estuaries can be used to generate electricity from tidal motion.
5. OTEC systems use temperature gradients to generate power.
6. Increasing the volume of the penstock in a hydroelectric power plant will always increase the power output.
7. The Three Gorges Dam is the world's largest hydroelectric dam.
8. Construction of the Three Gorges Dam had a significant financial, environmental and societal cost.
9. Water is used to turn a turbine in an OWC system.
10. A geothermal heat pump uses energy near the surface of the earth to heat and cool buildings.

Questions

1. What is hot, dry rock geothermal energy?
2. How is energy brought to the surface from a hot, dry rock geothermal source?
3. Can a drought affect the performance of a dam?
4. Is the energy associated with waves and tides mechanical energy or thermal energy?

5. What is the purpose of the United States Corps of Engineers?
6. What are four types of energy involved in hydroelectric power generation?
7. List three advantages and disadvantages associated with the Three Gorges Dam.
8. How do wave energy converters capture wave energy?
9. Can geothermal reservoirs be depleted by production?
10. What part of the world is most suitable for OTEC systems?

CHAPTER 9

RENEWABLE ENERGY – BIOENERGY AND SYNFUELS

After reading this chapter, you should be able to…

- Define biomass and describe its uses
- Identify multiple sources of biomass fuel
  - *P2P: Should farm land be used for food or fuel?*
- Describe biopower
  - *P2P: Is biomass a desirable alternative energy source?*
- Explain the significance of biofuels in Brazil
- Define synfuels and discuss how they are produced and their applications
  - *P2P: Are synfuels desirable alternative energy sources?*
- Summarize the environmental impact of primary energy sources

Biomass refers to wood and other plant or animal matter that can be burned directly or can be converted into fuels. We saw in Chapter 1 that wood was the energy source for virtually all of the energy consumed in the United States until the middle of the 19th century. Wood was replaced by other energy sources, notably fossil fuels, hydroelectricity, and nuclear energy until today biomass is a relatively small contributor to the energy mix. The United States consumed approximately 4.2 quads of biomass energy in 2010. To put this in perspective, the United States consumed approximately 97.4 quads of energy in 2010 (Appendix), which means that biomass consumption was approximately 4.3% of United States energy consumption in 2010.

Technologies exist to convert plants, garbage, and animal waste into natural gas. Synthetic fuels (synfuels) are used as substitutes for conventional fossil fuels such as natural gas and oil. Synfuels are produced by chemical reactions using such basic resources as coal or biomass. We discuss biomass and synfuels in this chapter.

9.1 Biomass

Biomass is biological material obtained from living or recently living organisms. People have used biomass for fuel ever since we learned to burn wood. Biomass includes land- and water-based vegetation, and waste materials that can be burned directly or can be converted into fuels. Vegetation growth often relies on photosynthesis, which is a process that converts solar energy into chemical energy. The process uses sunlight as an energy source to synthesize carbohydrates from carbon dioxide and water. Photosynthesis occurs in green plants, and some organisms such as bacteria and algae.

Some types of biomass used for energy include forest debris, agricultural waste, wood waste, animal manure, and the non-hazardous, organic portion of municipal solid waste. A major component of municipal solid waste is sewage. Developing countries are among the leading consumers of biomass because of their rapid economic growth and increasing demand for electricity in municipalities and rural areas.

Products derived from biological material are considered biomass. Methanol, or wood alcohol, is a volatile fuel that has been used in race cars for years. Another alcohol, clean burning ethanol, can be blended with gasoline to form a blended fuel (gasohol) and used in conventional internal combustion engines, or used as the sole fuel source for modified engines.

Availability is one advantage biomass has relative to other forms of renewable energy. Energy is stored in biomass in the solid, liquid, or gaseous state, and can be stored until needed. Biomass may be grown, harvested, stored, and converted into a more useful form. Other renewable energy forms, such as wind and solar energy, depend on environmental conditions that can vary considerably and are not always available when they are needed.

Biomass is typically transformed into useful energy using chemical reactions to produce heat during combustion. Technologies exist to convert plants, garbage, animal dung, and municipal solid waste (MSW) into natural gas. One example of a biomass to energy conversion is the decay of organic waste in a compost heap (Figure 9-1). The decay releases heat and methane. Another example of a biomass to energy conversion process is the production of biogenic gas from decay of organic waste in a landfill.

Figure 9-1. A Compost Heap on a Farm near Houston, Texas (© Fanchi, 2003)

A landfill is a pit filled with garbage. When the pit is full, we can cover it with dirt and insert a pipe through the dirt into the pit. The pipe provides a conduit for producing the natural gas that is generated from the decay of biological waste in garbage. The landfill gas is filtered, compressed, and routed to the main gas line for delivery to consumers. Figure 9-2 shows a landfill gas skid with a flare pipe. The ridge on the right hand side of the picture is the top of the landfill, and the dark features on the ridge are pieces of a material that are used to cover the landfill and prevent gas from escaping into the atmosphere.

Figure 9-2. Landfill Gas Skid with Flare near Denver, Colorado (© Fanchi, 2004)

9.1.1 Wood

Wood has historically been a source of fuel. We have already learned that wood was a primary energy source through much of history [Nef, 1977], but deforestation became such a significant problem in 16th century England that the English made the transition to an alternative source of fuel:

coal. The transition from wood to coal made it possible to slow deforestation. Today, many people rely on wood as a fuel source in underdeveloped parts of the world. Economic sources of wood fuels include wood residue from manufacturers, discarded wood products, and non-hazardous wood debris.

An increased reliance on wood as a fuel has environmental consequences, such as pollution, an increased rate of deforestation and an increase in the production of byproducts from wood burning. Figure 9-3 shows the afternoon sun as it was seen through smoke from a forest fire. One way to mitigate the environmental impact is to implement reforestation programs. Reforestation programs are designed to replenish the forests. Research is underway to genetically engineer trees that grow fast, are drought resistant, and easy to harvest. Fast-growing trees are an example of an energy crop, that is, a crop that is genetically designed to become a competitively priced fuel.

Figure 9-3. Afternoon Sun Obscured by Smoke near Denver, Colorado (© Fanchi, Summer 2002)

We know from observation and the chemistry of wood combustion that burning wood consumes carbon, hydrogen and oxygen. The combustion reactions produce water, carbon monoxide, and carbon dioxide, a greenhouse gas. Greenhouse gases are gases that capture sunlight reradiated as heat from the earth. Proponents of anthropogenic climate change believe an increase in the amount of greenhouse gases in the atmosphere will increase the capture of reradiated sunlight and lead to harmful changes in the climate. The emission of greenhouse gases produced by the combustion of biomass has an environmental impact that needs to be considered when comparing different energy sources.

9.1.2 Ethanol

Ethanol is an alcohol that can be added to gasoline to increase the volume of fuel available for use in internal combustion engines. A biofuel is a fuel derived from biomass. Ethanol is made from a biomass feedstock. A feedstock is the raw material that is used for processing or manufacturing. Residues that are the organic byproducts of food, fiber, and forest production are economical biomass feedstock. Examples of residue include sawdust, rice husks, corn stover, and bagasse. Corn stover is used to produce ethanol. It is a combination of corn stalks, leaves, cobs, and husks. Bagasse is the residue that is left after juice has been extracted from sugar cane. A biomass research facility is shown in Figure 9-4. A fermentation tank is in the foreground of the figure.

The process of ethanol production proceeds in several steps. First, the feedstock must be delivered to the feed handling area. The biomass is then conveyed to a pretreatment area where it is soaked for a short period of time in a dilute sulfuric acid catalyst at a high temperature. The pretreatment process liberates certain types of sugars and other compounds. The liquid portion of the mixture is separated from the solid sludge and washed with a base such as lime to neutralize the acid and remove compounds that may be toxic to fermenting organisms. A beer containing ethanol is produced after several days of anaerobic fermentation. The beer is distilled to separate ethanol from water and residual solids.

The production of ethanol from biomass requires distillation. During distillation, a mixture is boiled and the ethanol-rich vapor is allowed to condense to form a sufficiently pure alcohol. Distillation is an energy consuming process. Ethanol increases energy supply when it is used in petrofuels to increase the volume of transportation fuel. By contrast, the need to consume energy to produce ethanol decreases energy supply. A favorable energy balance is achieved if the energy required to produce ethanol is less than the energy provided by the consumption of ethanol.

Figure 9-4. Biomass Research Facility, National Renewable Energy Laboratory, Golden, Colorado (© Fanchi, 2003)

The energy in a gallon of ethanol is approximately 76,000 BTU. This is less than amount of energy in a gallon of gasoline, which is approximately 116,000 BTU. As a rule of thumb, the energy content in ethanol is about 2/3 the energy content in gasoline. If we blend ethanol and gasoline, the mixture will have an energy content that is greater than the energy content in ethanol but less than the energy content in gasoline.

Flex-fuel vehicles are designed to operate with gasoline or gasoline-ethanol blends up to 85% ethanol. The content of ethanol in the blend is denoted Exx where xx is the percentage of ethanol. For example, a gasoline-ethanol blend with 85% ethanol is denoted E85. According to the United States Department of Energy [DoE FFV, 2016], flex-fuel vehicles (FFV) "experience no loss in performance when operating on E85, and some generate more torque and horsepower than when operating on gasoline. However, since ethanol contains less energy per volume than gasoline, FFVs typically get about 15% – 30% fewer miles per gallon when fueled with E85."

Rudolf Diesel (1858-1913) was the first to develop an engine that could run on peanut oil. He demonstrated his engine at the 1900 Paris World Exhibition. Diesel's fuel is an example of a biofuel. Today, biodiesel is a relatively clean-burning renewable fuel that is commonly derived from vegetable oils.

Biodiesel can be blended with petroleum diesel to increase the supply of diesel fuel. Petroleum diesel is considered a middle distillate in the carbon number range C_{13} to C_{20} where C_x denotes a hydrocarbon molecule with x carbon atoms. For comparison, kerosene is a mixture of hydrocarbon molecules in the carbon number range C_{11} to C_{12}.

The adoption and widespread use of biofuels is hampered by cost when compared to the cost of using fossil fuels. One way to improve the economics of biofuels is to increase the cost of fossil fuel consumption. An increase in cost can be justified on environmental grounds by imposing a cost on the emission of greenhouse gases such as carbon dioxide. The cost can be imposed as a tax on the amount of greenhouse gases that are emitted by fossil fuel combustion, or by imposing a limit on the amount of greenhouse gases emitted.

Point to Ponder: Should farm land be used for food or fuel?
In recent years, biofuels have become increasingly popular as a means of extending the life of the world's oil supply. These fuel additives are renewable and reduce the amount of oil that must be used for transportation. This is particularly important in countries like the United States where there is a large transportation sector. Still, biofuels have an additional cost that must be considered.

Some of the biomass feedstock for biofuels comes from organic byproducts. These scraps do not satisfy the demand for biomass. Furthermore, the United States government is an example of a government that subsidizes biofuels to encourage farmers to use their land for biomass rather than food stuffs. As a consequence, this creates a conflict between two choices: use land to grow food, or use land to grow feedstock for biofuels.

Societies must recognize that the volume of water and land area suitable for growing biofuels and food are limited. The designation of land for one product can reduce the amount of land available for growing the other product. Governments seeking to increase energy supplies must balance their need for biofuels with their need for food. This concept is an example of opportunity cost, which is discussed further in Chapter 12.

9.1.3 Biopower

Biomass has historically been used to provide heat for cooking and comfort. Today, biomass fuels can be used to generate electricity and produce natural gas. The power obtained from biomass, called biopower, can be provided on a scale that is suitable for a single-family home up to a small city. Biopower is typically supplied by one of four different classes of systems: direct-fired, co-fired, gasification, and modular systems.

A direct-fired system is similar to a fossil fuel-fired power plant. High-pressure steam for driving a turbine in a generator is obtained by burning biomass fuel in a boiler. Biopower boilers presently have a smaller capacity than coal-fired plants. Biomass boilers provide energy in the range of 20 to 50 MW, while coal-fired plants provide energy in the range of 100 to 1500 MW. The technology exists for generating steam from biomass with an efficiency of over 40%, but actual plant efficiencies tend to be on the order of 20% to 25%.

Biomass may be combined with coal and burned in an existing coal-fired power plant. The process is called co-firing and is considered desirable because the combustion of the biomass-coal mixture is cleaner than burning only coal. Biomass combustion emits smaller amounts of pollutants, such as sulfur dioxide and nitrogen oxides. The efficiency of the co-firing system is comparable to the coal-fired power plant efficiency when the boiler is adjusted for optimum performance. Existing coal-fired power plants can be converted to co-firing plants without major modifications. The efficiency of transforming biomass energy to electricity is comparable to the efficiency of a coal-fired power plant and is in the range of 33% to 37%. The use of biomass that is less expensive than coal can reduce the cost of operating a coal-fired power plant.

Biomass gasification converts solid biomass to flammable gas. Cleaning and filtering can remove undesirable compounds in the resulting biogas. The gas produced by gasification can be burned in combined-cycle power generation systems. The combined-cycle system combines gas turbines and steam turbines to produce electricity with efficiency as high as 60%.

Modular systems use the technologies described above, but on a scale that is suitable for applications that demand less energy. A modular system can be used to supply electricity to villages, rural sites, and small industry.

Point to Ponder: Is biomass a desirable alternative energy source?

Figure 9-5 shows a car that uses 100% vegetable oil as its fuel. At the time the photo was taken, a gallon of gasoline in Honolulu, Hawaii cost approximately one third more than a gallon of gasoline in Denver, Colorado. Increases in the price of gasoline stimulate the search for alternative transportation fuels. Vegetable oil is an example of a biomass alternative energy source.

Biomass provides energy by generating heat using combustion. One byproduct of the combustion process is carbon dioxide, a greenhouse gas. The benefits that biomass is renewable and can serve as an energy source are offset by the production of carbon dioxide.

Figure 9-5. Car Powered by 100% Vegetable Oil, Waikiki, Oahu (© Fanchi, 2004)

9.2 Case Study: Biofuels in Brazil

Brazil is South America's leader in biofuel production. Brazil began its ethanol production program in the early 1980's when the country was facing a major debt crisis because of the amount of oil it was importing [Goldemberg, 2008]. Brazilian leaders decided to take advantage of their consistently warm temperatures and tropical climate to produce large quantities of sugarcane to use as biomass for biofuels. Approximately half of Brazil's 2008 sugarcane crop was used for biofuel production.

When it became clear that Brazil must find a way to reduce its need for oil imports, the government began a large incentive program to encourage the development of biofuel crops and production technologies. Almost one-third of the financing required to develop the technology was invested between 1980 and 1985, and over the following two and a half decades the technology advanced to the point that government incentives were reduced to less than half of what they were originally. In 1980, the Brazilian government subsidized ethanol production at US$700 per 1000 liters (US$2.65 per gallon). By 2004, the subsidy had fallen to US$200 per 1000 liters (US$0.76 per gallon) as the industry approached self-sufficiency. This is a result of the increasing cost of oil, advances in technology and improvements in economies of scale.

Anhydrous ethanol, which is 99.6% ethanol and 0.4% water, has approximately 67% of the volumetric energy desnity of gasoline. Hydrous ethanol (about 95.5% ethanol and 4.5% water) has a slightly lower volumetric energy density. The Brazilian government incentivized the use of ethanol as a stand-alone fuel in the 1980's, guaranteeing that the pump price of hydrous alcohol would be equivalent to 64.5% that of gasoline. This incentive led to an explosion in the popularity of ethanol-only cars and demand for ethanol-only fuel. Brazil experienced a shortage of ethanol-only fuel in the early 1990's which motivated a shift away from ethanol-only vehicles.

In 2003, Brazil introduced a flex-fuel motor that revolutionized the use of ethanol fuels. The flex-fuel motor is designed to use sensors to detect the ethanol level of the fuel in the tank. The motor then automatically adjusts the way it functions to reach maximum efficiency based on the fuel. This allowed for ethanol-gasoline mixtures of varying levels to be produced based on the best economics at the time and still be able to serve a large volume of motorists. While ethanol-only engines may be an option in the future, particularly if oil supplies are exhausted, the volume of ethanol needed to meet global demand is not available.

The success of the flex-fuel motor implies that the demand for ethanol in Brazil can increase dramatically in the future. The growth of Brazil's ethanol industry led to changes in the use of land that favored sugarcane production. The Brazilian government is enforcing regulations for clearing of land to curtail potentially adverse effects of land-use changes due to the expansion of sugarcane production. Some of the effects include transfer of livestock and crop (soybeans, cotton) production to biologically sensitive areas in the Amazons and Cerrados regions [USDA Brazil, 2011].

Demand for ethanol could grow elsewhere in the world as other nations seek to emulate Brazil's advanced ethanol production, especially those countries with the capability to support large sugarcane fields. The need for large regions of arable land is a significant limitation on the global adoption of the Brazilian model.

9.3 SYNFUELS

Synthetic fuels, or synfuels, are fossil fuel substitutes created by chemical reactions using basic resources such as coal or biomass. There are several ways to convert biomass into synfuels. Oils produced by plants such as rapeseed (canola), sunflowers and soybeans can be extracted and refined into a synthetic diesel fuel. Thermal pyrolysis and a series of catalytic reactions can convert the hydrocarbons in wood and municipal wastes into a synthetic gasoline.

Society can extend the availability of fossil fuels such as natural gas and oil by substituting synthetic fuels for conventional fossil fuels. Here we consider coal gasification, biomass conversion, gas-to-liquids conversion, coal-to-liquids conversion, and algae oil.

9.3.1 Coal Gasification

Large coal molecules are converted to gaseous fuels in coal gasification. Coal is thermally decomposed at temperatures on the order of 400° Centigrade (750° Fahrenheit) to 900° Centigrade (1650° Fahrenheit). The products of decomposition are methane and a carbon rich char. If steam is present, the char will react with the steam in the reaction to form carbon monoxide and hydrogen. The carbon monoxide-hydrogen mixture is called synthesis gas, or syngas. Carbon monoxide can react with steam in the reaction to form carbon dioxide and hydrogen. If the carbon dioxide is removed, the hydrogen-enriched mixture will react with carbon monoxide to produce methane and water vapor. The coal gasification process can be used to synthesize methane from coal. Methane is easier to transport than coal.

9.3.2 Biomass Conversion

Biomass, particularly biological waste, appears to be a plentiful source of methane. Methane production from biomass requires a conversion process. The simple methane digester shown in Figure 9-6 is a biomass conversion system that converts biological feed such as dung to methane. The screw agitator mixes the liquid slurry containing the feed. The mixing action facilitates the release of methane from the decaying biomass as a biomass conversion process. Methane trapped by the metal dome is recovered through the gas outlet.

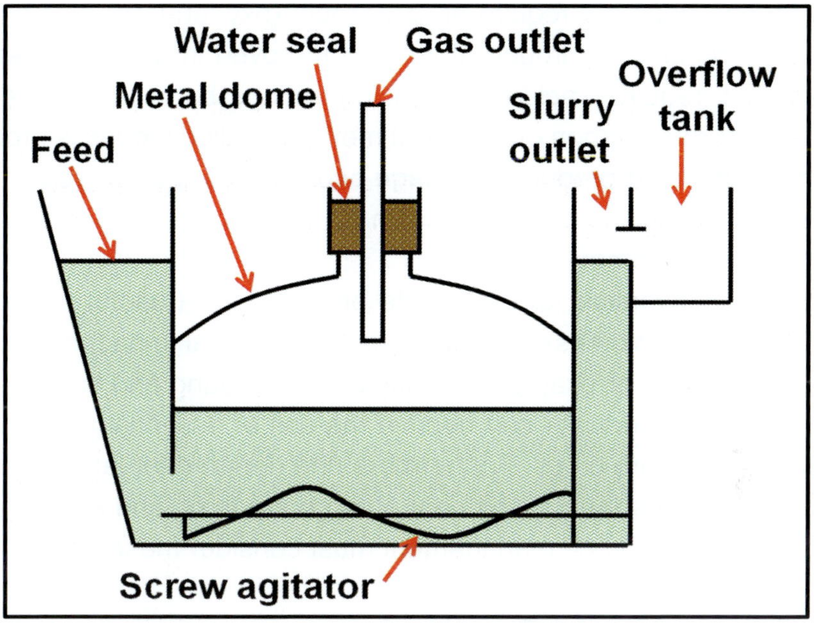

Figure 9-6. Methane Digester [after Cassedy and Grossman, 1998, page 298; and P.D. Dunn, Renewable Energies: Sources, Conversion and Application, 1986, P. Peregrinius, Ltd., London, United Kingdom]

Fermentation is another example of a biomass conversion process. Microorganisms are used in the fermentation process to convert fresh biological material into simple hydrocarbons or hydrogen. We illustrate fermentation processes by describing the anaerobic digestion process, a process that is well suited for producing methane from biomass.

The anaerobic digestion process, also known as anaerobic microbial degradation, proceeds in three stages. In the first stage, a complex biomass is decomposed by the first set of microorganisms. Common biomass materials used in anaerobic microbial degradation are polysaccharides known as cellulosic materials. Polysaccharides are materials that consist of chains of sugar (glucose) units. The decomposition of cellulosic material into simple sugar (glucose) molecules occurs in the presence of enzymes provided by microorganisms.

Stage one does not require an anaerobic (oxygen-free) environment. In the second stage, hydrogen atoms are removed in a dehydrogenation process that requires acidophilic (acid-forming) bacteria. In the third stage, a mixture of carbon dioxide and methane called biogas is produced from the acetic acid produced in stage two. The third stage requires the presence of anaerobic bacteria known as methanogenic bacteria in an oxygen-free environment.

Biomass conversion turns a waste product into a useful commodity. One difficulty with biomass conversion is the potential impact on the ecology of the region. For example, excessive use of dung and crop residues for fuel instead of fertilizer can deprive the soil of essential nutrients that are needed for future crops. Furthermore, the demand for large quantities of biomass requires the commitment of large areas of land and large volumes of water. Resource management must consider these issues.

9.3.3 Gas-to-Liquid Conversion

Synthetic liquid hydrocarbon fuels can be produced from natural gas by a gas-to-liquids (GTL) conversion process. Natural gas is composed of organic molecules such as methane and ethane which have a small number of carbon atoms. For example, methane has one carbon atom and ethane has two carbon atoms. The GTL process converts relatively small organic molecules into molecules with longer chains of carbon atoms. The primary product of the GTL process is a low sulfur, low aromatic diesel fuel. Aromatic molecules have multiple bonds between some of the carbon atoms and one or more carbon rings. Benzene, a carcinogen, is an example of an aromatic molecule.

Synthetic gas (syngas) can be used to synthesize liquid hydrocarbon fuels by the Fischer-Tropsch (F-T) process. The F-T process consists of four stages: syngas generation, gas purification, F-T synthesis, and product upgrading. F-T synthesis is the process of converting syngas to liquid hydrocarbons. Small hydrocarbon molecules such as methane are thermally decomposed to form carbon monoxide and oxygen. Catalysts are then used to produce hydrocarbons from the carbon monoxide and hydrogen. Some common catalysts are cobalt and iron. The F-T process produces a mixture of hydrocarbon molecules that range from low molecular weight methane to higher molecular weight products such as kerosene, diesel and wax.

9.3.4 Coal-to-Liquid Conversion

The first step in the coal-to-liquid conversion is coal gasification (Figure 9-7). Coal is fed into a gasifier where heat, pressure, and oxygen from an air separation plant convert coal into syngas. Syngas is composed of hydrogen and carbon monoxide.

A gas cleaning system removes impurities from syngas. Impurities include sulfur, mercury, and any carbon dioxide that may be present. Other elements of the gas cleaning system are designed to remove particulates, water, ammonia, and sulfuric acid. The syngas passes from the gas cleaning system to a Fischer-Tropsch (F-T) reactor where it is exposed to a catalyst such as iron or cobalt. The F-T process produces a mixture of hydrocarbon molecules as in the gas-to-liquids process.

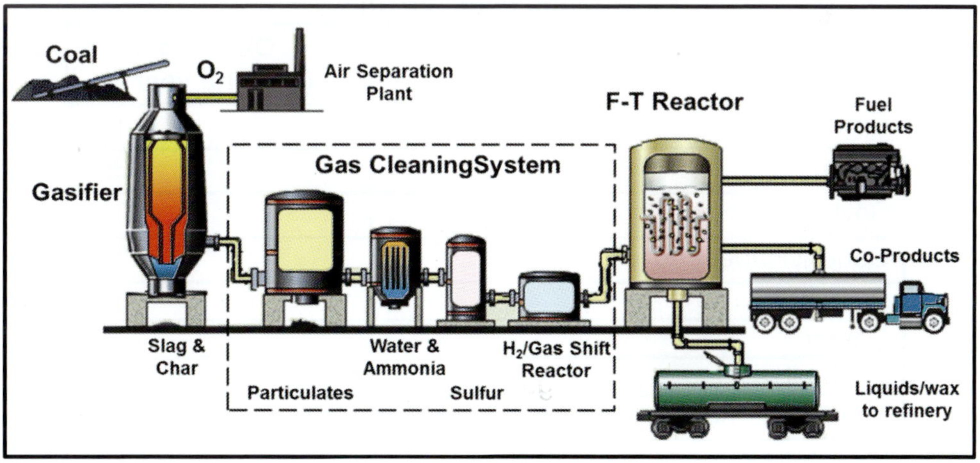

Figure 9-7. Coal-to-Liquid Conversion [after KCE CTL, 2016]

9.3.5 Algae Oil

Algae are a plentiful biomass that can be processed to synthesize oil and gas. The synthetic process begins with the selection of a strain of algae based on such factors as rate of growth, survivability in different environments, and ability to produce hydrocarbons. The growth of algae requires water, nutrients and access to sunlight. When exposed to sunlight, algae can reproduce quickly and use photosynthesis to convert water and carbon dioxide into sugar. Algae cells metabolize the sugar into oily lipids, an organic material that can be extracted from algae cells. The extracted oil is refined to produce diesel oil using a catalyst to replace oxygen with hydrogen. Further refinement can alter the length of carbon chains to produce gasoline and jet fuel from algae oil.

Table 9-1 compares oil yield for a selection of plant sources. The oil yields for microalgae with different oil contents are much larger than the other plant sources in the table. Mata, et al. [2010] define microalgae as "prokaryotic or eukaryotic photosynthetic microorganisms that can grow rapidly and live in harsh conditions due to their unicellular or simple multicellular structure." The advantages of microalgae are stimulating efforts to develop the resource as an energy crop.

Table 9-1
Oil Yield for a Selection of Plant Sources
[Mata, el al., 2010]

Plant Source	Oil Yield (L oil/ha year)
Corn/Maize	172
Hemp	363
Soybean	636
Canola/Rapeseed	974
Sunflower	1070
Castor	1307
Palm oil	5366
Microalgae (low oil content)	58,700
Microalgae (medium oil content)	97,800
Microalgae (high oil content)	136,900

Point to Ponder: Are synfuels desirable alternative energy sources?

Natural gas from the conversion of biological materials is renewable, but a byproduct of natural gas combustion is carbon dioxide, an undesirable greenhouse gas. Like biomass and fossil fuels, synfuels provide energy by generating heat using combustion. Once again the combustion process produces the greenhouse gas carbon dioxide as a byproduct.

Unlike biomass, coal gasification, gas-to-liquid conversion, and coal-to-liquid conversion depend on the fossil fuels coal and natural gas. The generation of greenhouse gases from consumption of biomass or synfuels by combustion adversely affects the desirability of using energy from biomass and synfuels.

9.4 Environmental Impact Summary

The fuel used to drive a power plant is called the primary fuel. Examples of primary fuels include oil, coal, natural gas, and uranium. Fossil fuels are the primary fuel for most power plants in the world today. Fossil fuel resources are finite, however, and power plants that burn fossil fuels emit greenhouse gases and gaseous pollutants. The extraction of some types of fossil fuels, such as coal and tar sands, can leave the land scarred and in need of reclamation (Figure 9-8).

Figure 9-8. Scarred Land near Banff, Canada (© Fanchi, 2002)

Fossil fuels are combustible fuels: energy is released from the fuel by combustion. Biomass and synfuels increase the supply of combustible materials. One important environmental consequence of using a combustible material is the production of greenhouse gases. The apparently obvious conclusion is to replace fossil fuels with renewable sources of energy that are environmentally benign. Before judging fossil fuels too harshly, however, it must be realized that every energy source has advantages and disadvantages.

The environmental impact of some clean energy sources is summarized below:

- Hydroelectric facilities, notably dams, can flood vast areas of land. The flooded areas can displace people and wildlife, and impact the ecosystems of adjacent areas with consequences that may be difficult to predict. Dams can change the composition of river water downstream of the dam and can deprive land areas of a supply of silt for agricultural purposes. A dam on a river can prevent the upstream migration of certain species of fish, such as salmon.
- Geothermal power plants can emit radiation and toxic gases such as hydrogen sulfide or greenhouse gases such as carbon dioxide. The produced water from a geothermal reservoir will contain dissolved solids that can form solid precipitates when the temperature and pressure of the produced water changes.
- Solar power plants have a relatively large footprint, can harm wildlife, interfere with air traffic, and may be visually offensive to some people.
- Wind farms have a relatively large footprint, can interfere with wildlife migration patterns, fling ice under suitable wintry conditions, catch fire, and be visually offensive, especially in areas where tourism depends on a pristine landscape.

Energy density, cost, and reliability are among the advantages that fossil fuels enjoy relative to other energy sources. A reliable energy source is a resource that is available almost all of the time. Some downtime may be necessary for facility maintenance or other operating reasons. The United States Department of Energy has defined a reliable energy source as an energy source that is available at least 95% of the time [DoE Geothermal, 2016]. The selection of an energy source should consider such factors as energy density, cost, reliability, social acceptability, and environmental impact. Improvements in large scale energy storage technology are an ongoing need for intermittent energy sources.

9.5 Activities

True-False

Specify if each of the following statements is True (T) or False (F).

1. Photosynthesis occurs in plants and bacteria.
2. Biological conversion of solar energy relies on photosynthesis.
3. Each stage of anaerobic digestion requires oxygen.
4. Biomass includes matter that was recently formed as a result of photosynthesis.
5. Biomass gasification converts solid biomass to flammable gas.
6. A methane digester produces Octane from a biological feedstock.
7. Distillation is a process that is used to extract pure ethanol from biomass.
8. Gas is generated in a landfill by the decay of organic material.
9. Synthetic diesel can be obtained from plants.
10. Ethanol can be made from sugarcane.

Questions

1. What are the 3 stages of anaerobic microbial degradation?
2. A landfill generates 4 million cubic feet of gas per day. What is the cash flow (in US$ per day) of the gas if the price of gas is US$5 per thousand cubic feet of gas?
3. What is the product X in the combustion reaction $C+O_2 \rightarrow X$?
4. What impact does carbon dioxide have on the environment? You should consider the atmosphere and the hydrosphere.
5. How is ethanol used in transportation?

6. The energy density of dry wood is 10,000 MJ/m^3. Assume that 5% of global energy consumption is due to burning of wood, and annual global energy consumption is approximately 5.8×10^{17} kJ. Estimate the volume of wood consumed?
7. What is the Fischer-Tropsch process?
8. Suppose 1 bushel of corn is needed to produce 3 gal of ethanol. How many bushels of corn per year are needed to provide 30 million gal ethanol per year?
9. A power plant generates 140 kWh electricity per ft^3 methane. How much methane must be produced at a wastewater treatment plant so the power plant can generate 70 million kWh electricity?
10. Medium oil content biodiesel from microalgae provides 98,000 liters of oil per hectare per year. How many square miles of microalgae would be needed to produce 7.3 billion barrels of oil per year?

CHAPTER 10

ENERGY CARRIER, ENERGY STORAGE AND HYBRID ENERGY SYSTEMS

After reading this chapter, you should be able to...

- Define different types of load
- Discuss hydrogen and its role as an energy carrier
    - *P2P: Is hydrogen the perfect fuel of the future?*
    - *P2P: Why is electrolysis important?*
    - *P2P: Where do we get the energy we need to produce hydrogen?*
- Describe the Hydrogen Economy
    - *P2P: Would people be willing to use hydrogen in their everyday lives?*
- Discuss large scale energy storage options
- Describe a hybrid energy system

Energy sources can be classified as renewable and nonrenewable. Some energy sources like coal, oil, and gas are relatively easy to transport and store. Other energy sources like wind and solar are intermittent energy sources that require backup to meet demand.

Demand for electric power is the load on a utility. Utilities need to have power plants that can meet three types of loads: base load, intermediate or cycling load, and peak load. The base load is the minimum baseline demand that must be met in a 24-hour day. Intermediate load is the demand that is required for several hours each day and tends to increase or decrease slowly. Peak load is the maximum demand that must be met in a 24-hour day. Today, most power plants that rely on fossil fuels and nuclear energy provide base load power, and wind farms and solar electric generating facilities provide supplemental power. If we want to rely on renewable energy in the future for all of our power needs, it will be necessary to find a way to provide power when intermittent energy sources are offline.

In addition, energy from nuclear, wind, and solar energy sources use electricity as an energy carrier, that is, an agent for distributing energy. Electricity has proven to be difficult to store in sufficient quantities for many commercial applications. An alternative energy carrier that has more useful transport and storage properties than electricity is hydrogen. To optimize the use of nuclear, wind, and solar energies, we need to resolve the energy transport and storage issues. The following possible solutions are discussed here: hydrogen as an energy carrier, large scale energy storage systems, and hybrid energy systems.

10.1 HYDROGEN

Hydrogen is found almost everywhere on the surface of the earth as a component of water. Each water molecule in the mountain reservoir in Summit County, Colorado (Figure 10-1) contains two atoms of hydrogen and one atom of oxygen. Hydrogen has many commercial uses, including production of ammonia for use in fertilizers, methanol production, hydrochloric acid production, and use as a rocket fuel. Liquid hydrogen is used in cryogenics and superconductivity. Hydrogen is important to us because it can be used as a fuel that is relatively easy to store and transport.

Hydrogen can be used as a fuel for a modified internal combustion engine or in a fuel cell. Fuel cells are electrochemical devices that directly convert hydrogen, or hydrogen-rich fuels, into electricity using a chemical process. Fuel cells do not need recharging or replacing and can produce electricity as long as they are supplied with hydrogen and oxygen.

Figure 10-1. Hydrogen in its Most Common Form on Earth (© Fanchi, 2000)

Hydrogen can be produced by steam reforming of hydrocarbons, electrolysis of water, and biological processes using plants like algae. Water electrolysis uses electrical energy to split water into its constituent elements. Electrolysis is a net energy consuming process in which more energy is consumed than produced. Hydrogen is not considered an energy source because more energy is required to produce hydrogen than can be obtained from hydrogen. Hydrogen is considered a carrier of energy because it can be used as a fuel to provide energy.

The properties of hydrogen make it a source of hope for a sustainable, global energy system. Here we first describe the properties of hydrogen before discussing the production of hydrogen and its use as the primary fuel in fuel cells.

10.1.1 Properties of Hydrogen

Hydrogen is the first element in the Periodic Table. The nucleus of hydrogen is the proton and hydrogen has only one electron. At ambient conditions on Earth, hydrogen is a colorless, odorless, tasteless, and nontoxic gas of diatomic molecules, that is, a molecule with two atoms represented by the symbol H_2. Selected physical properties of hydrogen, methane and gasoline are shown in Table 10-1. The properties of gasoline are illustrative since there are many grades of gasoline.

Table 10-1
Selected Physical Properties of Hydrogen, Methane, and Gasoline

Property	Hydrogen (gas)	Methane (gas)	Gasoline (liquid)
Molecular Weight (grams/moles) [a]	2.016	16.04	~110
Mass Density (kilograms/meters^3) [a,b]	0.09	0.72	720-780
Energy Density (megajoules/kilograms)	120 [a]	53[c,d]	46 [a,c]
Volumetric Energy Density (megajoules/meters^3)[a]	11 [a]	38[c,d]	35,000 [a,c]
Higher Heating Value (megajoules/kilograms) [a]	142.0	55.5	47.3
Lower Heating Value (megajoules/kilograms) [a]	120.0	50.0	44.0

a. Ogden [2002, Box 2, page 71]
b. at 1 atm and 0 °C
c. Hayden [2001, page 183]
d. Ramage and Scurlock [1996, Box 4.8, page 152]

A kilogram (2.2 pounds mass) of hydrogen, in either the gas or liquid state, has a greater energy density than the most widely used fuels today, such as oil and coal. The volumetric energy density of hydrogen is much less than gasoline however, which means that a larger volume of hydrogen is needed to provide the same energy as gasoline, or hydrogen must be compressed and stored under pressure.

Point to Ponder: Is hydrogen the perfect fuel of the future? Hydrogen is considered a clean, reliable fuel once it is produced because the combustion of hydrogen produces water vapor. Hydrogen can react with oxygen to form water in an exothermic combustion reaction. Hydrogen combustion does not emit greenhouse gases like carbon monoxide or carbon dioxide. When hydrogen is burned in air, it does produce traces of nitrogen oxides, which are pollutants. A particularly costly problem with hydrogen is that hydrogen is not readily available as free hydrogen; hydrogen is typically found in combination with other elements, such as oxygen in water. Consequently, elemental hydrogen must be freed before it can be used as a fuel. The production of hydrogen is discussed below.

10.1.2 Hydrogen Production by Electrolysis

Hydrogen can be produced by electrolysis. Electrical energy is used in electrolysis to split a molecule into smaller parts. In the case of water, electrolysis would decompose the water molecule into its constituent elements (hydrogen and oxygen) by the addition of electrical energy.

Electrolysis requires the presence of ions that can flow as an electrical current. Electrolyzing an aqueous salt solution enhances the production of hydrogen by the electrolysis of water. An aqueous salt solution is a mixture of ions and water. The addition of a small amount of a non-reacting salt such as sodium sulfate accelerates the electrolytic process. The salt is called an electrolyte and provides ions that can flow as a current.

We illustrate the electrolysis process in Figure 10-2 using the electrolysis of a water-salt solution. The salt in our example is table salt. Chemically, table salt is sodium chloride and is made up of one atom of sodium (Na) and one atom of chlorine (Cl). Sodium chloride (NaCl) separates into electrically charged Na^+ and Cl^- ions when dissolved in water. The ions respond to an electrical potential difference and make the water-salt solution electrically conductive.

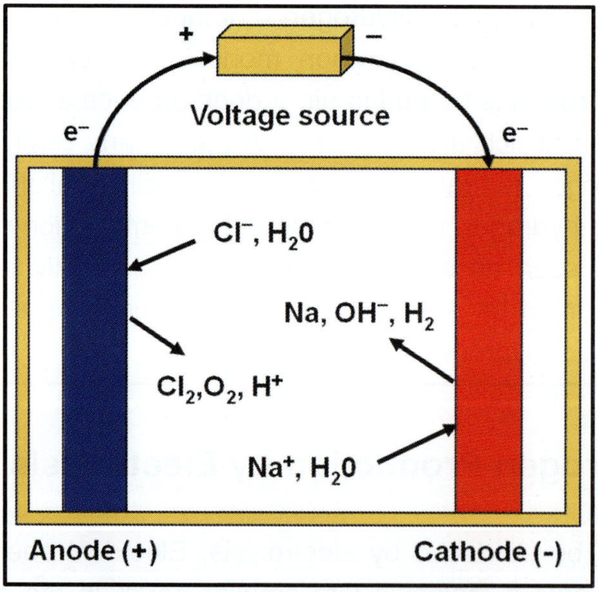

Figure 10-2. Electrolysis of Water-Salt (NaCl) Solution

The electrolytic cell in Figure 10-2 is a cell with two electrodes: the anode and the cathode. The anode is the positively charged electrode and attracts negatively charged ions (anions). By contrast, the cathode is the negatively charged electrode and attracts positively charged ions (cations). The voltage source (such as a battery) acts like an electron pump that pulls electrons from the anode and pushes them to the cathode. In the figure, a negatively charged chlorine ion with one extra electron loses its excess electron at the anode and combines with another chlorine atom to form the chlorine molecule (Cl_2). Electrons flow from the anode through an external circuit to the cathode where positively charged sodium ions combine with the electrons and become electrically neutral atoms of sodium.

Electrolysis is a reduction-oxidation (redox) reaction. Oxidation refers to the loss of electrons and occurs when electrons are released by a chemical species. Reduction refers to the gain of electrons and occurs when electrons are accepted by a chemical species. Oxidation occurs at the anode, and reduction occurs at the cathode. The electrolysis reaction can be viewed as two half-reactions: the oxidation half-reaction at the anode, and the reduction half-reaction at the cathode. A reducing agent or reductant is a chemical species that donates electrons in a reaction, and an oxidizing agent or oxidant is a chemical species that accepts electrons in a reaction.

The overall reaction for electrolysis of a water-salt solution in an electrolytic cell shows that hydrogen gas is produced at the cathode of the electrolytic cell and oxygen gas is produced at the anode when enough electrical energy is supplied. Hydrogen ions combine with hydroxyl ions to form molecules of water. Electrical energy is supplied in the form of an electrical voltage and must be greater than the threshold energy (activation energy) of the reaction.

The temperature of the electrolytic solution can affect the rate of hydrogen production by electrolysis. Conventional sources of heat such as hot plates or solar thermal devices can provide the heat. Electrolysis at elevated temperatures is called high temperature electrolysis. A related method of producing hydrogen is hydrogen production by infrared (HyPIR) electrolysis.

HyPIR electrolysis increases the rate of production of hydrogen gas by irradiating the electrolytic cell with infrared radiation of a specific frequency using a laser [Fanchi, 2012]. The optimum frequency for operation of the laser depends on the absorption frequency of the electrolytic solution. The photon energy excites the water molecule by increasing the vibration of the molecule. The band of frequencies that can be affected by the laser beam is broadened by thermal motion of the water molecule and collisions with other molecules. In HyPIR electrolysis, a laser emits a beam of infrared light that irradiates the electrolytic cell and facilitates the dissociation of water by stretching the hydrogen-oxygen bond. This excitation can appear as a heating effect since it increases the motion of the water molecule. The laser beam is more efficient than a conventional heat source at providing energy because it directly affects the hydrogen-oxygen bond.

Point to Ponder: Why is electrolysis important?
Hydrogen is a component of water. The distribution and amount of water on Earth implies that we have an abundant source of hydrogen for energy use if we can separate hydrogen from the water molecule, collect it and use it as a fuel.

Electrolysis can produce hydrogen for use as a fuel. Hydrogen production by electrolysis and hydrogen consumption can be environmentally benign. Hydrogen is a convenient fuel because it is relatively easy to store and it is portable. Hydrogen is not a source of energy, however, because the energy to break the hydrogen-oxygen bond is greater than the energy provided by the produced hydrogen. That is why electrolysis is considered a net energy consuming process and hydrogen is considered an energy carrier.

10.1.3 Hydrogen Production by Thermal Decomposition and Gasification

Electrolysis is one technique for producing hydrogen. Two other techniques are thermal decomposition and gasification. The process of decomposing the water molecule at high temperatures is called thermal decomposition. Heat in the form of steam can be used to reform hydrocarbons and produce hydrogen.

Steam reforming exposes a hydrocarbon such as methane, natural gas, or gasoline to steam at high temperature (700 to 1100° Centigrade) and high pressure [Sørensen, 2004]. Steam reforming reactions produce hydrogen. Steam reforming using hydrocarbons has limited long-term potential because hydrocarbons are not a renewable source of energy. The amount of methane in hydrates and shale can extend the usefulness of methane steam reforming. We can also produce a hydrocarbon like natural gas using a renewable technique such as methane digestion, but the rate of production is relatively slow compared to the current rate of consumption. The production of carbon dioxide, a greenhouse gas, is an undesirable characteristic of steam reforming. A process such as absorption, membrane separation or sequestration can be used to remove the carbon dioxide byproduct.

Some bacteria can produce hydrogen from biomass by fermentation or high-temperature gasification. The gasification process is similar to coal gasification.

Point to Ponder: Where do we get the energy we need to produce hydrogen?
One of the problems facing society is how to produce hydrogen using an environmentally acceptable process. Three possible sources of energy for producing hydrogen in the long-term are

wind energy, nuclear fusion, and solar energy. We have already seen that wind energy and solar energy are intermittent sources of energy. One way to enhance the usefulness of wind and solar energy is to use excess wind and solar energy to produce hydrogen during periods when the wind is blowing and the sun is shining. The produced hydrogen could then be used as a fuel when wind energy and solar energy are not being generated.

J.H. Ausubel [2000] has suggested that the potential of nuclear energy will be realized when nuclear energy can be used as a source of electricity and high-temperature heat for splitting water into its constituent parts. In this scenario, nuclear energy could be generated at sites with excess energy generating capacity and then used to produce hydrogen. The hydrogen could then function as a carrier of energy.

10.1.4 Hydrogen and Fuel Cells

Hydrogen is a carrier of energy that can be transported in the liquid or gaseous state by pipeline or in cylinders. Once produced and distributed, hydrogen can be used as a fuel for a modified internal combustion engine or as the fuel in a fuel cell. Fuel cells are electrochemical devices that directly convert hydrogen, or hydrogen-rich fuels, into electricity using a chemical rather than a combustion process.

A fuel cell consists of an electrolyte sandwiched between an anode and a cathode (Figure 10-3). The electrolyte is a substance that dissociates (or separates) into free ions so that the substance is electrically conductive. A water-salt solution is a common electrolyte because salt is composed of molecules that can separate into positively charged ions and negatively charged ions when dissolved in water. The presence of ions increases the electrical conductivity of the water-salt solution. The electrolyte in Figure 10-3 is a mixture of potassium hydroxide and water.

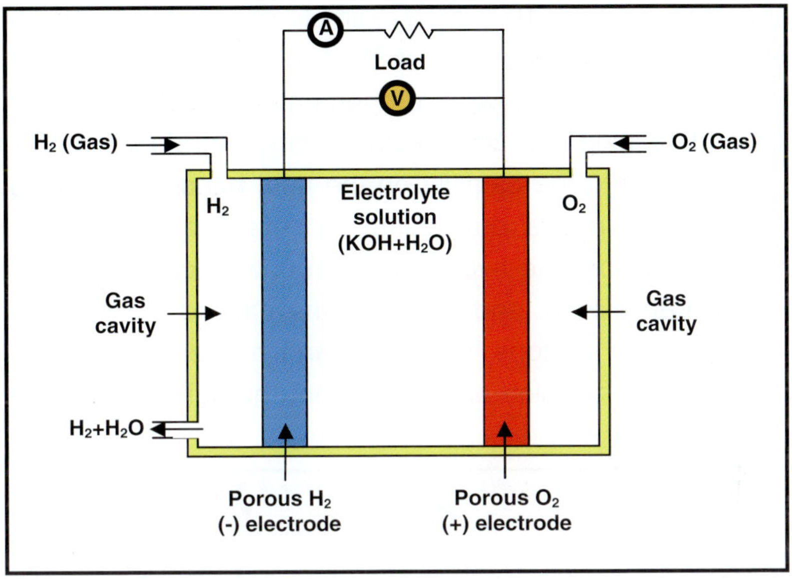

Figure 10-3. Schematic of a Fuel Cell [after Cassedy and Grossman, 1998, page 419; and Crowe, 1973]

The electrolyte solution is maintained at a lower pressure than the gas cavities on either side of the porous electrodes. The pressure gradient facilitates the separation of hydrogen and oxygen molecules. The type of electrolyte in a fuel cell distinguishes that fuel cell from other types of fuel cells. The load in the figure is a circuit with amperage *A* and voltage *V*. Hydrogen is fed to the anode and oxygen is fed to the cathode. When activated by a catalyst, hydrogen atoms dissociate into protons and electrons. The charged protons and electrons take different paths to the cathode. The electrons go through the external circuit and provide an electrical current, while the protons migrate through the electrolyte to the cathode. Once at the cathode, the protons combine with electrons and oxygen to produce water and heat. The heat is waste heat unless it can be captured for use in a co-generation process. Produced water must be removed from the fuel cell.

The fuel cell in Figure 10-3 is an example of a proton exchange membrane (PEM) fuel cell. A model of a PEM fuel cell is sketched in Figure 10-4. Part A of the figure shows hydrogen entering the catalyst on the left hand side and splitting into component electrons and hydrogen protons. Oxygen is flowing into the catalyst on the right hand side. Part B shows positively charged protons drawn into the membrane while negatively charged electrons from the hydrogen atom flow through an external circuit, pass through a load, and complete the path to the catalyst on the right hand side. The protons, electrons, and oxygen atoms combine to form water molecules, which exit the system through the exhaust shown in Part C.

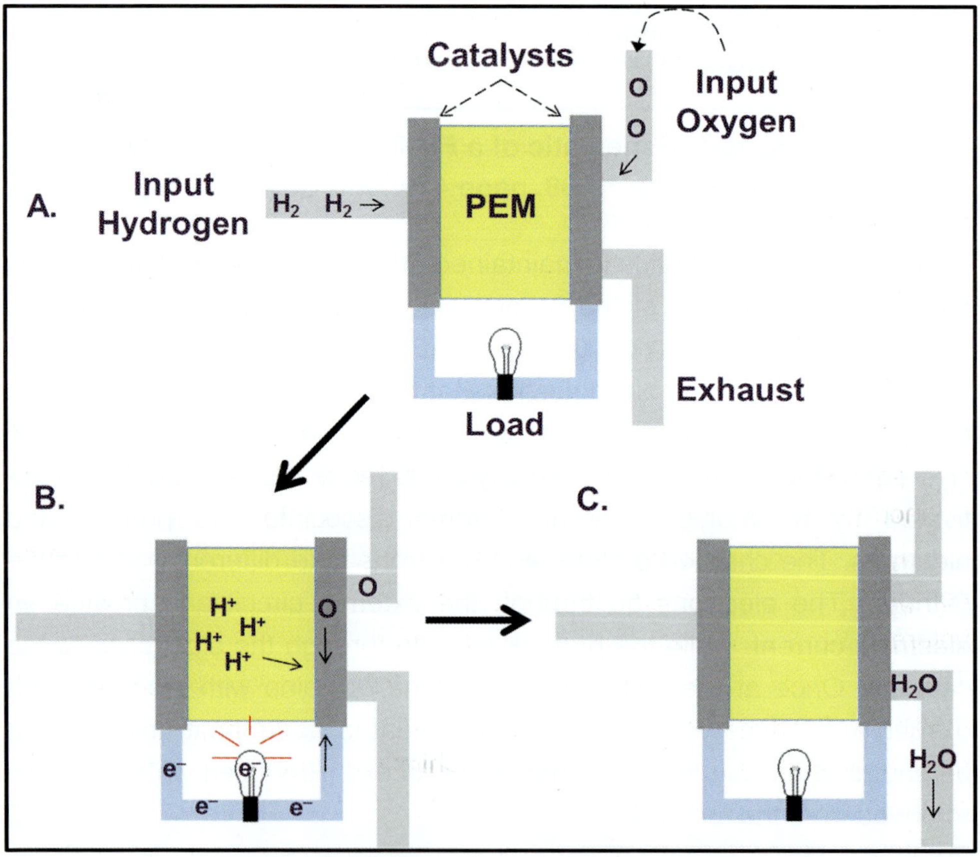

Figure 10-4. Model of a PEM Fuel Cell

Fuel cells produce clean energy in the form of electricity and heat as long as they are supplied with hydrogen and oxygen. Hydrogen and oxygen are fuels that must be provided from some suitable storage location.

10.2 The Hydrogen Economy

The historical trend toward decarbonization can be interpreted to imply that hydrogen will be the fuel of choice in the future. Power plants and motor vehicles will run on hydrogen, and the economies that emerge will be called hydrogen economies.

The concept of a hydrogen economy is not new. The use of hydrogen as a significant fuel source driving a national economy was first explored in the middle of the 20th century as a complement to the adoption of large scale nuclear electric generating capacity. Concerns about anthropogenic climate change and the desire to achieve sustainable development have renewed interest in hydrogen as a fuel.

A future that depends on hydrogen is not inevitable because many challenges remain. Hydrogen economies will require the development of improved technologies for producing, storing, transporting, and consuming hydrogen. We have already discussed some of the challenges involved in the production of hydrogen. Another example of a technological challenge that must be overcome in a transition to a hydrogen economy is the ability to store hydrogen.

Hydrogen can be stored in the liquid or gaseous state, but it must be compressed to high pressures or liquefied to achieve reasonable storage volumes because of the low density of the diatomic hydrogen molecule. The energy content of hydrogen gas is less than the energy contained in methane at the same temperature and pressure. The volumetric energy density of liquid hydrogen is about one third the volumetric energy density of gasoline. The relatively low volumetric energy density of hydrogen creates a storage problem if we want to store hydrogen compactly in vehicles.

Hydrogen can be stored effectively in the form of solid metal hydrides. A metal hydride is a metal that absorbs hydrogen. The hydrogen is absorbed into the spaces, or interstices, between atoms in the metal. According to Silberberg [1996, page 246], metals such as palladium and niobium "can absorb 1000 times their volume of H_2 gas, retain it under normal conditions, and release it at high temperatures." This form of storage may be desirable for use in hydrogen-powered vehicles. A model of a car that uses hydrogen fuel cells is shown in Figure 10-5.

Figure 10-5. Hydrogen Fuel Cell Car, Denver, Colorado Exhibit (© Fanchi, 2003)

Hydrogen can be hazardous to handle. A spectacular demonstration of this fact was the destruction of the German zeppelin *Hindenburg*. The *Hindenburg* used hydrogen for buoyancy. Hydrogen was contained onboard in a set of separate bags so that a leak in one bag would not lead to a collapse of the zeppelin. In 1937, the *Hindenburg* burst into flames while attempting a mooring in Lakehurst, New Jersey during a thunderstorm.

At the time of the disaster, people believed that the hydrogen in the *Hindenburg* was the cause of the explosion. Another possibility was sabotage. The *Hindenburg* was viewed as an example of leading edge technology in Nazi Germany. Destruction of the *Hindenburg* by sabotage would have been considered a political statement. A more modern explanation was proposed by Addison Bain at the end of the 20th century [Bain and Van Vorst, 1999]. Bain suggested that the chemical coating on the outside of the zeppelin was the cause of the explosion. When the chemical coating was ignited, perhaps by an electrical spark or lightning, the hydrogen began to burn. The one factor that is common to all explanations of the *Hindenburg* disaster is that hydrogen did ignite and burn. Today, a typical gas used in lighter-than-air ships is the inert gas helium.

Hydrogen, natural gas and gasoline are flammable in air. The two major chemical components in air are nitrogen and oxygen. Hydrogen forms an explosive mixture with air when the concentration of hydrogen in air is in the range of 4% to 75% hydrogen. For comparison, natural gas is flammable in air when the concentration of natural gas in air is in the range of 5% to 15% natural gas. Furthermore, the energy to ignite hydrogen-air mixtures is approximately one-fifteenth the ignition energy for natural gas-air or gasoline-air mixtures. The lower ignition energy of hydrogen in air makes it possible to consider hydrogen a more dangerous fuel than natural gas. On the other hand, the low density of hydrogen allows hydrogen to dissipate more quickly into the atmosphere than a higher density gas such as methane. Thus, hydrogen leaks can dissipate more rapidly than natural gas leaks. Adding an odorant to the gas can enhance the detection of gas leaks.

The environmental acceptability of hydrogen fuel cells depends on how the hydrogen is produced. If a renewable energy source such as solar energy is used to generate the electricity needed for electrolysis, vehicles powered by hydrogen fuel cells would be relatively clean since hydrogen combustion emits water vapor. Unfortunately, hydrogen combustion in air also emits traces of nitrous oxide compounds. Nitrogen dioxide contributes to photochemical smog and can increase the severity of respiratory illnesses.

Point to Ponder: Would people be willing to use hydrogen in their everyday lives?

The common conception of hydrogen depends on a person's cultural background. People in the United States tend to think of the *Hindenburg* on fire when they think of hydrogen as a fuel. One of the tasks of an evolving hydrogen industry is to educate the public about the misconceptions associated with hydrogen.

One example of the education effort is to re-educate the public about the *Hindenburg* disaster. Hydrogen economy proponents have presented Bain's explanation of the *Hindenburg* disaster at trade shows and public exhibitions. Videos of staged automobile accidents are another example of the education effort.

The videos show how a hydrogen-fueled car would burn following an accident compared to a gasoline-fueled car. Hydrogen from a breach in the storage tank of a hydrogen-fueled car is less buoyant than air and would rise, while gasoline from a breach in the storage tank of a gasoline-fueled car would pool on the ground around the vehicle. The gasoline could ignite and engulf the vehicle, while the hydrogen would form a vertical flare of flame escaping from the breach. The videos show that hydrogen could be used as a relatively safe fuel in transportation. These examples show that proponents of the hydrogen economy believe that an education effort is needed to overcome social resistance to adopting hydrogen as a primary fuel.

10.3 Large Scale Energy Storage

Solar and wind energy are two renewable energy sources that hold great promise for replacing fossil fuels as the dominant energy sources in the future energy mix. The amount of wind and solar energy that can be converted to useful power depends on location, time of day, season, and weather conditions. Consequently, both solar and wind energy are intermittent sources of energy.

The problem of intermittency is currently solved by using a more reliable source of energy, such as nonrenewable fossil fuels or nuclear energy, to provide energy as needed. In this power configuration, solar and wind energy are used to supplement other forms of energy. Our ability to rely on solar and wind energy as baseline sources of energy depends on our ability to store solar and wind energy in forms that can be readily accessed. One way to optimize the performance of intermittent energy sources is to use large scale energy storage, which is the subject of this section.

10.3.1 Energy Storage Concepts

We have seen that energy exists in many different forms, and that energy can be transformed from one form to another. The transformation property of energy is being used to store energy in a variety of forms, including chemical, biological, electrical, electrochemical, mechanical, and thermal. Each of these forms is introduced here.

Energy can be stored in chemical form as an energy carrier such as hydrogen or as a biofuel or synfuel. Chemical reactions like combustion release the stored chemical energy.

Energy can be stored in biological substances such as starch or glycogen. Starch is a carbohydrate that stores energy from photosynthesis. Glycogen stores energy for use in animal cells.

Energy can be stored in electrical form in devices such as capacitors and superconducting magnetic energy storage (SMES) systems. A capacitor stores electrical energy in the dielectric material sandwiched between two conductors. The SMES system stores energy in the magnetic field that is induced by the direct current in a superconducting coil. The coil has to be cooled to very low temperatures so that it can be superconducting. Superconductors are materials that offer no resistance to electron flow.

Dutch physicist Hans Kamerlingh Onnes (1853-1926) discovered the first superconductor in 1911. Onnes produced liquid helium by cooling helium to a temperature below 4.2° Kelvin (-269° Centigrade). He then observed that the resistivity of mercury vanished when mercury was cooled by liquid helium. Superconducting materials have been developed that operate at higher temperatures than 4.2° Kelvin, but superconductors can still only operate at temperatures that are far below room temperature. Once the coil is cool enough to operate as a superconductor, the current does not encounter any resistance and the energy in the magnetic field induced by the current can be stored indefinitely. It is important to note that energy is needed to refrigerate the superconducting coil, so the storage of magnetic energy requires the consumption of energy for refrigeration.

Energy is stored in electrochemical form in devices such as batteries and fuel cells. These devices use reduction-oxidation reactions to generate a flow of electrons.

Energy is stored in mechanical form using a variety of mechanisms. One mechanical energy storage mechanism is compressed air energy storage (CAES). CAES uses off-peak energy from a power plant to inject high pressure gas into a reservoir. The gas is withdrawn from the reservoir to drive a gas turbine when additional power is needed.

Another mechanical energy storage mechanism is hydroelectric energy storage. This method uses off-peak energy from a hydroelectric power plant to pump water from a lower elevation to a storage pool at a higher elevation. The elevated water is used to drive a water turbine when additional power is needed.

Energy is stored in thermal form using a variety of mechanisms. One thermal energy storage mechanism is molten salt. We discussed one application of molten salt thermal energy storage in Chapter 6. Molten salt was used to store solar energy. Solar energy can also be used to heat a saltwater pool. The pool is called a solar pond and can be used for applications such as heating and desalination.

Another thermal energy storage mechanism is a steam accumulator. Off-peak energy can be used to heat water, which is stored as hot water or steam under pressure in an insulated steel pressure tank. The hot water or steam is used to drive a turbine when additional power is needed. Steam accumulators are being used at the solar power complex near Seville, Spain.

The energy storage techniques discussed above are in various stages of development. Some of the techniques are not well suited for large scale energy storage, and others use too much energy or are too costly to be commercial. CAES and steam accumulators are suitable for large scale energy storage.

10.3.2 Compressed Air Energy Storage

Compressed Air Energy Storage (CAES) is an example of a large scale energy storage technology that has been implemented in Germany and Alabama. CAES is designed to provide off-peak energy from primary power plants during peak demand periods. Compressed air can be stored in solution-mined salt cavities, mined hard rock cavities, aquifers, geologic reservoirs, or storage tanks. The gas storage technique depends on location of the site, economics, commercial objectives and government regulations.

The Huntorf CAES facility in Germany and the McIntosh CAES facility in Alabama store gas in salt caverns. The CAES facility in Huntorf was built in 1978 and is the world's first CAES project [Crotogino, et al., 2001]. Gas was stored in the two salt caverns shown in Figure 10-6. The power plant in the figure can generate 290 MW. Off-peak energy at Huntorf is used to pump air underground and compress it in a salt cavern. The compressed air is produced during periods of peak energy demand to drive a turbine and generate additional electrical power.

Figure 10-6. Huntorf CAES Facility

The McIntosh power plant is a gas-fired power plant with approximately 340 MW capacity. The CAES facility associated with the plant, designated McIntosh unit 1, came online in 1991. During peak hours when additional energy production capacity is needed, the CAES facility produces compressed air that is added to a natural gas mixture for combustion.

A third commercial scale CAES facility was built in Gaines, Texas and commissioned in 2012 [CAES General Compression, 2016]. The 2.0 MW wind generation project is called the Dispatchable Wind Project. The project includes a wind turbine, a General Compression Advanced Energy

Storage (GCAES™) system, a storage cavern, and other ancillary facilities. During periods of low demand, a portion of the energy generated by the wind turbine is stored as high-pressure air and heated water. During periods of increased demand, high pressure air combines with heated water to generate electricity.

CAES systems are well-suited for areas where power plants and salt formations coexist. There are many other areas where current or proposed wind farms, solar power plants, and nuclear power plants, among others, overlie permeable formations that have or could contain gas. The potential for widespread implementation of CAES technology is significant.

10.3.3 Renewable Energy and CAES

Intermittent sources of energy such as wind and solar energy can function as the primary energy source for a utility if they can reliably provide baseline electricity directly to the electrical grid. One way to store solar and wind energy is to convert these energy forms to other forms of energy. For example, excess solar energy from a solar power plant could be used to heat and compress a fluid that could be stored in a reservoir. The hot, compressed fluid could be extracted from the reservoir and used to generate electrical power when the amount of energy from the solar power plant declines. Similarly, excess wind energy from a wind farm could be used to compress a fluid that could be stored in a reservoir.

A sketch of a wind energy system is shown in Figure 10-7, and a solar energy system is shown in Figure 10-8. In both cases, energy from a wind farm or solar power plant is stored in compressed air energy storage (CAES) systems. Off-peak energy is used to inject gas into the reservoir. A production well is used to withdraw gas to drive a turbine when wind or solar energy is not sufficient to meet demand.

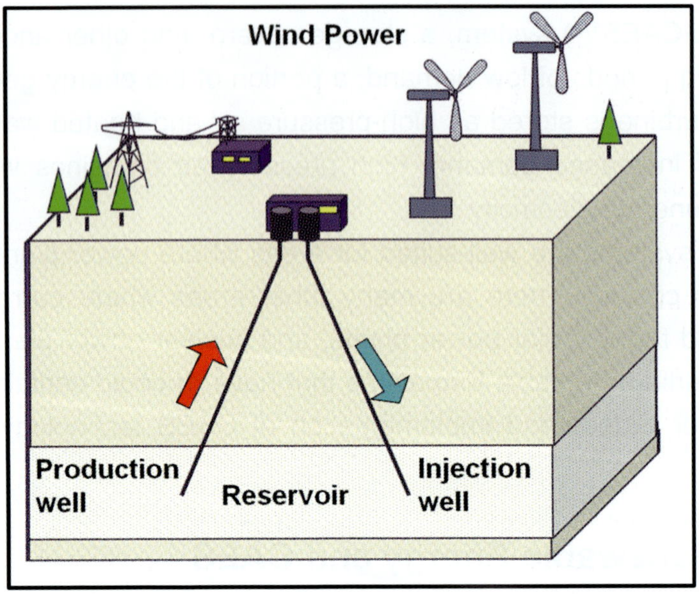

Figure 10-7. Sketch of CAES for a Wind Farm

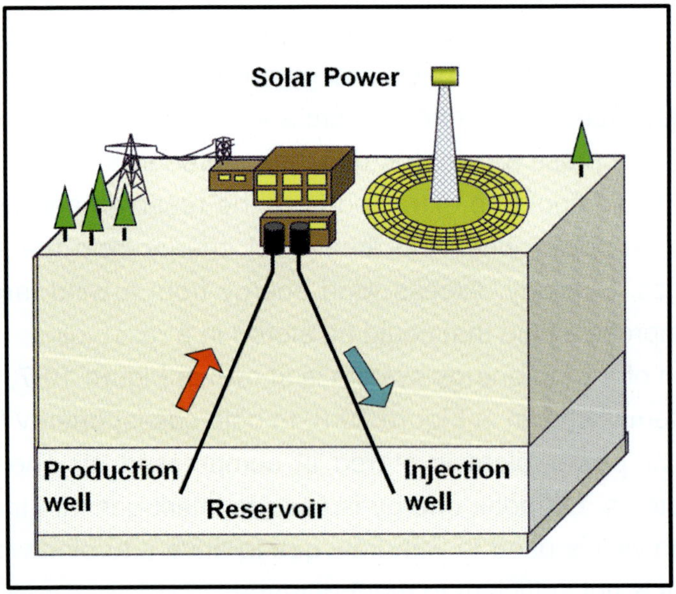

Figure 10-8. Sketch of CAES for a Solar Power Plant

10.4 A HYBRID ENERGY SYSTEM FOR THE HYDROGEN ECONOMY

A hybrid energy system is a system that combines two or more primary energy resources. For example, a wind farm supplemented by a natural gas fueled power plant is a hybrid energy system. Another example is the production and storage of hydrogen using an electricity-generating nuclear power plant.

Excess power capacity from sustainable energy systems such as nuclear plants, solar electric generating systems, and wind farms could be used to provide the energy needed to produce molecular hydrogen. The produced hydrogen would be suitable for a hydrogen-based economy, but the production of hydrogen requires a substantial amount of energy. Furthermore, hydrogen must be stored until it is needed. Figure 10-9 illustrates a hybrid system that could achieve the objectives of generating electricity, producing hydrogen, and storing hydrogen. The hybrid system includes a nuclear fission power plant and a hydrogen storage reservoir. Hydrogen is produced using off-peak power from the nuclear power plant and is stored in the reservoir until needed.

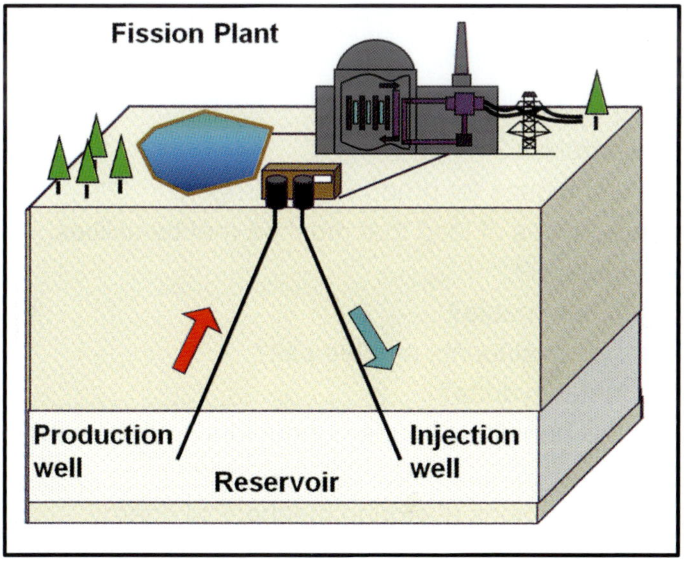

Figure 10-9. Sketch of a Nuclear Power Plant with Hydrogen Reservoir

10.5 Activities

True-False

Specify if each of the following statements is True (T) or False (F).

1. It is possible that the *Hindenburg* exploded because the chemical coating on the outside of the zeppelin caught on fire.
2. Molecular hydrogen can be produced from water by electrolysis.
3. The infrastructure exists for economically distributing and storing hydrogen.
4. Hydrogen is considered a source of energy.
5. The hydrogen atom is the first element in the Periodic Table.
6. Hydrogen is a common element found in sea water.
7. Thermal decomposition through steam reforming requires hydrocarbons and produces carbon dioxide.
8. Electrolysis is a net energy producing process.
9. Hydrogen gas has a higher energy density than gasoline.
10. CAES facilities are considered viable large-scale energy storage options.

Questions

1. What is the hydrogen economy?
2. What is electrolysis?
3. List three techniques for producing hydrogen.
4. List the three types of load that must be met by utilities.
5. What is a fuel cell?
6. Is hydrogen flammable?
7. What are the constituents of table salt?
8. What is a metal hydride?
9. What are the benefits of superconductors? What is the main drawback?
10. Describe two methods of storing mechanical energy.

CHAPTER 11

ELECTRICITY GENERATION AND DISTRIBUTION

After reading this chapter, you should be able to...

- Discuss the historical development of electric power
  - *P2P: Is society willing to adopt new technologies?*
- Describe electric power generation
- Explain how transformers work
- Explain how electric power is distributed
- Define distributed generation
- Discuss electric grid designs in use today
- Describe the Smart Grid
  - *P2P: Should we be concerned that the Smart Grid will infringe on our freedom?*
- Describe the European Supergrid concept

Power plants electrified the United States, and eventually the rest of the modern world, in only a century and a half. Although other types of energy are used around the world, electricity is the most versatile form for widespread distribution. The role of electric power plants is to generate electrical power for distribution through a transmission grid. Historical developments that led to the modern power generation and distribution system are described below. We then discuss some changes that are being implemented to prepare electric distribution systems for the emerging energy mix.

11.1 Historical Development of Electric Power

People first used muscle energy to gather food and build shelters. Muscle energy was used to grind grain with stones, chop wood with hand axes, and propel oar-powered ships. In many instances in history, conquered people became slaves and provided muscle energy for their conquerors.

Stones, axes, and oars are examples of tools that were developed to make muscle energy more effective. Water wheels and windmills replaced muscle power for grinding grain as long ago as 100 B.C. Wind and sails replaced muscle energy and oar-powered ships. Advances in the ability to use heat for power generation provided power for the industrial revolution.

11.1.1 Steam Generated Power

Early power stations were driven by wind and flowing water, and were built where wind and flowing water were available. The combination of heat and water produced steam, which could be used to drive a piston and provide mechanical energy (Figure 11-1). Steam generated power made it possible to move power plants away from sources of flowing water, but had to await the development of furnaces. Furnaces used heat to smelt ore and release metals such as copper, tin, iron, and uranium. Heat from fire frees the metal atoms and allows them to be collected in a purified state. Copper and tin were the first metals smelted and could be combined to form bronze. Purified metals could be used separately or as alloys to build devices that could be operated at elevated temperatures. This technological advance was an important prerequisite for generating power using steam.

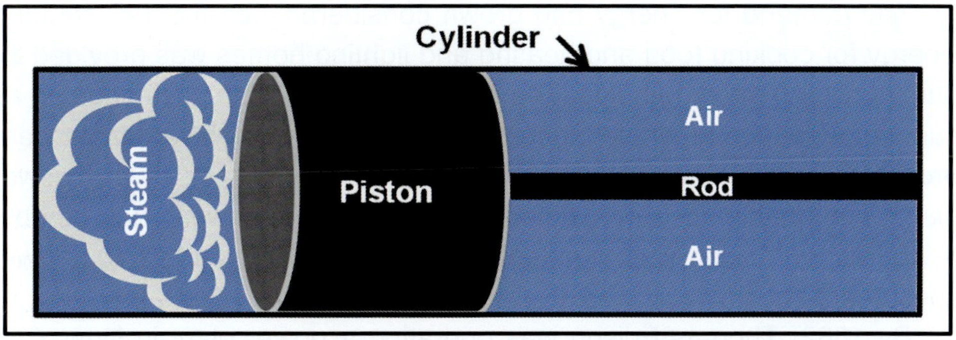

Figure 11-1. Schematic of a Simple Steam Engine

Scottish engineer James Watt improved the efficiency of the steam engine by introducing the use of a separate vessel for collecting and condensing the expelled steam. Watt's assistant, William Murdock, developed a gear system design in 1781 that enabled the steam engine to produce circular motion. The ability to produce circular motion made it possible for steam engines to provide the power needed to turn wheels. Steam engines could be placed on platforms attached to wheels to provide power for transportation. Thus was born the technology needed to develop steam-driven locomotives, paddle wheel boats, and ships with steam-driven propellers. The discovery that power plants did not have to be built near a particular fuel source meant that manufacturers had the freedom to build their manufacturing facilities in locations that optimized the success of their enterprise. If they chose, they could build near coal mines to minimize fuel costs, or near markets to minimize the cost of distributing their products.

Steam generated power was an environmentally dirty source of power. Burning biomass such as wood or a fossil fuel such as coal produced the heat needed to generate steam. Biomass and fossil fuels were also used in the home. Attempts to meet energy demand by burning primarily wood in 16th century Britain led to deforestation and the search for a new fuel source [Nef, 1977]. Fossil fuel in the form of coal became the fuel of choice in Britain and other industrialized nations. Coal gas, which is primarily methane, was burned in 19th century homes.

The demand for energy had grown considerably by the 19th century. Energy for cooking food and heating and lighting homes was provided by burning wood, oil, or candles. The oil was obtained from such sources as surface seepages or whale blubber. Steam generated power plants were providing power to consumers in the immediate vicinity of the power plant, but unable to serve potential customers that were not nearby. A source of power was needed that could be transmitted to distant consumers.

By 1882, Thomas Edison was operating a power plant in New York City. Edison's plant generated direct current electricity at a voltage of 110 volts. Nations around the world soon adopted the use of electricity. By 1889, a megawatt electric power station was operating in London, England. Industry began to switch from generating its own power to buying power from a power generating company. A fundamental inefficiency was present in Edison's approach to electric power generation, however. The inefficiency was not removed until the Battle of Currents was fought and won.

11.1.2 The Battle of Currents

The origin of power generation and distribution is a story of the Battle of Currents, a battle between two titans of business: Thomas Edison and George Westinghouse. The motivation for their confrontation can be reduced to a single, fundamental issue: how to electrify America.

Edison invented the first practical incandescent lamp and was a proponent of electrical power distribution by direct electric current. He displayed his direct current technology at New York City's Pearl Street Station in 1882. At the time, direct current could not be transmitted as far as alternating current without significant power loss. Late 19th century materials and technology enabled easier transformation of AC power between relatively low voltage and high voltage [Backhaus and Chertkov, 2013]. Electric power generation provides relatively low voltage power. High voltage power lines enable low-loss, long-distance power transmission. High voltage power must be transformed to lower voltage power for consumption.

Unlike Edison, Westinghouse was a proponent of alternating current because it could be transmitted over much greater distances than direct current. Alternating current could be generated at low voltages, transformed to high voltages for transmission through power lines, and then reduced to lower voltages for delivery to the consumer. Nikolai Tesla, a Serbian-American scientist and inventor who was known for his work with magnetism, worked with Westinghouse to develop alternating current technology. Westinghouse displayed his technology at the 1893 Chicago World's Fair. It was the first time one of the world's great events was illuminated at night and it showcased the potential of alternating current electricity.

The first large-scale power plant was built at Niagara Falls near Buffalo, New York in the 1890's. The power plant at Niagara Falls began transmitting power to Buffalo, less than 30 kilometers (20 miles) away, in 1896. The transmission technology used alternating current technology. The superiority of alternating current technology gave Westinghouse a victory in the battle of currents and Westinghouse became the father of the modern power industry. Westinghouse's success was based on the selection of a technology that was better at the time. Today, we know that high voltage direct current may be transmitted with less expense and less power loss than alternating current if it is transmitted over great distance. Direct current can also be used to connect alternating current transmission grids. The physical principles that are the basis of electric power transmission are discussed below.

A chronology of milestones in the development of electrical power is presented in Table 11-1 [after Brennan, et al., 1996, page 22; and Aubrecht, 1995, Chapter 6]. The milestones refer to the United States, which was the worldwide leader in the development of an electric power industry.

Table 11-1
Early Milestones in the History of the Electric Power Industry in the United States

Year	Event	Comment
1882	Pearl Street Station, New York	Edison launches the "age of electricity" with his DC power station
1893	Chicago World's Fair	Westinghouse displays AC power to the world
1898	Fledgling electric power industry seeks monopoly rights as regulated utilities	Chicago Edison's Samuel Insull leads industry to choose regulation over "debilitating competition"
1907	States begin to regulate utilities	Wisconsin and New York are first to pass legislation
1920	Federal government begins to regulate utilities	Federal Power Commission formed

11.1.3 Growth of the Electric Power Industry

The power industry started out as a set of independently owned power companies. Because of the large amount of money needed to build an efficient and comprehensive electric power infrastructure, the growth of the power industry required the consolidation of smaller power companies into a set of fewer but larger power companies. The larger, regulated power generating companies became public utilities and could afford to build regional electric power transmission grids. The ability to function more effectively at larger scales is an example of an economy of scale.

Utility companies were able to generate and distribute more power at lower cost by building larger power plants and transmission grids.

Power plants operated by utilities should be able to meet three types of loads: base load, intermediate or cycling load, and peak load. Base load and peak load are minimum baseline demand and maximum demand, respectively, in a 24-hour period. Intermediate load is the demand that is required for several hours each day.

Many power plants are built near major population centers to reduce distribution costs. Electric power for small towns and rural communities was an expensive extension of the power transmission system that required special support. The federal government of the United States provided this support when it created the Tennessee Valley Authority (TVA) and Rural Electric Associations (REA).

Point to Ponder: Is society willing to adopt new technologies?
The history of electric power development shows that society is willing to adopt new technologies. We have seen this recently in the adoption of the Internet and cell phones. When people understand the benefits of new technology and an appropriate infrastructure is created, people will adopt the technology. It can be inferred, then, that society will accept changes in its current energy system when people understand the benefits and repercussions of any changes.

11.2 Electric Power Generation

The first commercial-scale electric power plants were hydroelectric plants. The primary energy source of a power plant is the energy used to operate an electricity-generating power plant. For example, flowing water is the primary energy source for a hydroelectric plant. Today, most electricity is generated by one of the following primary energy sources: coal, natural gas, oil, water, or nuclear. Table 11-2 presents the supply of primary energy by type as a percentage of total primary energy supplied in the world for a selection of primary energy types. The statistics give us an idea of the relative importance of different primary energy sources. Fossil fuel was the dominant primary energy source at the beginning of the 21st century, and is the fuel of choice for transportation. Electricity is the leading energy carrier.

Table 11-2
World Total Primary Energy Supply by Energy Type
[IEA Key Statistics, 2015]

Primary Energy Type	World Total Primary Energy Supply	
	2009	2013
Oil	32.8 %	31.1%
Coal	27.2 %	28.9%
Natural Gas	20.9 %	21.4%
Biofuels and Waste	10.2 %	10.2%
Nuclear	5.8 %	4.8%
Hydroelectric	2.3 %	2.4%
Other*	0.8 %	1.2%
*Geothermal, solar, wind, etc.		

11.2.1 Hydroelectric Power Generation

An example of electric power generation is hydroelectric power generation. Many of the first commercial electric power plants relied on flowing water as their primary energy source. A schematic of a hydroelectric power plant is presented in Figure 11-2. Hydroelectric power generation was discussed in Chapter 8 where we pointed out that the mechanical energy of flowing water was converted to electrical energy. Water flows from an upper elevation to a lower elevation through a pipeline called a penstock. The water current turns a turbine that is connected to a generator. The turbine rotates the generator shaft. The mechanical energy of flowing water is transformed into the kinetic energy of rotation of the turbine.

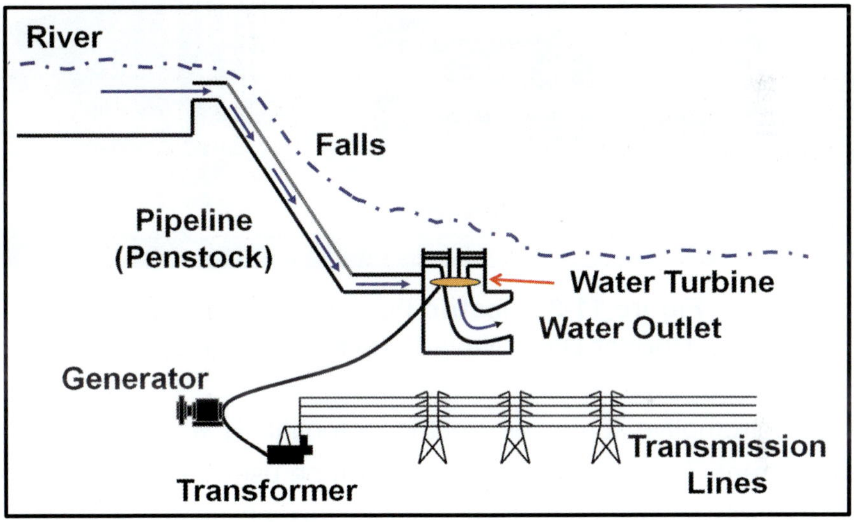

Figure 11-2. Principles of Hydroelectric Power Generation

An alternating current generator converts mechanical energy to electrical energy using the following principle: an electric current is induced in a loop of wire that is allowed to rotate inside a constant magnetic field. If flowing water drives a turbine attached to a coil of wire inside a magnet, the coil of wire will rotate inside the magnetic field and an electric current will be induced in the coil of wire. Figure 11-3 shows a row of electric power generators at Hoover Dam, Nevada.

Figure 11-3. Electric Power Generators, Hoover Dam, Nevada (© Fanchi, 2002)

An electromotive force, or electric potential, is associated with the electric current. The electromotive force can be used to transmit electrical power through transmission lines. To do this efficiently, a transformer must be included in the system.

11.3 TRANSFORMERS

Transformers perform the function of converting, or transforming, voltages. Voltage is the driving force for moving electrically charged particles. Current quantifies how many charged particles are moving. A transformer is a device that can convert a small alternating current voltage to a larger alternating current voltage, or vice versa. For example, it is desirable to work with relatively small voltages at the power plant and provide a range of voltages to the consumer. In between, in the transmission lines, a large voltage is required to minimize resistive heating losses in the line. A transformer that increases the voltage is referred to as a step-up transformer and is said to step-up the voltage; a transformer that decreases the voltage is referred to as a step-down transformer and is said to step-down the voltage. Transformer T_1 in Figure 11-4 is a step-up transformer from the lower voltage (L.V.) at the power station to a higher voltage (H.V.) in the transmission line. Transformer T_2 in the figure is a step-down transformer that converts the higher voltage in the transmission line to a lower voltage that is suitable for the consumer. The low voltages shown at opposite ends of the transmission line do not have to be the same. The actual voltages used in the transmission line depend on the properties of the transformer.

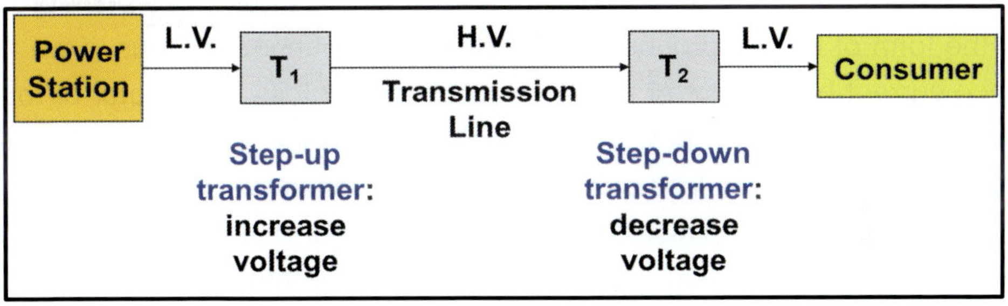

Figure 11-4. Power Transmission

Figure 11-5 shows that typical transmission line voltages can range from under 100,000 volts (100 kV) to over 750,000 volts (750 kV) [US-Canada Blackout Report, 2004, page 5]. A kilovolt (kV) is 1000 volts. The figure shows voltage step-up transformers between the generating station and transmission lines, and voltage step-down transformers between transmission lines and customers.

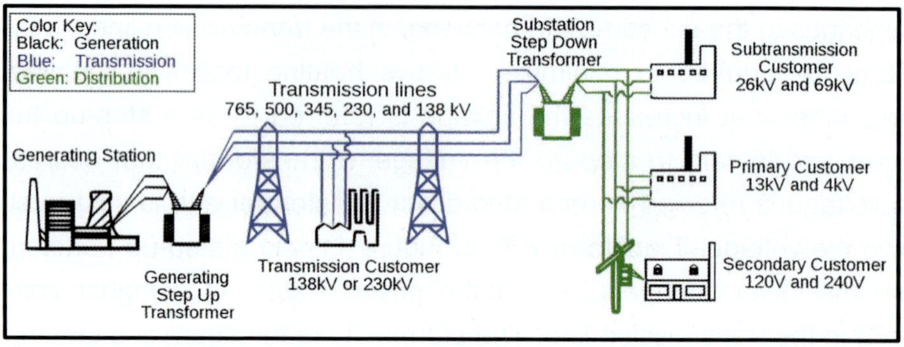

Figure 11-5. Schematic of a Power Transmission System

Advances in high voltage direct current (HVDC) technology have made it possible to transmit electrical power on HVDC power lines. Direct current can be transmitted at high voltages over great distances with relatively low power loss and cost compared to alternating current power transmission. In general, power is proportional to current, and power loss in the form of heat is proportional to the square of current. Thus, if we reduce current by a factor of two, we reduce power loss by a factor of four. For a given amount of power, an increase in voltage lets us reduce both the current and power loss associated with its transmission.

HVDC power transmission is a desirable transmission technology, yet HVDC power transmission is not widely used. The infrastructure does not exist for many applications to operate using HVDC power. Transmission voltage needs to be reduced to be compatible with end-use applications such as lighting and motors. Modern semiconductor technology can be used to transform HVDC, but the existing infrastructure is designed with voltage transformers that work with alternating current electricity rather than direct current electricity. Converting the existing electric power distribution system to HVDC for all applications is a significant economic cost. Consequently, HVDC power transmission has been limited to select applications, such as connecting the electricity grids described below.

11.4 ELECTRIC POWER DISTRIBUTION

Transmission lines are used to distribute electric power. As a rule, society would like to minimize power loss due to heating to maximize the amount of primary energy reaching the consumer from power plants. We can reduce power loss by reducing the current or by decreasing the distance of transmission. In most cases, it is not a viable option to decrease the transmission distance. It is possible, however. For example, we could choose to build a manufacturing facility near a power station to minimize the cost of transmission of power. One consequence of that decision is that the manufacturer may incur an increase in the cost of transporting goods to market. Another possibility is to install smaller, lower capacity power plants closer to more consumers. The result is a broader distribution of generating capacity known as distributed generation and is discussed in more detail below.

A more viable option for reducing power loss is to reduce the current that must be transmitted through transmission lines. Power loss increases faster with an increase in current than with an increase in transmission distance. The transmission of direct current incurs resistive power losses based on the direct current produced at the power plant. The purpose of transformers is to reduce the current in the transmission line.

11.4.1 Transmission and Distribution System

Electric power generating stations are connected to loads (consumers) by means of a transmission system that consists of transmission lines and substations. The substations are nodes in the transmission grid that route electric power to loads at appropriate voltages. Figure 11-6 shows the basic elements of an electric power transmission system. Typical transmission voltages in the United States range from 69 kilovolts to 765 kilovolts, and the alternating current frequency is 60 hertz. The infrastructure for providing electric power to the loads from the substations is the distribution system.

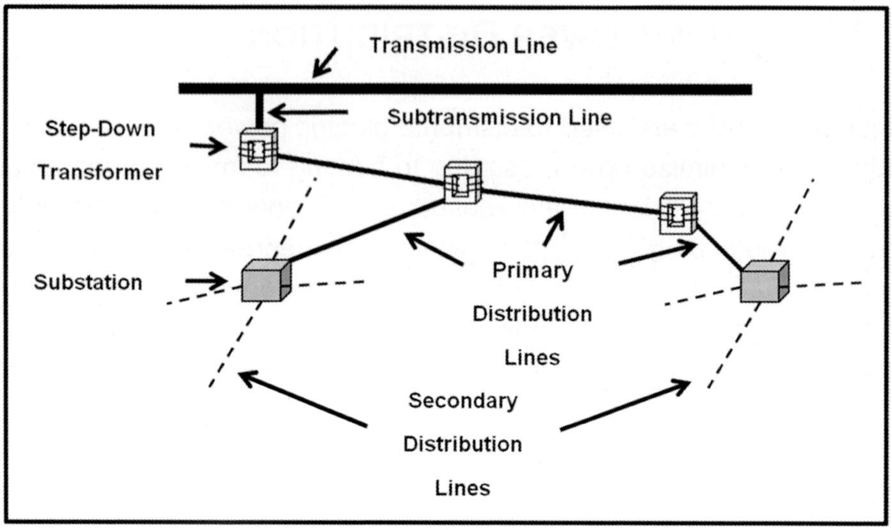

Figure 11-6. Power Transmission System

Electrical power can be transmitted over great distances using complex transmission systems. A failure in power transmission can affect large geographic areas and many people, as illustrated in the satellite images shown in Figure 11-7. The images are courtesy of the United States National Oceanic and Atmospheric Administration [NOAA Black Out, 2010]. They show the North American power blackout in August 2003. Power transmission failures can result in the loss of electrical power for millions of people.

Power losses in transmission lines limit the distance that electrical power can be transmitted. An option for the future is to use superconductors as transmission lines. At low temperatures, resistance to electrical current vanishes in superconducting materials. Electrical power losses are minimized when resistance to an electric current vanishes. Superconductors are not yet feasible for widespread use in power transmission because they require costly refrigeration.

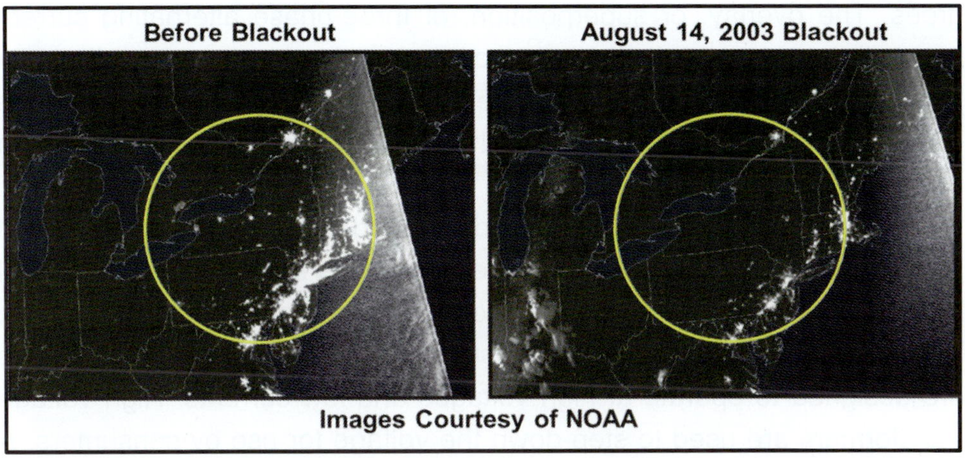

Figure 11-7. North American Power Disruption in August 2003

Electric power is cost-effectively transmitted and distributed by operating a three-phase system. Three-phase electricity refers to alternating current and voltages that are out of phase with each other by 120 de-

Figure 11-8. Transmission of Three-Phase Electricity (© Fanchi, 2003)

grees. The overlay, or superposition, of three-phase alternating current and voltage makes the alternating current and voltage more uniform. A three-phase alternating current generator is used to provide three-phase electricity. Transmission lines for three-phase electricity are shown in Figure 11-8.

Three-phase electricity can be distributed to three one-phase loads using three separate conductors. The conductors are designed with small areas to minimize the amount of conducting material, such as copper, that must be bought and put into position. Three-phase transmission lines are designed to operate at high voltages and low currents. High voltage transformers are used to step-down the voltage for use by consumers. A typical pole-mounted transformer is shown in Figure 11-9.

Figure 11-9. Pole-Mounted Transformer (© Fanchi, 2003)

A three-phase electrical transformer can convert three-phase electrical power at a distribution voltage of 13.8 kilovolts to single-phase voltages that are suitable for consumers. Residential consumers in the

United States typically use single-phase voltages of 120 volts for small appliances and 240 volts for larger appliances such as clothes dryers. Large, industrial consumers can use three-phase voltages on the order of 2160 volts or higher.

11.4.2 Household Circuits

Households are among the most common consumers of electricity in the modern world. Electricity is delivered as alternating current to a typical house in the United States using either a two-wire line or a three-wire line. The potential difference, or root mean square voltage, between the two wires in the two-wire line in the United States is 120 volts, and is 240 volts in many parts of Europe. One of the wires in the two-wire line is connected to a ground at the transformer and the other wire is the "live" wire. The three-wire line has a neutral line, a line at +120 volts, and a line at −120 volts.

A meter is connected in series with the power lines to measure the amount of electricity consumed by the household. Figure 11-10 shows series and parallel connections for three electrical devices. The series connection is designed so that the total current flows through each electrical device. The amount of current flowing through each electrical device in the parallel connection depends on the resistance of each device.

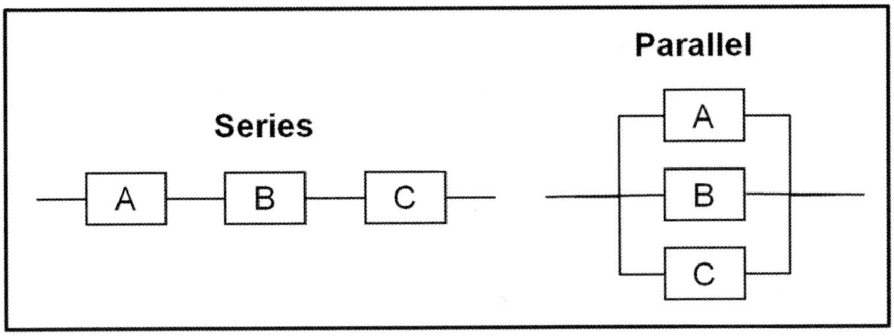

Figure 11-10. Schematic of Series and Parallel Connections

In addition to the meter, a circuit breaker is connected in series with the power lines to provide a safety buffer between the house and the power line. A fuse may be used instead of a circuit breaker in older homes. The fuse contains a metal alloy link, such as a lead and tin alloy, with a low melting temperature. If the alloy gets too hot because of resistive heating, the link will melt and break the circuit. More modern circuit breakers are electromagnetic devices that use a bimetallic strip. If the strip gets too hot, the bimetallic strip will curl based on the coefficients of thermal expansion of the two metals that make up the bimetallic strip. The change in shape of the strip when it curls breaks the circuit.

The power line and circuit breaker are designed to handle the current in the circuit, which can be as high as 30 amps. Many household applications require 15 amps of current. Lamps and appliances such as microwaves and toasters operate at 120 volts, while electric ranges and clothes dryers operate at 240 volts. Each circuit in the house has a circuit breaker or fuse to accommodate different loads.

The circuit in the house is connected in parallel to the power lines. The parallel connection makes it possible to turn on and shut off an electrical device without interfering with the operation of other electrical devices. If a series connection were used, the circuit would be broken whenever one of the electrical devices was turned off. That is why circuit breakers are connected in series between the household circuit and the power lines; the circuit breaker is designed to disconnect all electrical devices in the house from the power lines in the event of an overload, such as a power surge. An open circuit occurs when the continuity of the circuit is broken. A short circuit occurs when a low resistance pathway is created for the current.

Electricity can be harmful if a person touches a live wire while in contact with a ground. Electric shocks can cause burns that may be fatal and can disrupt the routine functioning of vital organs such as the heart. The extent of biological damage depends on the duration of contact with the current and the magnitude of the current. A current in excess of 100 milliamps (0.1 amp) can be fatal if it passes through a body for a few seconds.

Electrical devices and power lines should be handled with care. Three-pronged power cords for 120-volt outlets provide two prongs that are grounded and one prong that is connected to the live wire. The grounded prongs are provided for additional safety in electrical devices designed to use the three-pronged cords. One of the ground wires is connected to the casing of the appliance and provides a low resistance pathway for the current if the live wire is short-circuited.

11.4.3 Distributed Generation

Practical considerations limit the size of power plants. Most large-scale power plants have a maximum capacity of approximately 1000 MW. The size of the power plant is limited by the size of its components, by environmental concerns, and by energy source. For example, the area occupied by a power plant, called the plant footprint, can have an impact on land use. Conventional power plants that burn fossil fuels such as coal or natural gas can produce on the order of 1000 MW of power. Power plants that depend on nuclear reactors also produce on the order of 1000 MW of power. By contrast, power plants that rely on solar energy presently can produce on the order of power. Power from collections of a few hundred MW wind turbines in wind farms can vary from one megawatt to hundreds of megawatts. If we continue to rely on nuclear or fossil fuels, we can expect to work with reactors that generate on the order of 1000 MW of power. Nuclear reactors with a capacity on the order of 100 MW have been developed for naval propulsion. Lower power nuclear reactors could be used commercially if public support is available. If we switch to power plants that depend on solar energy or wind energy, the power generating capacity of each plant is less than 1000 MW of power and we must generate and transmit power from more plants to provide for existing and future power needs.

One power distribution strategy is to build more power plants with less power generating capacity and more widespread distribution. The federal government of the United States passed a law in 1978 called the Public Utilities Regulatory Policies Act (PURPA) that allows non-utilities to generate up to 80 MW of power and requires utilities to purchase this power. PURPA was the first law passed in decades to relax the monopoly on power generation held by utilities and reintroduce competition in the power generating sector of the United States economy.

Historically, distributed generation was the first power generation technology. The distributed generation of energy is the generation of energy where it is needed and on a scale that is suitable for the consumer. Examples of distributed generation include a campfire, a wood stove, a candle, a battery-powered watch, a car, a residential wind turbine, and a solar panel. Each of these examples generates its own power for its specific application.

The modern electricity grid transmits electric power to distant locations using a system that relies on large-scale power generating plants and extensive transmission capability. Some people believe that the future of energy depends on a renaissance in distributed generation. In this view, large-scale power plants will be replaced by many smaller-scale power-generating technologies. Borbely and Kreider define distributed generation as "power generation technologies below 1 MW of electrical output that can be sited at or near the load they serve." [Borbely and Kreider, 2001, page 2] This definition does not include small-scale power-generating technologies whose ideal locations depend on the locations of their energy source. For example, hydropower, wind farms, and solar power plants may not be considered distributed generation technologies according to Borbely and Kreider's definition because they depend on the availability of water, wind and solar energy that may not be near consumers. Consequently, hydropower, wind farms, and solar power plants are located near their energy sources. According to Borbely and Kreider's definition, a wind turbine or solar panel that provides power to a household would qualify as distributed generation if it provides less than 1 MW power.

11.5 Developments in Electric Grid Design

We have seen that power from wind and sunshine is not always produced when it is needed. The sun does not shine on a particular spot on the surface of the earth all day long, but the sun does shine all the time. We can use sunlight 24 hours per day if we can collect sunlight at different locations around the globe, or from space. Similarly, the wind does not usually blow all the time at a particular location, but wind is always blowing somewhere. Distributed generation can be used to harvest both wind energy and solar energy. Improved power distribution and transmission systems could be used to produce energy from wind and sunshine in different parts of the world and transmit it to places where it is needed. A regional or global energy distribution system would have to be developed and maintained to achieve this capability. Advances in grid design are discussed here.

11.5.1 North American Grid

The contiguous United States electrical grid is part of the North American grid. The North American grid consists of three alternating current subgrids connected by high voltage direct current (HVDC) lines. The three subgrids include the Eastern Interconnect region, the Western Interconnect region, and the Texas Interconnect region that includes most of Texas. The interconnect regions are shown in Figure 11-11 [DoE Power Grid, 2010]. The Texas and Western Interconnect regions are linked with the Mexican grid, and the Eastern and Western Interconnect regions are connected with the Canadian grid.

The Eastern, Western and Texas Interconnect regions are subdivided into 10 North American Electric Reliability Council (NERC) regions. The NERC regions are ECAR (East Central Area Reliability Coordination Agreement), ERCOT (Electric Reliability Council of Texas), FRCC (Florida Reliability Coordinating Council), MAAC (Mid-Atlantic Area Council), MAIN (Mid-America Interconnected Network), MAPP (Mid-Continent Area Power Pool), NPCC (Northeast Power Coordinating Council), SERC (Southeastern Electric Reliability Council), SPP (Southwest Power Pool), and the WSCC (Western Systems Coordinating Council). The NERC regions are shown in Figure 11-11. Alaska and Hawaii are separate systems. The Hawaiian Islands, in particular, use minigrids to provide power to each island.

Electric grids connect consumers to electrical power. In addition to connecting power stations with consumers, transmission lines provide a level of redundancy that allows demand to be met by complementary power sources. This does not mean that system failure cannot occur, as we saw with the 2003 blackout described above.

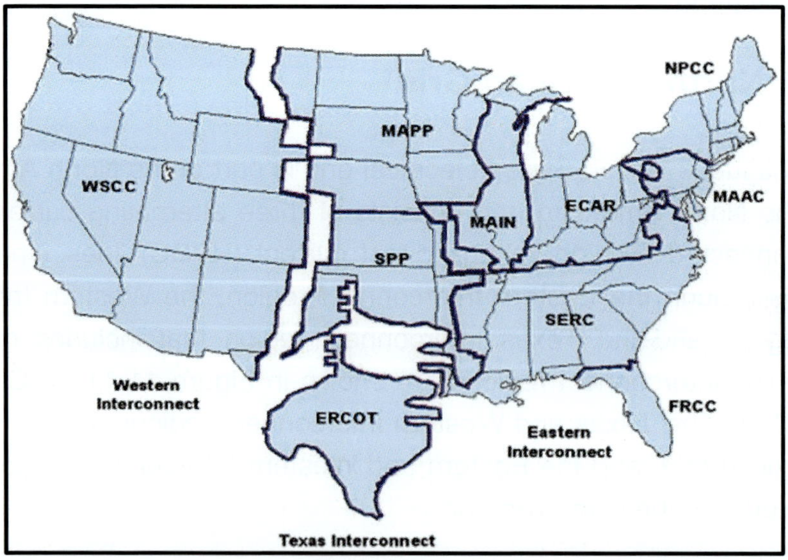

Figure 11-11. North American Electricity Grid

11.5.2 Smart Grid

The United States Federal Energy Regulatory Commission (FERC) has authority over the interstate transmission of electric power. It is charged with modernizing the United States electric system under the auspices of the 2007 Energy Independence and Security Act [US FERC, 2016]. FERC is developing a Smart Grid which is expected to apply digital technologies to the grid, and enable real-time coordination of information from generation supply resources, demand resources, and distributed energy resources.

Title XIII of the Energy Independence and Security Act of 2007 specifies characteristics of a Smart Grid [US FERC, 2016]:

1. Increase use of digital information and controls technology to improve reliability, security, and efficiency of the electric grid.
2. Dynamically optimize grid operations and resources, with full cybersecurity.
3. Deploy and integrate distributed resources and generation, including renewable resources.
4. Develop and incorporate demand response, demand-side resources, and energy efficiency resources.
5. Deploy "smart" technologies (real-time, automated, interactive technologies that optimize the physical operation of appliances and consumer devices) for metering, communications concerning grid operations and status, and distribution automation.
6. Integrate "smart" appliances and consumer devices.
7. Deploy and integrate advanced electricity storage and peak-shaving technologies, including plug-in electric and hybrid electric vehicles, and thermal storage air conditioning.
8. Provide consumers timely information and control options.
9. Develop standards for communication and interoperability of appliances and equipment connected to the electric grid, including the infrastructure serving the grid.

10. Identify and lower unreasonable or unnecessary barriers to adoption of smart grid technologies, practices, and services.

Point to Ponder: Should we be concerned that the Smart Grid will infringe on our freedom?
Some people are concerned that they will lose some of their freedom when the Smart Grid is fully implemented. The source of this concern is the FERC charge to provide "real-time coordination of demand resources." If demand resources include businesses and households, then FERC could regulate how businesses and households use energy for such applications as heating and cooling. The Smart Grid could ration power or limit your use of power at inconvenient times without special authorization. For example, if you need power for a special event such as a conference, you might have to obtain authorization from a utility or governmental agency. The need to obtain authorization can infringe on existing liberties and gives some people authority to allocate resources. This increases the possibility that corruption will occur. Items 8 and 10 of the Smart Grid characteristics listed above can be used to alleviate these concerns, but are unlikely to eliminate corruption.

11.5.3 European Supergrid

The Union for the Coordination of Transmission of Electricity (UCTE) began unifying the electricity grids of the European Union member states in the 1950's after World War II. The UCTE was formed in 1951 as the Union for the Coordination of Production and Transmission of Electricity (UCPTE). In 1999, UCPTE became an association of transmission system operators (TSOs) called the UCTE when the EU deregulated electricity markets. Members of UCTE include most of continental Europe. Electricity grids in the North African states of Morocco, Algeria and

Tunisia are synchronized with the UCTE grid through the Gibraltar AC link.

UCTE is now a member of the European Network of Transmission System Operators for Electricity (ENTSO-E). Membership in ENTSO-E includes more TSOs and more countries than the UCTE, including TSOs from Great Britain, which was not a member of UCTE. The mission of ENTSO-E embraces the following challenges [ENTSO-E, 2016]:

- Security - pursue coordinated, reliable and secure operations of the interconnected electricity transmission network, while anticipating the decision to cope with upcoming system evolutions.
- Market - provide a platform for the market by proposing and implementing standardized market integration and transparency frameworks that facilitate competitive and integrated continental wholesale and retail markets.
- Sustainability - facilitate secure integration of new generation sources, particularly renewable energy, as well as significantly contributing to the EU's greenhouse gases reduction and renewable energy supply goals.
- Network Adequacy - promote the adequate development of the interconnected European grid and investments for a reliable, efficient and sustainable power system.

Plans exist for altering the EU electricity grid to a Supergrid that encompasses all EU member states. The Supergrid is advocated by DESERTEC Foundation, an initiative of the Club of Rome. The Club of Rome was founded in 1968 by Italian industrialist Aurelio Peccei and Scottish scientist Alexander King. It is a global not-for-profit organization with the mission of acting “as a global catalyst for change through the identification and analysis of the crucial problems facing humanity and the communication of such problems to the most important public and private decision makers as well as to the general public.” [Club of Rome, 2010, see Organization]. The concept of a Supergrid is also supported by the

European Wind Energy Association (EWEA) since this grid should help connect large planned North Sea wind farms to European consumers.

The Supergrid planned by DESERTEC is expected to transmit electrical power generated by offshore wind farms in the North Sea and electrical power generated by Spanish and North African solar power plants [Desertec Red Paper, 2010; Desertec White Paper, 2010]. The plan includes construction of desalination plants to provide potable water to North African and Middle Eastern nations. The Supergrid will connect the EU with Saharan Africa. Friends of the Supergrid (FOSG), an organization that supports development of the Supergrid, has pointed out that the Supergrid has several advantages [FOSG, 2016]. The Supergrid will help the EU and participating nations decarbonize 20% of the power sector by 2020 and 90% by 2050; integrate renewable energy into the continent's energy mix; transmit power from renewable resources to load centers throughout the EU; balance the EU electricity network and enhance security of supply. In addition, the Supergrid will enhance the EU electricity market, create highly skilled jobs, and create opportunities for participating companies to market sustainable energy technology. Critics of the plan express concern that the project will require significant expenditures by EU member states, make the EU dependent on energy from non-member states, and will be vulnerable to attack. Brexit, the British vote on June 23, 2016 to exit the EU, shows that EU membership may change and could impact implementation of the EU Supergrid.

11.6 Activities

True-False

Specify if each of the following statements is True (T) or False (F).

1. George Westinghouse was a proponent of alternating current.
2. The load on a utility is the demand for electric power.
3. A generator can convert mechanical energy to electrical energy.
4. Thomas Edison proposed using alternating current to distribute electric power.

5. Steam engines were developed to convert thermal energy into mechanical energy.
6. Thomas Edison launched "the age of electricity" at the Pearl Street station in 1882 with his DC power station.
7. Electrical power can be transmitted over great distances without loss of power using the existing transmission grid.
8. The first commercial-scale electric power plants were hydroelectric plants.
9. A short circuit occurs when the continuity of the circuit is broken.
10. The European Supergrid will include wind power from the North Sea region and solar power from Spain and North Africa.

Questions

1. Suppose the minimum daily power provided by a utility is 100 MW and the maximum daily power is 50% greater than the minimum. Express the base load and peak load of the utility in MW.
2. Is electrical power transmitted over transmission lines using low voltage or high voltage? Explain your answer.
3. What does an alternating current transformer do?
4. Is the voltage used in household circuits in the United States the same as the voltage used in Europe?
5. What is distributed generation?
6. What is the efficiency of a power station that uses 120 MW power to generate 100 MW power?
7. How many phases are used to transmit and distribute electric power in the United States?
8. What two systems are connected in series to power lines that enter households? What are their purposes?
9. What are the three alternating current subgrids called in the North American grid?
10. How might a "smart" grid interfere with personal and business freedom?

CHAPTER 12

ENERGY ECONOMICS

After reading this chapter, you should be able to...

- Explain basic principles of economics including supply and demand, and elasticity
- Define CAPEX and OPEX
- Discuss costs and benefits of energy
  - *P2P: Is the military budget a hidden cost of importing oil and gas?*
- Define economies of scale
- Explain how management decisions in the energy sector are made
- Describe life cycle analysis
- Describe levelized energy costs and their importance
  - *P2P: Can we bias cost comparisons?*
- Define the First Oil Crisis and discuss its geopolitical significance
- Describe the impact of fluctuating oil prices
  - *P2P: Why would a country want to keep reserves of fossil fuels secret?*

Energy may be the most important factor that will influence the shape of society in the 21st century. The cost and availability of energy significantly impacts our quality of life, the health of national economies, the relationships between nations, and the stability of our environment. What kind of energy do we want to use in our future? Will there be enough? What will be the consequences of our decisions?

The selection of an energy source depends on such factors as availability, accessibility, environmental acceptability, capital cost, and ongoing operating expenses. We discuss how economic factors affect decision-making in this chapter. Many of the concepts presented here are discussed in standard economics textbooks, such as Varian [2009] and Dahl [2004].

12.1 Principles of Economics

Economic decisions are made on a microeconomic or macroeconomic scale. Microeconomics is the study of how individual businesses or households decide to allocate limited resources. Macroeconomics is the study of the behavior of economic units that are aggregated into community, national or global economies. Both the decision-making of individual economic units and the policy-making of governments depend on the supply of and demand for energy.

Prices of goods and services in a market economy are determined by supply and demand. For comparison, in a planned economy prices of goods and services are set by a central government. A market is a venue where goods and services can be exchanged for other goods and services or for currency. The rate of exchange, or the value of the goods or services being exchanged, is considered the price and can be quantified using a currency such as US dollars. The difference between the price of a unit and the cost of a unit is profit.

When a country or other decision-making body considers which energy sources are best suited for them, it is important to not only examine the price today, but the potential price in the future. Historically for most countries, the determining factor for deciding which energy sources to use is price. Factors such as environmental impact, political impact, and national security are growing in importance and can be factored into price.

12.1.1 Supply and Demand

Supply and demand are expressed by the relationship between price and quantity of units. Figure 12-1 illustrates supply and demand curves. The plot of supply and demand is a snapshot in time. The demand curve slopes down from left to right because normal market behavior is to buy more of a product when the price of a unit is low and buy less when the price is high. The supply curve tends to slope up from left to right because normal market behavior is to provide less of something when the price is low and provide more when the price is high. The point on the curves where supply and demand intersect is called market equilibrium. This is the point that markets seek to achieve, where the amount produced (supplied) is equal to the amount consumed (demanded).

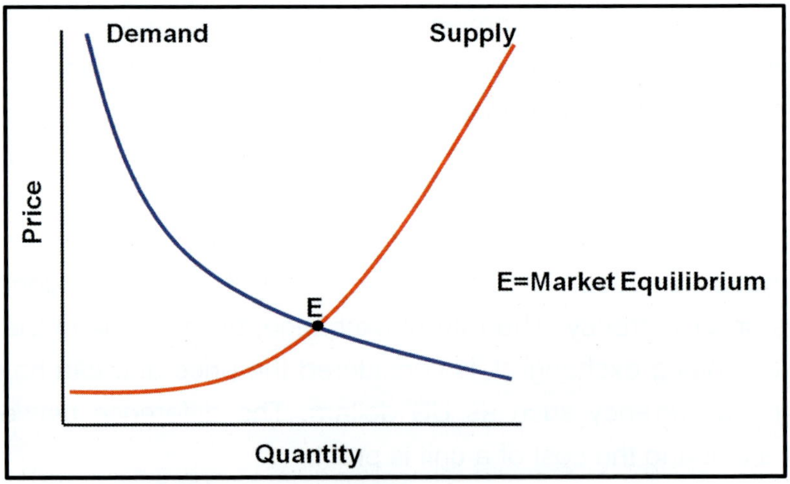

Figure 12-1. Supply and Demand Curves

The economic theory of supply and demand tells us how quantity of units is affected by price. Figure 12-2 shows how a change in quantity influences price with regard to demand. The base case labeled 12-2a relates price and a specified quantity of units. An increase in price in case 12-2b is accompanied by a decrease in quantity sold. On the other hand, a decrease in price in case 12-2c is accompanied by an increase in quantity sold. A similar discussion can be applied to supply. In that case, an increase in price is accompanied by an increase in quantity available for sale. A decrease in price is accompanied by a decrease in quantity available for sale.

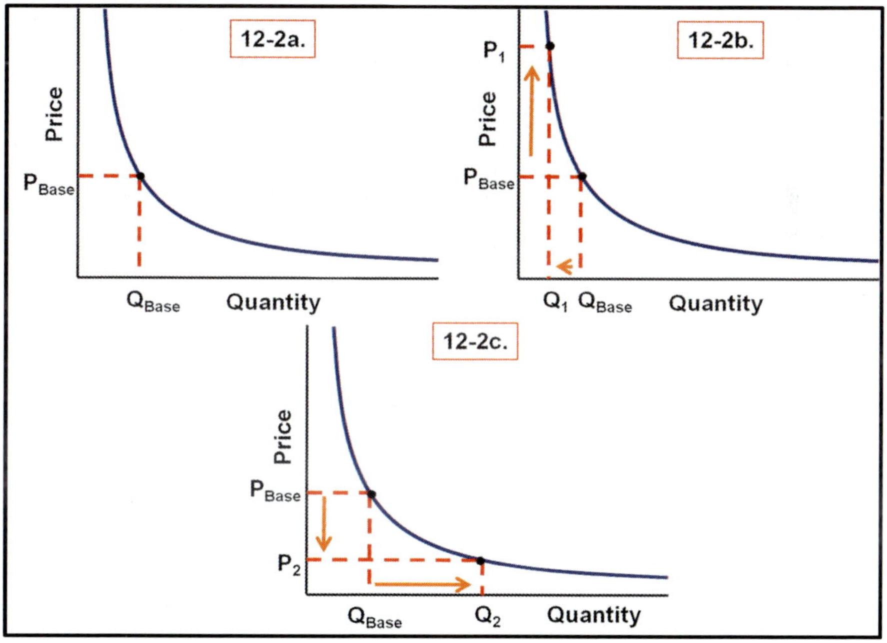

Figure 12-2. Demand and Price Relationship

Demand does not need to come entirely from private consumers and businesses. Governments can be consumers of goods and services as well, and this is included in the demand curve. Demand also does not necessarily imply that all of a good will be consumed. The demand for goods often varies with time. Consequently, the supply and demand curves for many goods and services change significantly as a function of time. For example, in the United States, the demand for heating oil increases during the winter, and the demand for gasoline increases during the summer vacation season.

12.1.2 Elasticity

The slope of the demand curve is a measure of the elasticity of demand. Elasticity of demand refers to the sensitivity of demand to changes in price and can be viewed as elastic or inelastic. Elastic demand and inelastic demand are compared in Figure 12-3. Demand is elastic when a price change is accompanied by a corresponding change in quantity sold. Demand is inelastic when a price change has little or no impact on quantity sold.

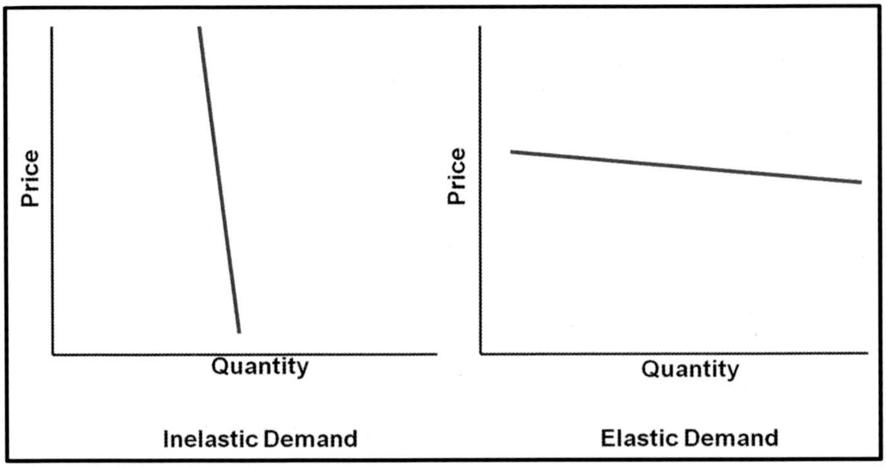

Figure 12-3. Elasticity of Demand

There are several possible reasons that demand for a product may be elastic. For example, if a product has a very close substitute or many potential substitutes, then an increase in price of that product will motivate consumers to switch to one of the other products. This is called product substitution. Another explanation for elastic demand is that a product may be a discretionary purchase, that is, the product has a limited number of specific uses that are not considered necessities. Thus an increase in price will cause a willingness to delay the purchase or a large drop in consumer interest.

Inelastic demand occurs when price can change significantly without having a significant impact on demand. For example, in recent years, the price of oil in the United States had seen its greatest percentage increase since the first oil crisis of the 1970's. From May 2003, when the average price of a barrel of crude oil was roughly US$23, to July 2008 when the average price per barrel was US$135, crude oil prices rose roughly 586%. While the price increased significantly, the demand for petroleum products in the United States did not significantly decline. United States petroleum use (demand) was roughly 600 million barrels in May 2003, and was 606 million barrels in July 2008, a 1% increase. With oil prices affecting most Americans at the gas pump, many changed their driving habits as much as they could, but the country still required vast quantities of oil to function due to the immense size of the transportation sector and the lack of a competitive substitute for oil. In this example, demand for oil in the price range from US$23 per barrel to US$135 per barrel was inelastic, that is, demand for oil was largely unaffected by increases in price.

The elasticity of a curve can vary depending on price. This was illustrated in Figure 12-2 above, where the slope of the demand curve was steepest at the top, meaning that demand at that quantity level is inelastic. As price drops, the demand became more elastic. This figure illustrates demand for oil in the United States. When the price is highest, demand decreases but there is a quantity of oil that will be demanded with minimal sensitivity to price. As the price gets lower, however, Americans tend to use oil in increasing amounts because there are very few substitutes for oil, particularly for transportation, and oil has many different uses.

The existence of substitutes is one of the largest determining factors for price elasticity. That is why each oil crisis over the past 40 years has been met with a call to decrease dependence on foreign oil. It can be said that the cost of imported energy is a tax on a nation's economy; the greater the amount of energy imported, the more it negatively affects the economic health of the nation.

The cost of energy in oil importing nations is sensitive to price-setting decisions by foreign bodies. One of the first demonstrations of this was the 1973 oil crisis when OPEC enacted an oil embargo against the United States and caused an unprecedented increase in oil price globally.

12.1.3 CAPEX and OPEX

The cost of commercializing wind energy and solar energy is increased by the significant cost of constructing wind farms, power plants and related infrastructure to distribute the energy to consumers. These costs are known as capital expenditures, or CAPEX. They are typically incurred at the beginning of the project and are expected to create future benefits.

Once the production facility has been built, the cost of fuel for wind and solar energy facilities is negligible and long term costs are operational expenses, or OPEX. OPEX includes fuel, personnel, and maintenance costs. Although some alternative power sources would be considered too costly today, over the life of the equipment, they may be cheaper than other energy options.

Figure 12-4 illustrates a typical cost curve for a project. At the beginning, the cost is highest because the basic facility must be built. Cost then stabilizes at a lower level when the facility is completed and the principal costs are OPEX. At the end of the life of the facility, the cost to decommission the facility and repair or mitigate any environmental damage leads to an increase in cost.

When making economic decisions regarding energy, it is important to consider substitute products such as alternative energy. Many countries are making decisions today to construct expensive alternative energy facilities such as wind farms, solar power plants and hydroelectric dams. As we have seen, some people claim that peak oil has already been reached, while others argue that it may be decades before peak oil is reached. Access to oil, environmental impact, energy independence and security are factors that need to be considered by a country that is deciding whether to implement policies advocating the development of alternative energy sources.

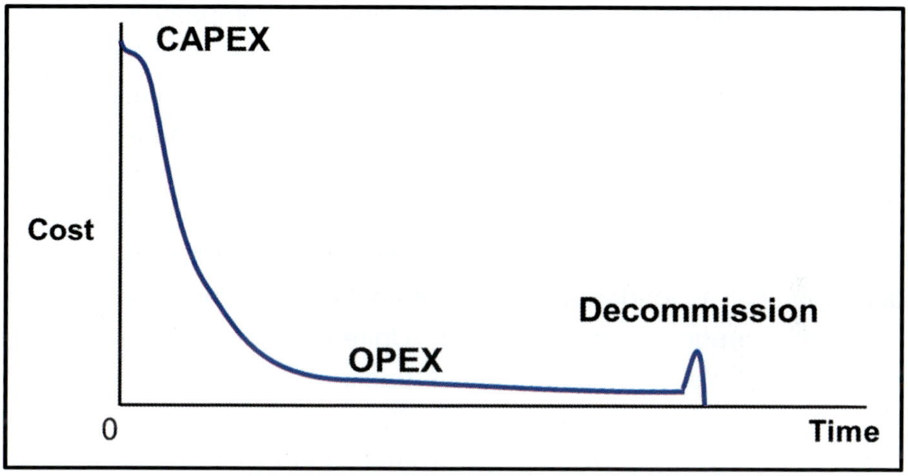

Figure 12-4. Cost Curve for a Typical Project

12.2 Costs and Benefits

Energy is not free. The financial cost of energy comes in two forms, direct and indirect. The societal cost of energy can be more difficult to quantify. Direct and indirect costs, and tangible and intangible costs and benefits are explained here.

12.2.1 Direct and Indirect Costs

Costs to energy producers come in two forms: direct and indirect costs. Direct costs are the costs directly associated with producing and transporting energy to consumers. This includes the cost of materials and labor to construct the facility, and operating costs. Fuel costs are direct costs that are significant costs for coal, oil, gas, and nuclear power plants, but are negligible for solar, wind, and hydroelectric plants. Many power plants use large volumes of water that must be purchased as another direct cost.

A significant up-front direct cost of new alternative energy projects is the cost of financing for these projects. When a company is seeking to build a power facility using a relatively new technology, that technology is considered to be riskier than more established technologies. Thus, the company typically pays a higher interest rate on any money borrowed from the private sector to build the facility because of the lender's perception that the risk of failure to function or meet expectations is greater than building the facility with established technology. Some governments may choose to serve as lender to reduce the cost of financing a new technology and encourage its development. United States President Barack Obama signed the economic stimulus package known as the American Recovery and Reinvestment Act in 2009 to encourage development of renewable energy by providing government loan guarantees to several private sector companies. These loans typically had lower rates for larger sums of money than were available from private lenders. Some of the subsidized companies failed. For example, the solar company Solyndra declared bankruptcy in 2011.

Direct costs are costs incurred for activities or services that provide benefits to specific projects. Indirect costs are costs that benefit multiple projects. Overhead is an indirect cost. The term "overhead" typically refers to expenses required for a business to function, but do not directly generate profits. Overhead includes expenditures for managing facilities, paying administrative personnel, and paying rent and utilities. The following example illustrates the concept of overhead.

Suppose an energy company owns a wind farm, a coal-fired power plant, and a natural gas power plant. Such a company would typically employ a number of people to account for the costs and revenues related to the wind farm and each of the power plants. The expense for this accounting work can be proportionally divided among the three energy systems, but it is not feasible to determine exactly how much accounting time is dedicated to each system. The cost of accounting is an indirect cost. Office buildings that contain office support services such as accounting and human resources, as well as front office services such as customer service, are also considered indirect costs.

Indirect costs can include permitting and other governmental costs that a company must pay. When a company chooses to build a new power plant, they must receive authorization from local and national government agencies. The cost of the permits and the costs of preparing any related paperwork are considered indirect costs. An indirect cost for companies operating in some countries is the cost of influencing officials who have the power to grant permits or mineral extraction rights.

Another type of cost is opportunity cost. Opportunity cost is the concept that the selection of one opportunity that requires the use of scarce resources means another opportunity is not selected. The value of the opportunity that was not selected is the opportunity cost. For example, suppose a company has enough money to either build several wind turbines or drill for natural gas, but they cannot do both. If the company chooses to drill for gas, the wind turbines that go un-built and the related revenues from those turbines are the opportunity cost to the company for choosing the natural gas project.

12.2.2 Tangible and Intangible Costs and Benefits

Economic decisions should consider tangible financial costs as well as intangible costs. Intangible costs are the costs that tend to be difficult to quantify and can be financial or societal. For example, the environmental impact of an energy source can be measured with both tangible and intangible costs. Tangible costs would include the cost of installing environmental protection measures such as scrubbers in coal plants, radiation barriers in nuclear power plants, or sound dampening walls around drilling rigs in urban areas. Tangible environmental costs also include the projected costs of cleaning up a site, whether that clean-up is the result of a chemical spill, or the cost of reclaiming a drill site or power plant site after it is decommissioned.

Intangible costs include factors such as carbon emissions and their impact on the atmosphere. Energy producers can estimate the financial cost to decrease emissions, but the damage to the environment can be difficult, if not impossible, to quantify. Society as a whole must consider the societal cost of using an energy source that produces these emissions. Another intangible cost can be the aesthetics related to an energy source.

Oil refineries, coal power plants, and even wind farms are considered unattractive by some people. The installation of industrial facilities can impact tourism if it adversely affects the ambiance of an area. An intangible cost that governments and energy companies must pay close attention to is the public perception of an energy source. For example, fossil fuels have received significant negative publicity because of the growing concern about greenhouse gas emissions, and nuclear energy has always had to fight negative public perception about its safety. An unquantifiable cost to countries continuing to use fossil fuels or nuclear fission power is the negative publicity that is associated with these fuels domestically and internationally.

Energy sources also have tangible and intangible benefits that must be considered. Tangible benefits include something as simple as price if one energy source is cheaper than another. Another tangible benefit includes effectiveness. Gasoline, for example, is effective as a transportation fuel. This is a tangible benefit for oil relative to other sources of energy. Intangible benefits include the environmental impact of the energy source. Unlike fossil fuels, which produce undesirable carbon emissions, many alternative energy sources have a good public image because they have lower carbon emissions. The lack of carbon emissions is a tangible benefit for wind farms, solar power plants, and hydroelectric plants because it eliminates the cost of mitigating greenhouse gas emissions, and the associated image of being a "clean" energy source is an intangible benefit.

Point to Ponder: Is the military budget a hidden cost of importing oil and gas?

Most industrialized nations rely on oil and gas to support their energy needs. Some of these countries, such as the United States, Great Britain, Russia and China maintain significant military capabilities for national defense and global influence. Part of this military expenditure is used to maintain access to resources and some of these costs could be considered part of the cost of continued reliance on fossil fuels. For example, if US$10 billion is spent to keep open supply lines to a region in one year, and you import 500 million barrels of oil from that region in a year, that would add US$20.00 to the cost of each barrel of imported oil, or about US$0.48 to each gallon of gasoline. This cost assumes that no lives are lost and, if they are, that they have no monetary value. Each society must determine the value they want to place on the lives of people who are being asked to protect their sources of energy (Figure 12-5).

Figure 12-5. Military Cemetery in Washington, D.C. (© Fanchi, 2002)

12.3 Economies of Scale

The cost of wind and solar energy is relatively high compared to fossil energy. This is due in part to the higher cost of production using technology that is still developing. As the demand for a product increases, producers can expand production and increase their profit by improving the technology and reducing the average cost of the product. The reduction of average cost of the product is achieved primarily by distributing fixed costs, such as overhead and administrative costs, over a larger quantity of products. The reduction of the average cost of a product as the scale of production expands is referred to as an economy of scale. Economies of scale are achieved if the average cost of a product decreases when a firm expands the production of the product. The history of the modern wind turbine industry illustrates the concept of economies of scale.

Interest in renewable energy, including wind turbines, was stimulated in the United States and Europe when people realized that oil supply could be used as a political weapon following the 1973 oil crisis. The most important result of that oil crisis in the context here is that the price of oil increased to the point that leaders in the United States and Europe began to understand that dependence on a foreign energy supplier carried a significant cost. Domestic access to energy resources has considerable political value. This realization has been the impetus for renewable energy research since the 1973 oil crisis. For example, the United States National Renewable Energy Laboratory (NREL) in Golden, Colorado was established by the Solar Energy Research Development and Demonstration Act of 1974 during the Gerald R. Ford Administration. Originally called the Solar Energy Research Institute, the facility began operating in July 1977 and received significant funding for solar research during the Jimmy Carter Administration. The facility was elevated to national laboratory status during the George H.W. Bush Administration in September 1991 and became NREL. The laboratory has been continuously funded by administrations representing both American political parties.

The cost of renewable energy was prohibitively high and interest in renewable energy waned when the price of oil stabilized after the end of the crisis in early 1974. Several oil supply disruptions have occurred since 1973, and the demand for renewable energy has continued to grow. Wind turbine technology in particular developed quickly and is now economically competitive. Demand for wind turbines has increased, wind turbine technology has improved, and production capacity has expanded. Today, the number of commercial wind turbines is increasing and the cost for manufacturing and installing wind turbines has decreased. This is an example of economies of scale. Wind energy has become economically competitive with fossil energy and is a viable technology for energy providers to consider adding to their energy portfolio.

Recognition that energy from oil can be replaced by energy from another energy source can be expressed in terms of the concept of fungibility. We use a specific example to clarify the concept. Electrical energy can be provided using several different energy sources. If the price of one energy source gets too expensive, a cheaper source can be substituted for the more expensive source. This is product substitution, and is possible because electrical energy is fungible, that is, electrical energy produced by one type of energy source, such as burning oil or coal, is interchangeable with electrical energy produced by another type of energy, such as harvesting wind or solar energy.

A unit of a product is fungible if individual units of the product are interchangeable. Thus a kilowatt-hour of electricity from the burning of oil is interchangeable with a kilowatt-hour of electricity from wind harvested by a wind turbine. Electricity is a fungible commodity; therefore it can be produced from any of a large number of energy sources and offered to the consumer at a price that optimizes the cost to the producer. Consumers can use electricity without knowing how their electricity was produced or where it came from.

12.4 Management Decisions in the Energy Sector

Decision makers at energy companies and within governments must consider many economic factors when selecting energy projects and formulating policy. Energy projects range from drilling a well to building expensive, offshore platforms or solar power plants, or even venturing into a new energy type entirely. Some of the concepts involved in making management decisions are presented here.

12.4.1 Cash Flow and Economic Indicators

An economic analysis of competing investment options usually requires preparing cash flow predictions. The value of money can change with time. The time value of money is included in the economic analyses by applying a discount rate to adjust the value of money in future years to the value of money during a base year. The resulting cash flow is called the discounted cash flow because future cash is less valuable than present cash. The cash flow of an investment option is the net cash generated by the investment option or expended on the investment option as a function of time. The net present value (NPV) of the discounted cash flow is the value of the cash flow at a specified discount rate.

Net present value is the difference between revenue and expenses. Net present value, revenue, and expenses depend on the time value of money. We can account for the time value of money by introducing a discount rate in the calculation. Revenue can be calculated from the price per unit produced times the quantity produced. The quantity produced can be volume of oil or gas, kilowatt-hours of electricity, or any other appropriate measure of resource production. Expenses include capital expenditures, operating expenditures, and taxes.

The time dependence of NPV is illustrated in Figure 12-6. This figure shows that NPV can be negative. A negative NPV says that the investment option is operating at a loss. The loss is usually associated with initial capital investments and operating expenses that are incurred before the investment option begins to generate revenue. For example, mirrors have to be installed in a solar electric generating system and a generator built before electricity can be sold to the transmission grid. The installation of mirrors and a generator are capital expenses that occur before revenue is generated by the sale of electricity. Similarly, large investments in the design and construction of offshore platforms must be made before wells can be drilled and oil produced. The eventual growth in positive NPV occurs when revenue exceeds expenses. The point in time when NPV equals zero is the payout time. In the example shown in Figure 12-6, payout time is approximately 1.8 years. The concept of payout time can be applied to discounted cash flow or undiscounted cash flow.

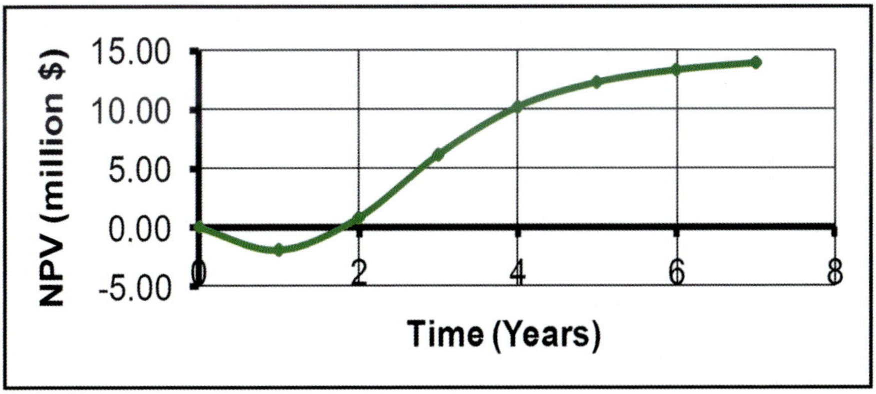

Figure 12-6. Typical Cash Flow

Several commonly used indicators of economic performance are listed in Table 12-1. The discount rate at which the maximum value of NPV is zero is called the discounted cash flow return on investment (DCFROI) or Internal Rate of Return (IRR). Payout time, NPV, and DCFROI account for the time value of money. An indicator that does not account for the time value of money is the profit-to-investment (PI) ratio. The PI is the total undiscounted cash flow without capital investment divided by total investment. It is often useful to prepare a variety of plots, such as NPV versus time and NPV versus discount rate, to show how economic indicators perform as functions of time.

Table 12-1
Indicators of Economic Performance

Discount Rate	Factor to adjust the value of money to a base year.
Net Present Value	Value of cash flow at a specified discount rate.
DCFROI or IRR	Discount rate at which net present value is zero.
Payout Time	Time when net present value equals zero.
Profit-to-Investment Ratio	Undiscounted cash flow without capital investment divided by total investment.

Economic analyses using indicators of economic performance provide information about the relative performance of different investment options. The economic viability of an investment option is usually decided after considering a combination of economic indicators. For example, an investment option may be considered economically viable if the DCFROI is greater than 30% and payout time is less than two years. It should be remembered, however, that quantitative indicators provide useful information, but not complete information.

Economic viability is influenced by both tangible and intangible factors. The tangible factors, such as building a generator or drilling a well, are relatively easy to quantify. Intangible factors such as environmental and socio-political concerns are relatively difficult to quantify, yet may be more important than tangible factors.

12.4.2 Life Cycle Analysis

The analysis of the costs associated with an energy source should take into account the initial capital expenditures and annual operating expenses for the life of the system. This analysis is called life cycle analysis and the costs are called life cycle costs. Life cycle costing requires the analysis of all direct and indirect costs associated with the system for the entire expected life of the system. According to Bent Sørensen [2004, page 767], life cycle analysis includes analyzing the impact of "materials or facilities used to manufacture tools and equipment for the process under study and it includes final disposal of equipment and materials, whether involving reuse, recycling or waste disposal." A list of life cycle costs is presented in Table 12-2.

Table 12-2
Life Cycle Costs of an Energy System
[Goswami, et al., 2000, page 528]

Capital equipment costs
Acquisition costs
Operating costs for fuels, etc.
Interest charges for borrowed capital
Maintenance, insurance, and miscellaneous charges
Taxes (local, state, federal)
Other recurring or one-time costs associated with the system
Salvage value (usually a cost) or abandonment cost

It is important to recognize that the future cost of some energy investment options may change significantly as a result of technological advances. The cost of a finite resource can be expected to increase as the availability of the resource declines, while the cost of an emerging technology will usually decline as the infrastructure for supporting the technology matures.

The initial cost of one energy system may be relatively low compared to competing systems. If we only consider initial cost in our analysis, we may adopt an energy option that is not optimum. For example, the annual operating expenses for an option we might choose based on initial cost may be significantly larger than those of an alternative option. In addition, projections of future cost may be substantially in error if the cost of one or more of the components contributing to an energy system changes significantly in relation to our original estimate. To avoid making less-than-optimum decisions, we should consider all of the life cycle costs of each investment option. We also need to evaluate the sensitivity of cash flow predictions to plausible changes in cost as a function of time.

Inherent in life cycle analysis is an accurate determination of end use efficiency. End use efficiency is the overall efficiency of converting primary energy to a useful form of energy. It should include an analysis of all factors that affect the application. As a simple example, consider the replacement of a light bulb.

The simplest decision is to choose the least expensive light bulb. On a more sophisticated level, we need to recognize that the purpose of the light bulb is to provide light and some light bulbs can provide light longer than other light bulbs. In this case we need to consider the life of the light bulb in addition to its price. But there are still more factors to consider. If you live in an equatorial region, you might prefer a light bulb that emits light and relatively little heat, so you can reduce air conditioning expenses. On the other hand, if you live in a cooler northern climate, you might desire the extra heat and choose a light bulb that can also serve as a heat source. Once you select a light bulb, you want to use it where it will do the most good. Thus, if you chose a more expensive light bulb that has a long life and generates little heat, you would probably prefer to use the light bulb in a room where the bulb would be used frequently, such as a kitchen or office, rather than a closet where the bulb would be used less frequently. All of these factors should be taken into account in determining the end use efficiency associated with the decision to select a light bulb.

One of the goals of life cycle analysis is to make sure that decision makers in industry, government, and society in general, are aware of all of the costs associated with a system. In the context of energy resource management, Sørensen [2004, Section 7.4.3] has identified the following impact areas: economic, environmental, social, security, resilience, development, and political. Some typical questions that must be answered include the following:

1. Does use of the resource have a positive social impact, that is, does resource use provide a product or service without adversely affecting health or work environment?
2. Is the resource secure, or safe, from misuse or terrorist attack?
3. Is the resource resilient, that is, is the resource relatively insensitive to system failure, management errors, or future changes in the way society assesses its impact?
4. Does the resource have a positive or negative impact on the development of a society, that is, does the resource facilitate the goals of a society, such as decentralization of energy generating facilities or satisfying basic human needs?
5. What are the political ramifications associated with the adoption of an energy resource?
6. Is the resource vulnerable to political instability or can the resource be used for political leverage?

A thorough life cycle analysis will provide answers to all of these questions. Of course, the validity of the answers will depend on our ability to accurately predict the future.

12.4.3 Risk Analysis and Real Options Analysis

A characteristic of natural resource management is the need to understand the role of uncertainty in decision making. The information we have about a natural resource is usually incomplete. What information we do have may contain errors. Despite the limitations in our knowledge, we must often make important decisions to advance a project. These decisions should be made with the recognition that uncertainty and risk are present and can influence investment decisions. Here, risk refers to the possibility that an unexpected event can adversely affect the value of an asset. Uncertainty is not the same as risk. Uncertainty is the concept that our limited knowledge and understanding of the future does not allow us to predict the consequences of our decisions with 100% accuracy. Risk analysis is an attempt to quantify the risks associated with investing under uncertainty.

One of the drawbacks of traditional risk analysis is the limited number of options that are considered. The focus in risk analysis is decision making based on current expectations about future events. For example, the net present value analysis discussed above requires forecasts of revenue and expenses based on today's expectations. Technological advances or political instabilities are examples of events that may significantly alter our expectations. We might overlook or ignore options that would have benefited from the unforeseen events. An option in this context is a set of policies or strategies for making current and future decisions. Real Options Analysis attempts to incorporate flexibility in the management of investment options that are subject to considerable future uncertainty.

The best way to incorporate options in the decision making process is to identify them during the early stages of analysis. Once a set of options has been identified for a particular project, we can begin to describe the uncertainties and decisions associated with the project. By identifying and considering an array of options, we obtain a more complete picture of what may happen as a consequence of the decisions we make. Real Options Analysis helps us understand how important components of a project, particularly components with an element of uncertainty, influence the value of the project.

12.4.4 Levelized Energy Cost

A useful measure for evaluating the feasibility of an energy generating technology and comparing different energy generating systems is levelized cost of energy (LCOE). LCOE is total life cycle cost divided by total lifetime energy production. It is an estimate of the cost of generating energy, typically in the form of electricity, for a particular energy generating system. The LCOE includes all costs incurred over the expected life of the system, and can be used to compare the cost per unit of energy for different energy systems. A net present value calculation is performed to calculate the break-even price for energy generated by the system. Average levelized cost of electricity generation for plants entering service in 2020 are illustrated in Table 12-3.

Table 12-3
Average Levelized Cost of Electricity Generation
[US EIA Levelized Costs, 2015]

Plant Type	2013 US$/kWh 2013 data	Normalized 2013 data
Geothermal	0.0478	0.5021
Gas, Advanced Combined Cycle	0.0726	0.7626
Wind	0.0736	0.7731
Gas, Conventional Combined Cycle	0.0752	0.7899
Hydroelectric	0.0835	0.8771
Conventional Coal	0.0951	0.9989
Advanced Nuclear	0.0952	1.0000
Gas, Advanced CC with CCS	0.1002	1.0525
Biomass	0.1005	1.0557
Advanced Coal	0.1157	1.2153

Table 12-3
Average Levelized Cost of Electricity Generation
[US EIA Levelized Costs, 2015]

Plant Type	2013 US$/kWh 2013 data	Normalized 2013 data
Solar PV	0.1253	1.3162
Advanced Coal with CCS	0.1444	1.5168
Wind – Offshore	0.1969	2.0683
Solar Thermal	0.2397	2.5179

The second column of Table 12-3 presents average national levelized costs for electricity generating plants that rely on different energy sources. The levelized costs are calculated using 2013 data and refer to plants that will enter service in 2020. The acronym CC refers to combined cycle, and CCS refers to carbon control and sequestration. In some cases, CCS refers to carbon capture and storage. A combined cycle power plant combines two or more thermodynamic cycles to improve overall efficiency. For example, a combined cycle gas turbine power plant burns natural gas to drive one turbine, and uses excess heat from the exhaust to drive a steam turbine. Advanced technologies improve power generation efficiency in new power plants when compared with conventional technologies.

The levelized costs in Table 12-3 are normalized in the third column using the levelized cost of a nuclear power plant as the reference cost for normalization. Levelized costs in Table 12-3 do not include financial incentives such as tax breaks or government subsidies. More details about the levelized costs are provided in the 2015 Annual Energy Outlook [US EIA Levelized Costs, 2015]. It is important to note that LCOE can change based on a variety of factors including market conditions, technological advances, and government policies.

Figures 12-7 and 12-8 graphically display the information in Table 12-3. Imposing environmental regulations that further restrict greenhouse gas (GHG) emissions increases the cost of fossil fuel-fired plants because technology designed to mitigate GHG emissions must be installed. The comparison in the figures shows that the levelized costs of some renewable energy systems are comparable to the levelized costs of some systems that consume fossil fuels, especially when the levelized cost of CCS technology is included.

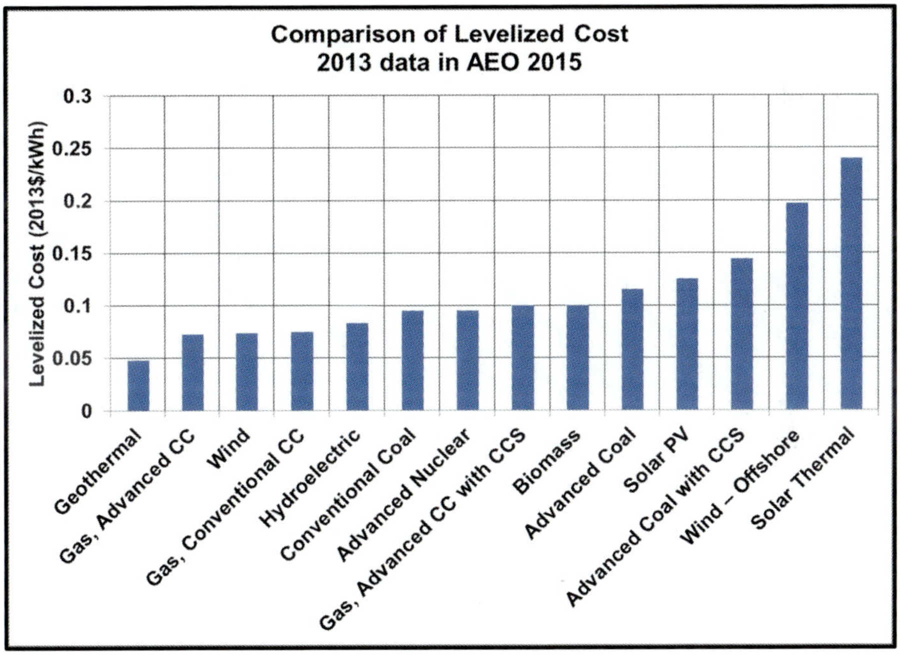

Figure 12-7. Comparison of Levelized Cost of Electricity Generation

Another way to make renewable technologies more competitive economically is to impose a carbon tax. For example, solar PV is a few cents per kWh more expensive than fossil fuels when expressed in 2013 US$. One way to make solar PV more competitive economically is to add a carbon tax to combustible fuels, such as natural gas or coal, to account for the environmental impact of greenhouse gas emissions. A carbon tax that is the difference between the cost of solar PV and the cost of fossil fuel power would have to be a few cents per kWh to make solar PV competitive with fossil fuel power.

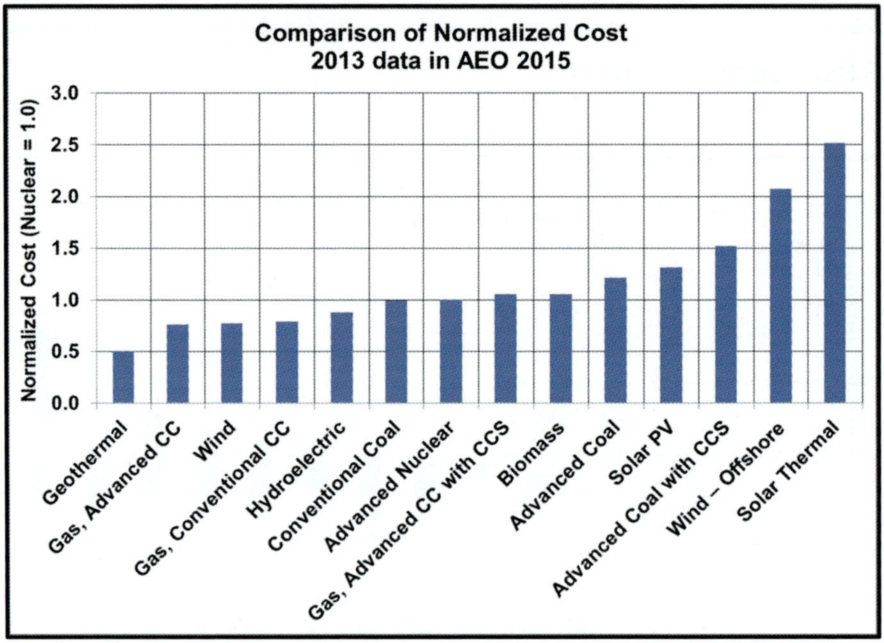

Figure 12-8. Comparison of Normalized Cost of Electricity Generation

Normalized cost for 2009 and 2013 are compared in Figure 12-9. The 2013 data show a substantial decline relative to 2009 data for geothermal and Solar PV, while fossil fuel-fired plants show an increase. These changes can be attributed to changes in government regulations, improvements in technology, and economies of scale.

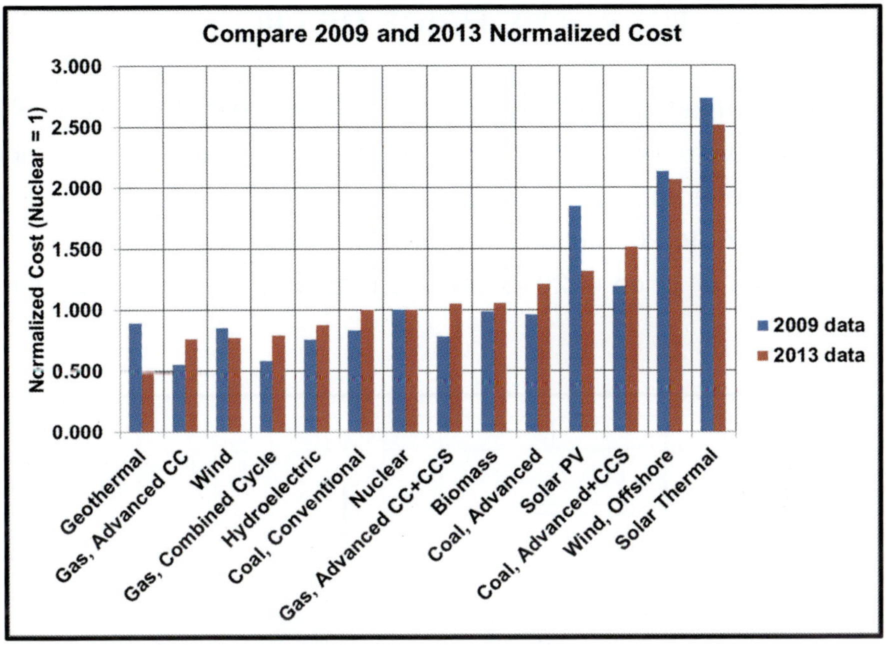

Figure 12-9. Comparison of Normalized Cost of Electricity Generation using 2009 Data and 2013 Data

Point to Ponder: Can we bias cost comparisons?

Every cost comparison relies on assumptions. It is possible to bias a comparison of costs by selecting assumptions that either increase or reduce the cost of a particular technology. For example, if we include the cost of carbon capture and sequestration (CCS) technology in the cost comparison, we can increase the cost of fossil fuel-fired power plants. On the other hand, we can reduce the cost of electricity generation using renewable energy technology by assuming that tax credits or subsidies will be applied. The bias can be magnified if we combine assumptions that favor one technology relative to another. If the bias is significant, it can change the ranking of technologies.

12.5 Case Study: The Price of Oil

Alarmists have made dire predictions in the media that the price of oil will increase with virtually no limit since the first oil crisis in 1973. Figure 12-10 shows average annual imported crude oil price in 2015US$ per barrel from 1974 to 2015 [US EIA Oil Price, 2015]. The abrupt decline in imported crude oil price in 2015 showed that oil price can fall as quickly as it can rise.

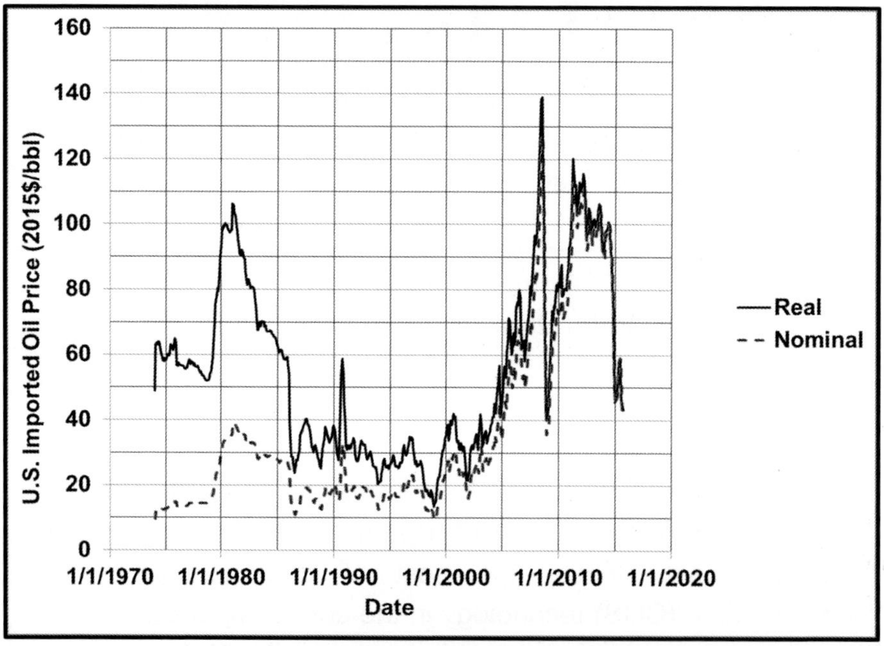

Figure 12-10. United States Average Annual Imported Crude Oil Price

Historical events provide insight into the volatility of oil price. In this section, we describe an event that was set in motion by the ongoing conflict in the Middle East and is known as the first oil crisis. We then discuss how oil price is affected by oil supply.

12.5.1 Events Leading Up to the First Oil Crisis

The Middle East has been embroiled in conflict since the formation of Israel in 1947. The British occupied Palestine in 1917, which was near the end of World War I. After World War I ended in 1918, the League of Nations approved the Mandate of Palestine, which authorized British administration of Palestine.

The United Nations partitioned Palestine into a Jewish state, an Arab state, and a UN administered Jerusalem at the end of World War II in 1945. The partitioning took effect in November, 1947 and the British left Palestine in 1948. The partitioning motivated two conflicting responses.

The response of Arab leaders was to reject the partitioning. By contrast, Israeli leaders declared independence and asserted national sovereignty when Britain left the region in 1948. The initial result of the partitioning was a civil war between Palestinian Jews and Palestinian Arabs. When Israel declared itself a sovereign state, the civil war escalated to an international war between Israel and nearby Arab countries. The war ended in July 1949, but it was not resolved. The conflict in the Middle East set the stage for a future war that would launch the first oil crisis.

On September 14, 1960, the Organization of Petroleum Exporting Countries (OPEC) was founded at a conference in Baghdad, Iraq. Thirteen countries including Iran, Indonesia, Angola, Nigeria, Ecuador, Venezuela and seven Arab countries joined the group with the goal of creating a neutral bargaining unit for oil producing countries. At the time, these countries were facing pressure from oil companies to lower prices and separately had little influence.

Initially OPEC sought to increase the share of oil revenues given to the oil producing countries and to increase their control over the supply of oil. By the start of the 1970s, OPEC had become a unified force and was gaining influence over world oil supply and price.

In May, 1967 Egypt expelled United Nations forces from the Sinai region separating Egyptian and Israeli territory. Israel, threatened by the presence of Egyptian forces on their border, called up reservists in preparation for armed conflict. On June 5, 1967, Israel attacked the Egyptian forces. Jordan and Syria joined Egypt in the struggle against Israel. The war lasted from June 5 to June 10, 1967 and is called the Six-Day War. Israel won a decisive victory and tripled its size by taking control of the Sinai Peninsula, the Gaza Strip, the West Bank of the Jordan River including East Jerusalem, and the Golan Heights.

Four years later, in a seemingly unrelated event, the United States and several other countries decided in 1971 to discontinue using gold as a standard for the value of their currencies. This led directly to depreciation of the value of the dollar and other currencies that oil producing nations were receiving for their goods.

The new currency valuation system had an adverse effect on the purchasing power of oil producing countries. Change in the value of currencies was accompanied by price increases in goods sold to these countries from the major oil consumers (goods such as wheat, grain, sugar, cement, even refined petroleum products) while the price of oil remained the same. Oil producing countries, including the members of OPEC, were receiving less real income for their exported oil at the same time that they were paying higher prices for the goods they purchased. Members of OPEC decided to begin pricing oil against the value of gold instead of the value of the dollar. The result was the first stage of the "oil shock" felt in the United States and throughout the world as oil prices suddenly increased.

Meanwhile, in the Middle East, tensions remained high as Arab nations viewed the existence of the Israeli state as an affront to the region. The Egyptians and Syrians, beaten in the Six-Day War, sought to recover the land they had lost as well as to defend their national honor after the defeat in 1967. On October 6, 1973, on the Jewish holiday of Yom Kippur, Egypt and Syria attacked Israel in what became known as the Yom Kippur War.

Israel's closest ally, the United States, expected the war to be short-lived based on the length of previous wars, but the Egyptians and Syrians were receiving assistance from Arab allies and support from the Union of Soviet Socialist Republics (U.S.S.R.). In the first 24 to 48 hours of the war, the attacking nations made positive advances, but shortly thereafter momentum began to shift to the defending Israelis, although losses were unexpectedly high.

Concerned that Israel was suffering significant losses, United States Secretary of State Henry Kissinger and President Richard Nixon initiated Operation Nickel Grass, an airlift operation designed to bring supplies to Israel to help in the war. There were also reports that the United States had flown reconnaissance missions over Egypt to provide intelligence to Israel about Egyptian military movements and placements in preparation for an Israeli counterattack, launched on October 14, 1973.

The United States' involvement in the Yom Kippur War would turn out to have enormous implications. Virtually the entire Arab world considered Israel an adversary and sought its destruction. On September 1, 1967, at the Khartoum Arab Summit, participating Arab nations signed the Khartoum Resolution which included the infamous "three no's": no peace with Israel, no recognition of Israel, and no negotiation with Israel. By supporting Israel in the war, the United States not only enraged Egypt and Syria, but the entire Arab world. The final straw came on October 17, 1973, when President Nixon requested $2.2 billion in emergency aid from Congress for Israel. Two days later, on October 19, thirteen days after the start of the Yom Kippur War, Libya initiated an oil embargo against the United States and the rest of the Arab nations followed suit.

12.5.2 The Arab Oil Embargo

The purpose of the Arab oil embargo was two-fold. First, it was important that Arab countries supporting the war against Israel provide an incentive for the United States to withdraw its support from Israel. By eliminating the supply of oil to the United States from OPEC members, the price of oil for the United States and the price of oil products within the United States were expected to increase. The second purpose of the embargo was to punish the United States for taking sides in the conflict, and particularly for taking the side of Israel. The United States believed it had little choice in the matter, however, since the United States and Britain were largely responsible for the existence of the Israeli state. In addition, the United States knew that the U.S.S.R. was supporting the Arab side of the conflict, effectively making the Yom Kippur War a battlefield in the Cold War.

The Yom Kippur War ended a week after the oil embargo began, but the embargo did not end until March 17, 1974. The ceasefire for the war was largely negotiated by the United States and U.S.S.R. on October 22, 1973, although the fighting continued for several days and skirmishes continued even longer.

The impact of the oil embargo was exacerbated by the collapse of the United States stock market in January 1973. The 1973-74 market crash was in large part a result of the destabilization of the United States' currency following President Nixon's decision to move the country off the gold standard. The price of a barrel of crude oil increased significantly from January 1973 to January 1974.

The significant oil price increase in 1973 shows the impact of the first oil crisis. Americans were forced to ration gasoline, with customers lining up at gas stations and accusations of price gouging. Arab leaders learned that oil could be used as leverage over the economy of a global superpower, and United States leaders realized that the energy destiny of the United States had become dependent on the actions of other nations. The Arab oil embargo prompted nations around the world to begin seriously considering a shift away from a carbon-based economy.

The first oil crisis is an important modern example of the interconnection between politics, economics and energy policy. Another example is the steep decline in oil price in 2014 and 2015 when Saudi Arabia chose to defend its market share by keeping the supply of oil greater than the demand for oil. The decline in oil price can force competitors to defer or shut-in high cost oil production.

12.5.3 How Does Oil Price Affect Oil Supply?

The economic viability of producing oil from conventional and unconventional reservoirs depends on cost and the price of oil. The supply of oil increases as the price of oil increases since it becomes possible to economically produce more difficult resources such as unconventional oil. In addition, many older fields contain oil that was not produced because it could only be extracted using more wells or more sophisticated, and expensive, extraction methods. Most oil producing companies choose to seek and produce less expensive oil so they can compete in the international marketplace.

Table 12-4 illustrates the sensitivity of oil producing techniques to the price of oil. Oil prices in the table include prices in the original 1997 prices and inflation-adjusted prices to 2016. The inflation rate used in the table is 5%. The actual inflation rate for oil prices depends on a number of factors, such as size and availability of supply and demand.

Table 12-4
Sensitivity of Oil Recovery Technology to Oil Price

Oil Recovery Technology	Oil Price Range	
	1997US$/bbl	2016US$/bbl 5% Inflation
Conventional	15 – 25	38 – 63
Enhanced Oil Recovery (EOR)	20 – 40	51 – 101
Extra Heavy Oil (e.g. tar sands)	25 – 45	63 – 114
Alternative Energy Sources	40 – 60	101 – 152

Table 12-4 shows that more sophisticated technologies can be justified as the price of oil increases. It also includes a price estimate for alternative energy sources, such as wind and solar. Wind energy and solar PV are increasingly competitive with oil as an energy source for generating electricity. In some cases there is overlap between one technology and another. For example, steam flooding is an Enhanced Oil Recovery (EOR) process that can compete with conventional oil recovery techniques such as water flooding, while chemical flooding is an EOR process that can be as expensive as many alternative energy sources.

According to Table 12-4, alternative energy sources become cost competitive when the price of oil rises above 2016US$101 per barrel. If the price of oil stays at 2016US$101 per barrel or higher for an extended period of time, energy consumers will begin to switch to less expensive energy sources. This switch is known as product substitution and has already begun in many countries. For example, consumers in European countries pay much more for gasoline than consumers in the United States. Countries such as Denmark, Germany and Holland are developing wind energy as a substitute to fossil fuels for generating electricity. France has opted to rely on nuclear fission energy.

Historically, oil exporting countries have tried to maximize their income and minimize competition from alternative energy and expensive oil recovery technologies by supplying just enough oil to keep the price below the price needed to justify product substitution. The increase in shale oil production in the United States led to competition to preserve market share among oil exporting nations in 2014 and 2015. In particular, Saudi Arabia chose not to reduce oil production to establish a balance between supply and demand. Instead, the Saudis chose to compete for market share by maintaining production in an attempt to drive out high-cost producers. This position was not endorsed by all members of OPEC because OPEC members were relying on oil revenue to support their national budgets. Global oversupply of oil resulted in a substantial decline in oil price. The price of oil cannot drop without limit, however, because oil exporting nations rely on oil income to pay their bills.

Oil importing countries can attempt to minimize their dependence on imported oil by developing technologies that reduce the cost of alternative energy. If an oil importing country contains mature oil reservoirs, the development of relatively inexpensive technologies for producing oil remaining in mature reservoirs or the imposition of economic incentives to encourage domestic oil production can be used to reduce the country's dependence on imported oil.

Point to Ponder: Why would a country want to keep reserves of fossil fuels secret?

Every country depends to some extent on fossil fuels. As we have seen, fossil fuels are an important fuel source for commercial, industrial, and transportation systems. The amount of fossil fuels in a country has an impact on that country's ability to function as a geopolitical power. A country may choose to keep the size of its resource base secret to limit information to its adversaries and introduce uncertainty in negotiations with businesses and other countries.

Recent history has illustrated why some countries would want to mislead people about their fossil reserves. Knowing that fossil fuels are presumed to be nonrenewable, this resource has high value to many countries that have few other resources of value. Keeping their actual reserves secret lets countries, such as those in the Middle East, control the supply and price of their resources as long as possible, and gives them a bargaining advantage over foreign nations seeking those resources.

In the late 1980's, members of OPEC had an incentive to mislead others about their national reserves. At the time, OPEC decided to tie production from member countries to their stated reserves. Thus, if a country claimed it had greater reserves, it would be allowed to produce more oil and generate greater revenue. Kuwait was among the first, and the other OPEC nations followed suit, to claim their reserves were suddenly much greater than previously reported. Looking at the numbers, it could be argued that these countries simply switched from reporting their reserves to reporting their total resources which is typically a significantly larger volume. This move inflated their reported reserves, and could impact predictions relating to a future oil peak. [Campbell, 2005]

Many of the "facts" relating to oil and gas reserves internationally are simply estimates based on the amount that each country chooses to report. Reported reserves can be compared to previous reports and estimates to determine if they are reasonable, but this has limited value. It is this uncertainty that makes it difficult for anyone to accurately predict how much oil and gas remains in the world, and when it might run out. Much of the information must be taken at face value from countries that have a motivation to mislead outsiders.

Further, countries are not the only entities that have an interest in keeping their reserves secret, or at least misleading outsiders about their reserves. In 2004, Royal Dutch/Shell stock declined significantly when it was discovered that the company had been overstating its proved reserves by including fields that had not yet been fully analyzed. These reserves were shifted to the "probable" category or even a lower category, meaning that their value to the firm decreased measurably. Why did Shell overstate the proved reserves? Simply, because it made the company's balance sheet look better. An improvement in the balance sheet can make the company look more valuable, can lead to an increase in the company's stock price, and can improve the company's apparent financial position for receiving loans.

The implications of misleading reserves resonated throughout the industry. It cast doubts on the integrity of reserves reports by other companies, including national oil companies. Shell's top executive resigned in the wake of the scandal, and brought the company under increased scrutiny by the United States Securities and Exchange Commission (SEC).

12.6 ACTIVITIES

True-False

Specify if each of the following statements is True (T) or False (F).

1. Microeconomics is the study of how individual businesses or households decide to allocate limited resources.
2. An increase in price leads to an increase in demand for a product.
3. Demand for a product is inelastic if it stays the same regardless of price.
4. OPEX for wind and solar power are zero.
5. The Yom Kippur War was the catalyst for the Arab oil embargo on the United States.
6. Electricity is fungible because one unit of electricity can be interchanged with another, regardless of the source.
7. The net present value of a project is always positive.
8. Salvage value can be either a revenue or a cost and should be included in a life cycle analysis.
9. Uncertainty is the concept that our limited knowledge and understanding of the future does not allow us to predict the consequences of our decisions with 100% accuracy.
10. The goal of LCOE is to compare energy costs across energy systems.

Questions

1. What does it mean if a project has a negative NPV?
2. What is life cycle analysis?
3. What is the purpose of Real Options Analysis?
4. Suppose a nuclear power plant can provide electrical energy at a cost of 6.7 cents/kWh. The corresponding cost for pulverized coal is 4.2 cents/kWh and for natural gas averages 4.7 cents/kWh. If a carbon tax of 3 cents/kWh is imposed, will the cost of nuclear energy be more expensive or less expensive than the cost of energy from pulverized coal or natural gas?

5. Which renewable energy systems in Table 12-3 have a lower levelized cost than power plants that burn gas and use CCS technology?
6. Explain the elasticity of demand.
7. What is the difference between OPEX and CAPEX?
8. What is end use efficiency?
9. How did the United States' decision to remove its currency from the gold standard affect the oil market?
10. What were two reasons for imposing an Arab oil embargo on the United States in 1973?
11. Why do oil companies want to prevent oil prices from getting "too high"?
12. Why would a country want to keep its oil and gas reserves secret?

CHAPTER 13

FUTURE ISSUES – GEOPOLITICS OF ENERGY

After reading this chapter, you should be able to...

- Discuss the importance of energy issues in a geopolitical context
- Define sustainable development
- Differentiate between "lifeboat ethics" and "spaceship ethics"
    - *P2P: Do we need to consider the ethics of energy distribution?*
- Discuss different geopolitical systems including the Clash of Civilizations
    - *P2P: Does one country have the right to direct another country's technological development?*
- Discuss the significance of key resources
- Describe the relationship between political systems and energy policy
- Discuss the merits of regulating global carbon emissions

Decisions are being made today by companies and governments around the world based on the assumption that fossil fuels will be replaced as the primary source of energy. Some governments of oil consuming nations are maintaining military forces capable of keeping oil supply lines open between producers and consumers. Oil exporting nations are trying to optimize their revenues by influencing the market price of oil. Every one of us has a stake in the decisions that are being made to provide energy.

One of the problems facing society is the need to develop and implement a strategy that will provide energy to meet future global energy needs and satisfy environmental objectives. The development of strategies depends on our view of the future. At best, we can only make educated guesses about what the future will bring. The quality of our educated guesses depends on our understanding of the situation and the quantity and quality of information we have available.

The concept of sustainable development is a vision of what the future should be and is a road map for how we should prepare for the future. It has been adopted by the United Nations and helps explain evolving business practices in the energy industry. We discuss sustainable development as a road map for the future and present some forecasts that recognize the need to replace fossil fuels. Before we consider options to shape the future energy mix, we review how we arrived at this point, and why a change may be needed.

13.1 Sustainable Development

An energy mix is emerging to meet energy demand in the 21st century. The demand for energy is driven by factors such as increasing trends in population and consumption. The ability to meet the demand for energy depends on such factors as price volatility, supply availability, and efficiency of energy use. One measure of how efficiently a country is using its energy is energy intensity.

In the context of energy policy, energy intensity is defined as total primary energy consumption divided by gross domestic product. Countries that have low energy consumption and high domestic productivity will have relatively low energy intensities. Countries with high energy consumption and low domestic productivity will have relatively high energy intensities. By considering the change in energy intensity as a function of time, we can see if a country is improving its efficiency of energy consumption relative to its domestic productivity.

Figure 13-1 presents energy intensity (in annual BTU consumption per US$ in 2005 dollars) as a function of time for a select group of countries and the world. If one of our goals is to maintain or improve quality of life with improved energy efficiency, we would like to see energy intensity decrease as a function of time. Energy intensity was increasing for Saudi Arabia up to 2011, and energy intensity was decreasing for the other countries in the sample, including the United States. China's energy intensity increased early in the 21st century and was leveling off by 2011. India's energy intensity increased in the latter part of the 20th century and was declining by 2011. World energy intensity is approximately unchanged for the entire period.

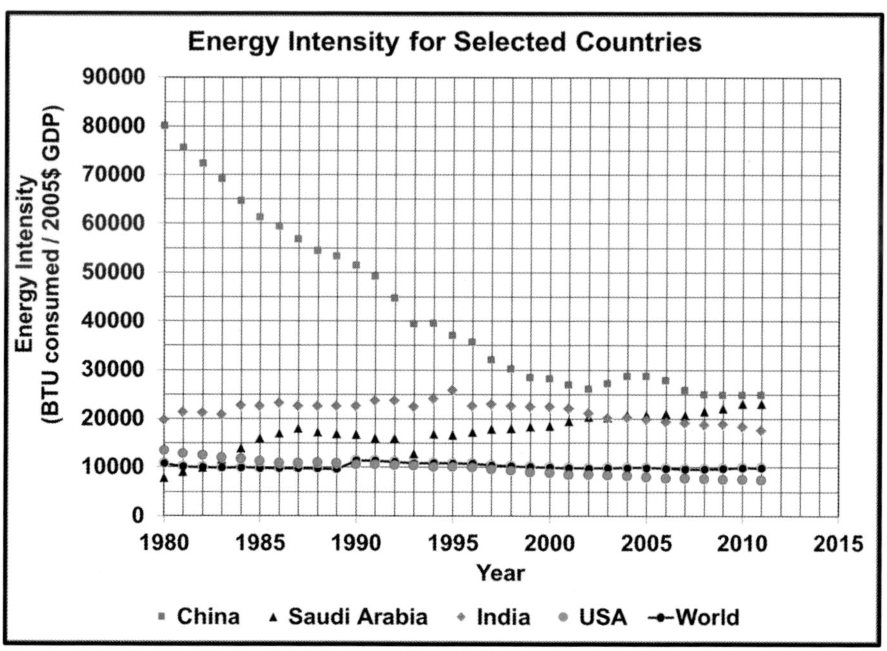

Figure 13-1. Energy Intensity for Selected Countries [US EIA Energy Intensity, 2016]

The emerging energy mix is expected to rely on clean energy, that is, energy that is generated with minimal environmental impact. The goal is to achieve sustainable development which includes the consideration of the rights of future generations. The concept of rights is a legal and philosophical concept. It is possible to argue that future generations do not have any legal rights to current natural resources and are not entitled to rights. From this perspective, each generation must do the best it can with available resources. On the other hand, many societies are choosing to adopt the concept of preserving natural resources for future generations. National parks are examples of natural resources that are being preserved.

Elkington [1994] advocated a strategy that would advance the goal of sustainability in business practices by introducing an accounting framework that would simultaneously benefit the company, its customers, and the environment. This is now commonly called the 'triple bottom line' or the 3Ps for profit, people, and the planet.

Whittaker [1999] identified the three components of sustainable development and the three goals of the triple bottom line (TBL) as economic prosperity, social equity, and environmental protection. From a business perspective, the focus of TBL is the creation of long-term shareholder value by recognizing that corporations are dependent on licenses provided by society to do business. If business chooses not to comply with sustainable development policies, society can enforce compliance by imposing additional government regulation and control. Critics point out that profit can measure the economic value of business activity, but it is difficult to account for the economic value of business activities on the planet (environment), and people (labor and the community).

13.2 Energy and Ethics

One issue that must be considered in the context of sustainable development is the distribution of energy. Should energy be distributed around the world based on need, ability to pay, or some other value? This question is an ethical issue because the answer to the question depends on the values we choose to adopt.

The distribution of energy in the future will depend on whether or not a nation has a large per capita energy base. Should nations with energy resources help those in need? If so, how should they help? Two of the more prominent ethical positions for helping nations without energy resources are identified as *lifeboat ethics* and *spaceship ethics*. These positions are considered here for two reasons: they are diametrically opposed ethical positions that apply to the global distribution of energy; and they illustrate that people of good will can take opposite positions in a significant debate.

Proponents of *lifeboat ethics* oppose the transfer of wealth by charitable means. In this view, the more developed industrial nations are considered rich boats and the less developed, overcrowded nations are poor boats. The rich boats should not give the poor boats energy because their help would discourage the poor boats from making difficult choices such as population control and investment in infrastructure. Lifeboat ethics is a "tough love" position; it encourages nations to seek self-sufficiency. On the other hand, it might make some nations desperate and encourage the acquisition of energy resources by military means.

Proponents of *spaceship ethics* argue that everyone is a passenger on spaceship Earth. In this view, some passengers travel in first class while others are in steerage. A more equitable distribution of energy is needed because it is morally just and it will prevent revolts and social turmoil. Thus, the wealthy should transfer part of their resources to the poor for both moral and practical reasons. On the other hand, nations that receive charitable donations of energy may be unwilling to make the sacrifices needed to become self-sufficient.

It may be that a synthesis of these two ethical positions would be the optimum policy for distributing energy. Countries with resources could help other countries develop their own resources with the clear understanding that the developing countries must become self-sufficient.

Point to Ponder: Do we need to consider the ethics of energy distribution?

The issue of energy distribution is really an issue of energy access. If people do not have access to essential resources, they may feel compelled to fight for those resources. Ethical positions such as *lifeboat ethics* and *spaceship ethics* attempt to address these concerns. Can access to energy be provided without discouraging self-reliance? Can the amount of useful energy be increased so that the need for sacrifice is lessened and the cost of energy is decreased? These are questions facing us today.

13.3 Energy and Geopolitics

Quality of life, energy, and the distribution of energy are important components of global politics. Readily available, reasonably priced energy is a critical contributor to the economic well-being of a nation. We have already seen that deforestation in England motivated the search for a new primary fuel. The need for oil encouraged Japanese expansion throughout Asia in the 1930's and was one of the causes of World War II. The 1973 Arab-Israeli war led to the first oil crisis with a significant short-term increase in the price of oil. This oil price shock was followed by another in 1979 after the fall of the Shah of Iran. These oil price increases are considered shocks because they were large enough to cause a significant decline in global economic activity [Verleger, 2000, page 76]. Our ability to correctly forecast energy demand depends on our understanding of technical and socio-political issues. In this section, we give a brief introduction to global politics and then discuss its implications for our understanding of future energy demand.

13.3.1 Clash of Civilizations

The world has been undergoing a socio-political transition that began with the end of the Cold War and is continuing today. Huntington [1996] provided a view of this transition that helps clarify historical and current events, and provides a foundation for understanding the socio-political issues that affect energy demand.

Huntington argued that a paradigm shift was occurring in the geopolitical arena. A paradigm is a way of thinking about a subject. It is a model that is realistic enough to help us make predictions and understand events, but not so realistic that it tends to confuse rather than clarify issues. A paradigm shift is a change in paradigm.

A geopolitical model can be used to order events and make general statements about reality. We can use the model to help us understand causal relationships between events and groups of people in communities ranging in size from organizations to national alliances. A geopolitical model lets us anticipate future developments and, in some instances, make predictions. It helps us establish the importance of information in relation to the model and it shows us paths that might help us reach our goals.

The Cold War between the Soviet Union and the Western alliance led by the United States established a framework that allowed people to better understand the relationships between nations following the end of World War II in 1945. When the Cold War ended with the fall of the Berlin Wall and the break-up of the Soviet Union in the late 1980's and early 1990's, it signaled the end of one paradigm and the need for a new paradigm. Several geopolitical models have been proposed. Huntington considered four possible paradigms for understanding the transition (Table 13-1).

Table 13-1
Huntington's Possible Geopolitical Paradigms

1	One Unified World
2	Two Worlds (West versus non-West; us versus them)
3	Anarchy (184+ Nation-states)
4	Chaos

The paradigms in Table 13-1 cover a wide range of possible geopolitical models. The One Unified World paradigm asserts that the end of the Cold War signaled the end of major conflicts and the beginning of a period of relative calm and stability. The Two Worlds paradigm views the world in an "us versus them" framework. The world was no longer divided by political ideology (democracy versus communism); it was divided by some other issue. Possible divisive issues include religion, and rich versus poor (generally a North-South geographic division). The world could also be split into zones of peace and zones of turmoil. The third paradigm, Anarchy, views the world in terms of the interests of each nation, and considers the relationships between nations to be unconstrained. These three paradigms, One Unified World, Two Worlds, and Anarchy, range from simple (One Unified World) to complex (Anarchy).

The Chaos paradigm says that post-Cold War nations are losing their relevance as new loyalties emerge. In a world where information flows freely and quickly, people are forming allegiances based on shared traditions and value systems. The value systems are notably cultural and, on a more fundamental level, religious. The new allegiances are in many cases a rebirth of historical loyalties. New alliances are forming from the new allegiances and emerging as a small set of civilizations. The emerging civilizations are characterized by ancestry, language, religion, and way of life.

Huntington considered the fourth paradigm, Chaos, to be the most accurate picture of current events and recent trends. He argued that the politics of the modern world can be best understood in terms of a model that considers relationships between the major contemporary civilizations shown in Table 13-2. The existence of a distinct African civilization has been proposed by some scholars, but is not as widely accepted as the civilizations identified in the table.

Table 13-2
Huntington's Major Contemporary Civilizations

Civilization	Comments
Sinic	China and related cultures in Southeast Asia
Japanese	The distinct civilization that emerged from the Chinese civilization between 100 and 400 A.D.
Hindu	The peoples of the Indian subcontinent that share a Hindu heritage.
Islamic	A civilization that originated in the Arabian peninsula and now includes subcultures in Arabia, Turkey, Persia, and Malaysia.
Western	A civilization centered around the northern Atlantic that has a European heritage and includes peoples in Europe, North America, Australia, and New Zealand.
Orthodox	A civilization centered in Russia and distinguished from Western Civilization by its cultural heritage, including limited exposure to Western experiences (such as the Renaissance, the Reformation, and the Enlightenment).

Table 13-2
Huntington's Major Contemporary Civilizations

Civilization	Comments
Latin America	Peoples with a European and Roman Catholic heritage who have lived in authoritarian cultures in Mexico, Central America and South America.

Each major civilization has at least one core state [Huntington, 1996, Chapter 7]. France and Germany are core states in the European Union, which is viewed as part of Western Civilization. The United States is also a core state in Western Civilization. Russia and China are core states, perhaps the only core states, in Orthodox Civilization and Sinic Civilization, respectively. Core states are sources of order within their civilizations. Stable relations between core states can help provide order between civilizations.

It can be argued that the lack of a stable relationship between core states in the Islamic Civilization is a key contributor to the turmoil within that civilization. OPEC was the framework for a stable relationship, but the resurgence of oil production in the United States resulted in a breach in OPEC solidarity and the oil price drop discussed in Chapter 12.

The growth of multiculturalism in some states has established communities within those states that may not share the values and allegiances of the host state. A multicultural state in this context is a member state of one civilization that contains at least one relatively large group of people that is loyal to many of the values of a different civilization. For example, Spain is a member state of Western Civilization with a sizable Islamic population. After the collapse of the Soviet Union, some multicultural states (e.g. Yugoslavia and Czechoslovakia) that were once bound by strong central governments separated into smaller states with more homogeneous values.

Within the context of the multi-civilization geopolitical model, the two World Wars in the 20th century began as civil wars in Western Civilization and engulfed other civilizations as the hostilities expanded. Those wars demonstrate that civilizations are not monolithic: states within civilizations may compete with each other. Indeed, the growth of multiculturalism and large migrant populations in some states are making it possible for states within a civilization to change their cultural identity as cultures within member states compete for dominance. The change in cultural identity can lead to a change in allegiance to a civilization.

The Cold War and the oil crises in the latter half of the 20th century were conflicts between civilizations. Western Civilization has been the most powerful civilization for centuries, where power in this context refers to the ability to control and influence someone else's behavior. The trend in global politics is a decline in the political power of Western Civilization as other civilizations develop technologically and economically. Energy is a key factor in this model of global politics. This can be seen by analyzing the energy dependence and relative military strength of core states. For example, consider the relationship between Western and Islamic Civilizations.

Western Civilization is an importer of oil and many states in Islamic Civilization are oil exporters. The result is the transfer of wealth from oil importing states of Western Civilization to oil exporting states in Islamic Civilization. By contrast, the United States, a leading core state in Western Civilization, is the leading military power in the world with a large arsenal of nuclear weapons. Most core states in Islamic Civilization are relatively weak militarily and do not have nuclear weapons. The wealth being acquired by Islamic Civilization is being used to alter the balance of military power between Western Civilization and Islamic Civilization.

Iran, a core state in Islamic Civilization, is developing nuclear technology. The possibility of acquiring nuclear weapons is raising concerns in other countries in Islamic Civilization such as Saudi Arabia. Iran (formerly Persia) and Saudi Arabia (Arabic) are both Islamic nations, but have different religious beliefs within the Islamic faith. The increased stress in

Islamic Civilization could be a pre-cursor to civil war within the civilization. Another factor adding stress to relationships between OPEC nations, many of which are member states of Islamic Civilization, is the competition for market share in a world where expanded use of renewable energy and shale oil production are decreasing the demand for imported oil.

Turmoil resulting from stress in the heart of Islamic Civilization fostered a migration of Middle Easterners, many Muslims, to Europe and the United States. The migration surged in 2015. If the migration is large enough, it can change the cultural identity of host nations. The possibility of a major change in culture can raise concerns among citizens of host nations, and leads to the question: is the migration a humanitarian gesture by the host nation, or is it an incursion by another civilization?

Point to Ponder: Does one country have the right to direct another country's technological development?

The United States, backed by United Nations resolutions, has attempted to block countries such as Iran and North Korea from developing nuclear weapons technologies. The goal was to prevent the proliferation of nuclear weapons, particularly to states that have governments that are deemed dangerous to Western Civilization. However, the policy of containing the distribution of nuclear technology also blocks countries from acquiring nuclear power which could help them improve quality of life for their people. This raises the question: should countries that have advanced technology be allowed, or even morally obligated, to control the distribution of technology to less-advanced societies? On the other hand, should a technologically advanced country be prohibited from sharing technology that can be used to build advanced weapons with a country that has shown signs of political instability or military aggressiveness? Answers to these questions could lead to international agreements. Who is going to enforce these agreements? The United Nations seems

> like an obvious answer, but how effective has the United Nations been as a peacekeeper? Is it capable of enforcing agreements between nations, especially if powerful nations disagree on enforcement strategy and tactics?

13.3.2 Clash Over Resources

Ideological differences between civilizations can lead to a struggle for global influence between core states. The battlefields in this struggle can range from economic to ideological to military. The outcome of this struggle is influenced by access to energy.

Energy importing states in one civilization rely on access to energy sources from energy exporting states in other civilizations. If the relationship between energy trading states is hostile, energy becomes a weapon in the struggle between civilizations. For example, the growth of non-Western civilizations, such as Sinic and Hindu Civilizations, has increased demand for a finite volume of oil. This increases the price of oil as a globally traded commodity and increases the flow of wealth between oil importing civilizations and oil exporting civilizations. Oil importing nations may try to reduce their need for imported oil by finding energy substitutes or by conservation. The social acceptability of energy conservation varies widely around the world. In some countries such as Germany, energy conservationists and environmentalists are a political force (the Green Party). In other countries, such as the United States, people may espouse conservation measures but may be unwilling to participate in or pay for energy conservation practices, such as recycling, carpooling, or driving energy-efficient vehicles.

Energy production depends on the ability of energy producers to have access to natural resources. Access depends on the nature of relationships between civilizations with the technology to develop natural resources and civilizations with territorial jurisdiction over the natural resources. Much oil production technology was developed in Western Civilization and gave Western Civilization the energy it needed to become the most powerful civilization in the world. As Western Civilization consumed its supply of oil, it became reliant on other civilizations to provide enough oil to maintain its oil-dependent economies. This dependence in times of stress between civilizations can lead to social turmoil and conflict between states that are members of different civilizations. Impending turmoil is motivating member states to develop alternatives to fossil fuels in an effort to achieve energy independence. The EU Supergrid discussed previously is a program designed to give the EU energy independence by 2050.

Huntington's Clash of Civilizations provides one perspective on modern geopolitical events. Another perspective was advocated by Klare [2004], who argued that modern geopolitics is driven by a struggle for resources. Klare summarized his thesis as follows [Klare, 2004, page xii]: "After examining a number of recent wars in Africa and Asia, I came to a conclusion radically different from Huntington's: that resources, not differences in civilizations or identities, are at the root of most contemporary conflicts." In **Resource Wars** [Klare, 2001], Klare made the case that oil, water, land and minerals were each important enough to be a source of contention, but he narrowed the list to petroleum in **Blood and Oil** [Klare, 2004]. Inexpensive and abundant petroleum is essential to modern economies.

13.4 Case Study: Government and Energy

Politics plays a crucial role in the development of energy resources and systems in today's world. The form of government in place for a given country can have a large impact on policy decisions that affect the use of energy resources. A government is a group of people that controls and directs the making and administration of public policy. Policies can include licensing of power facilities, licensing the production of natural resources, making public lands available for energy purposes, and regulation of industries that generate power or consume energy. Government is often responsible for developing infrastructure that is needed for the public good, including power transmission systems. Government can also play a role in the research and development of technologies, although the extent of government involvement depends on the governing body. We begin by defining different forms of government, than discuss competing energy visions associated with political differences.

13.4.1 Political Primer: Forms of Government

The terms "left" and "right" in politics are typically attributed to the seating arrangement in the French National Assembly during the French Revolution in 1789. The social order at the time included three estates: the First Estate consisted of priests and clergy; the Second Estate consisted of nobility and warriors; and the Third Estate consisted of workers or the common people. The Third Estate together with some aristocrats and low clergy sat in seats on the left hand side of the King, whereas most aristocrats and high clergy sat on the right hand side of the King [Rosas and Ferreira, 2013]. In general, supporters of the monarchy and aristocracy sat on the right, while people who opposed the old order sat on the left.

Today, if we stand at the podium of the European Parliament debate chamber in Strasbourg, France and look out over the members in the audience, we would see members with liberal, socialist, green, or communist views on the left side of the podium, and conservatives, Eurosceptics, and anarchists on the right side of the podium. Modern liberalism tends to seek social justice by supporting the use of government taxation and social programs to redistribute wealth. Socialism and Communism seek greater control of the means of production by increasing central planning and regulation. Green party members support government regulation of activities that can affect the environment. By contrast, modern conservatives support capitalism and limited government regulation. Eurosceptics are critical of the European Union governance system and believe that EU governance threatens national sovereignty. Brexit may be viewed as a manifestation of Euroscepticism. Anarchists support minimal or no government control with political power vested in the individual.

Figure 13-2 illustrates the range of political beliefs in terms of government control. There are different interpretations of political belief systems. Here, the "far left" represents total government control while the "far right" represents no government control. A ruling class tends to accumulate political power as you move to the left while political power tends to be vested in the individual as you move to the right.

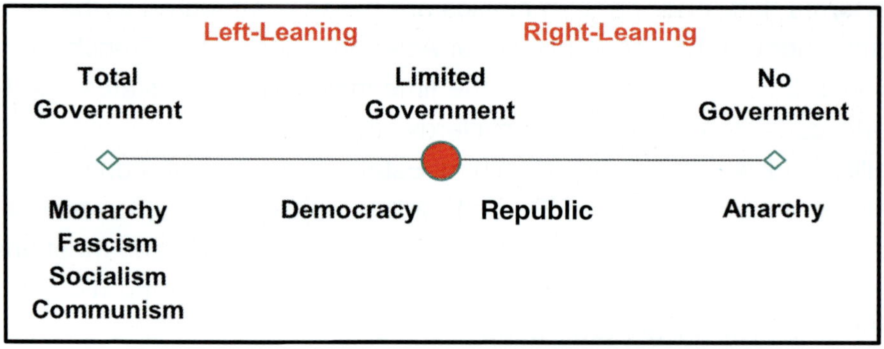

Figure 13-2. Spectrum of Governments

The role of government depends on the form and size of government. Citizens exercise political power directly in a democracy. Representatives are selected to exercise political power in a republic. The United States was founded as a constitutional republic, that is, a republic with the powers of government enumerated in a written constitution. In a monarchy, a hereditary chief of state serves for life and exercises power ranging from nominal to absolute. Political authority is centralized under a dictator in fascism. The dictator can impose stringent government controls. Political power is centralized in a ruling class in socialism, communism, and a modern form of government controlled by bureaucracies.

The size of government can be quantified by expressing government spending in terms of the size of the economy. For example, government spending by the United States Federal government ranged from approximately 18% of GDP to 25% of GDP during the first decade of the 21st century. As a rule of thumb, governments to the left in Figure 13-2 tend to spend a larger percentage of national GDP than those to the right. Government expenditures during times of war can alter this observation. The amount of money spent directly by the government is not the same as the amount of resources controlled by the government through policies, laws and regulations. It is possible in centrally planned economies for government to control the activities of privately held firms.

13.4.2 The Politics of Competing Energy Visions

Governments influence the energy sector. The political right would argue that the government should play a limited role in the development of clean energy. Their argument is that the free market will determine which technologies are best used and when transitions between energy sources should occur. As fossil fuel resources are depleted, they argue, the price will increase and the market will be inclined to seek alternatives such as wind and solar. The counter-argument is that without government attaching a cost to pollution and greenhouse gas emissions, the market would not account for them and the environment would suffer.

The political left, on the other hand, believes that government should tax and regulate industries that pollute and emit carbon. This factors environmental cost into the cost of the resource. In addition, the left believes that the government can play a role in supporting clean technologies in their infancy when they are too expensive and risky for private enterprises to research and develop. This is where direct government investment (Solyndra solar panel) and indirect subsidies (Spanish solar power) come in. The counter-argument to this approach is that government intervention in the free market will distort prices, tend to artificially inflate or suppress prices in a way that is unsustainable, and select winners and losers. Additionally, the government has limited resources, and those resources could potentially be better spent elsewhere. Finally, government assistance invariably creates the potential for fraud, corruption and political favoritism.

13.5 Global Regulation of Carbon Emissions

Many countries have been evaluating different methods for minimizing carbon emissions into the atmosphere. The primary motivation for reducing greenhouse gas emissions is to control anthropogenic climate change. Several policies have been suggested for assigning a cost to carbon emissions. An option that has been implemented in the European Union (EU) is called "Cap and Trade."

The premise behind Cap and Trade is that the central government of a country will be given a carbon emission goal by an international organization such as the European Union or the United Nations. For example, EU countries subject to the Kyoto Protocol agreed to reduce their carbon emissions by 8% relative to 1990 output levels by the year 2012. This is the "cap" for the country. The country can decide how to implement the policy. One implementation strategy proceeds as follows.

The government determines the amount of emissions that can be produced and gives companies, including utilities that run heavily-emitting power plants, a certain amount of carbon credits that they can use. This is their allowed amount of carbon emissions. If a company emits less gas than it is allowed by its credits, the company may "trade" the unused credits to other companies that exceed their cap. Trade, in this case, usually means selling credits to the highest bidder.

The EU instituted a Cap and Trade program in January 2005. The Cap and Trade program, called the Emission Trading Scheme (ETS), began as a trial. In 2007, ETS progressed from the Phase I trial stage to Phase II. Phase II links the EU ETS to that of other countries involved in similar carbon trading plans in accordance with the Kyoto Protocol. EU Phase III is scheduled to be in effect from 2013 to 2020.

The Cap and Trade system can be designed to allow a country to phase in its emissions cap. In other words, the country may reduce the number of carbon credits available each year to force a reduction in carbon emissions. The idea is that a gradual reduction would not require drastic, costly changes in a short period of time. Proponents of Cap and Trade point out that the system minimizes government intervention in the free market since companies may freely buy and sell credits and the market sets the price. The only responsibility the government has is to set the number of credits available each year. Critics of Cap and Trade express concern that Cap and Trade will not be effective at reducing greenhouse gas emissions, and that Cap and Trade is amenable to political favoritism and corruption. In the EU system, credits are allocated on an industry by industry basis, which creates an incentive for industries to lobby for preferential treatment [Reyes, 2011].

13.5.1 Effect of Cap and Trade on Energy Supply and Demand

An example of a supply and demand curve offered by proponents of Cap and Trade is given in Figure 13-3. P_{Base} and Q_{Base} are the base price and quantity at equilibrium before any added legislation. In the figure, energy supply decreases due to the added cost of carbon-emitting products, but energy demand also decreases as consumers "do their part" to limit carbon emissions by decreasing energy use and as consumers change their purchasing habits because of the increased price of carbon-emitting products. The overall result is that prices remain stable and the amount of carbon-emitting products on the market decreases to $Q_{C\&T}$. The result is a win-win as consumer prices remain the same, but carbon emissions decrease.

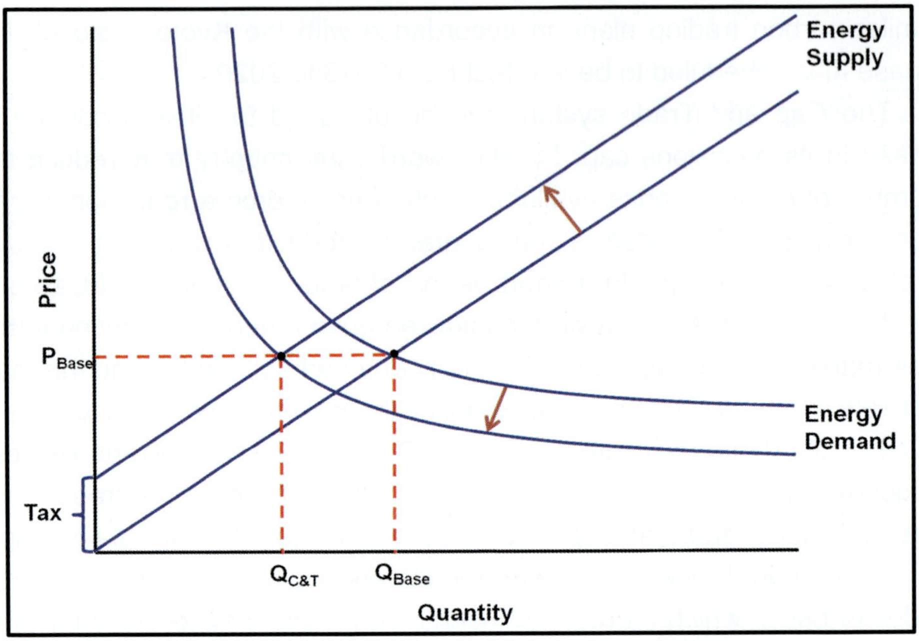

Figure 13-3. Proponents of Cap and Trade

Figure 13-4 presents an alternative scenario. It presumes that consumer behavior will not change with the additional cost of carbon emissions. While this may be overly simplified, it serves to illustrate the possibility that the quantity of energy that will be demanded will only decrease a small amount, but the price of energy will increase significantly.

Under Cap and Trade, energy providers are forced to buy carbon credits and bear the direct burden of the increase in cost. To the extent possible, energy providers will shift the cost of carbon credits to consumers through price increases on consumer goods such as gasoline and electricity. This is why opponents of Cap and Trade programs argue that Cap and Trade programs are a regressive tax on the middle and lower classes, that is, a tax on people that spend the largest percentage of their income on energy.

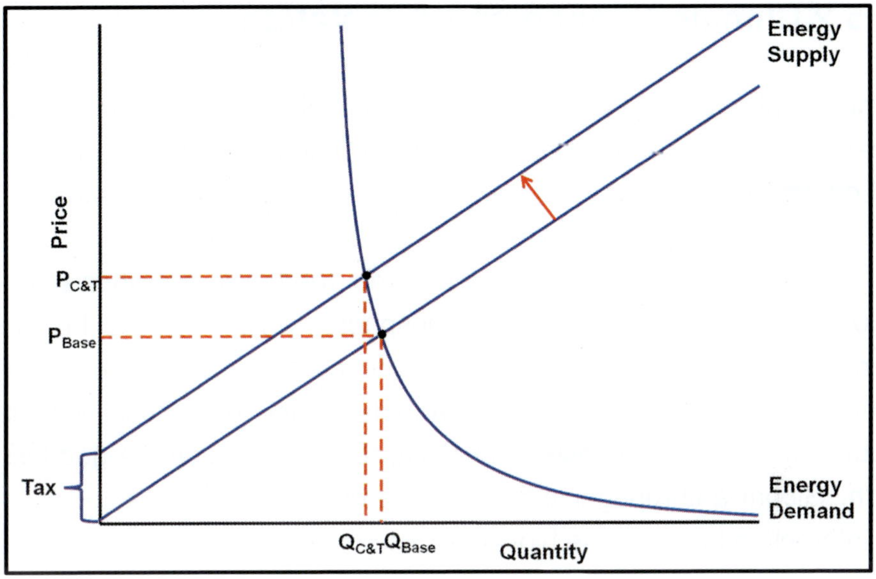

Figure 13-4. Opponents of Cap and Trade

We also need to consider the opportunity costs of Cap and Trade programs. If financial resources within a nation are being disproportionately spent to decrease carbon emissions, those same financial resources cannot be used for other purposes. This has a major impact on energy companies. While they can pass much of the cost on to consumers, they are not likely to be able to maintain the profit margins that existed prior to the Cap and Trade going into effect. Critics of the energy industry may argue that the industry does not need as much profit as it has made historically, but critics of Cap and Trade argue that the reduction in profit also means less money available for exploration, research, and projects that are risky and expensive. In this sense, it can be argued that Cap and Trade will tax everyone in the economy, including the poor, and it will stifle technological progress.

13.5.2 Climate Change and Adaptation

Some people believe that greenhouse gas emissions are having a major impact on climate, while others believe the impact will not be significant. Still others believe that anthropogenic climate change has already begun and that it is too late to do anything about it.

James Lovelock introduced the earth feedback hypothesis in the 1960's. He hypothesized that the biosphere, which includes humanity, interacts with the earth's physical environment to maintain an environment that can sustain life. The earth feedback hypothesis was later dubbed the Gaia hypothesis after Gaia, the Greek supreme goddess of Earth. Global anthropogenic climate change could be expected to alter the interaction between the biosphere and the physical components of the earth.

In **The Vanishing Face of Gaia** [Lovelock, 2009, page 68], Lovelock questioned the validity of climate forecasts and said the most important question in climate change is "How much and how fast is the earth heating?" Lovelock did not deny that anthropogenic climate change was occurring. On the contrary, Lovelock said that sea level change was a trustworthy indicator of the earth's heat balance. He pointed out that sea level was rising because ice was melting on land and the ocean was expanding because it was getting warmer. Evidence of ice melting is represented by the retreat and disappearance of glaciers. Figure 13-5 shows water flowing from a melting glacier near Saas Fe, Switzerland in June 2004. According to Lovelock, humans must accept that anthropogenic climate change is inevitable and prepare to adapt. He said that "Until we know for certain how to cure global heating, our greatest efforts should go into adaptation, to preparing those parts of the earth least likely to be affected by adverse climate change as the safe havens for a civilized humanity" [Lovelock, 2009, page 68].

Figure 13-5. Water Flowing from a Melting Glacier near Saas Fe, Switzerland (© Fanchi, 2004)

13.6 ACTIVITIES

True-False

Specify if each of the following statements is True (T) or False (F).

1. Energy intensity is energy consumption divided by gross national income.
2. A goal of lifeboat ethics in the global distribution of energy is to encourage nations to seek energy self-sufficiency.
3. Political power is the ability to control and influence someone else's behavior.
4. A paradigm shift is a change in paradigm from one model to another.
5. The world is obligated to implement a policy of sustainable development.
6. Huntington considered the end of the Cold War the catalyst for a geopolitical paradigm shift.
7. In Huntington's model, core states are sources of order within their civilizations.

8. Many oil fields that have already been abandoned still contain viable, recoverable oil.
9. Cap and Trade is amenable to corruption.
10. Cap and Trade systems could be considered a tax on the lower and middle classes.

Questions

1. Define spaceship ethics in the context of the global distribution of energy.
2. Why would wealthy nations be willing to implement spaceship ethics when developing policies for distributing energy?
3. Why would a country with energy resources be willing to share with a country without energy resources?
4. Suppose $100 billion is spent on the military in a year to protect the delivery of 20 million barrels of oil per day to the global market. How much does the military budget add to the cost of a barrel of oil?
5. What is energy intensity on the national level?
6. List Huntington's four potential geopolitical paradigms.
7. How has the rise of multiculturalism affected civilizations in Huntington's paradigm?
8. How does Klare's perspective on modern geopolitics differ from Huntington's?
9. Describe the concept behind Cap and Trade.
10. What is the Gaia hypothesis?

CHAPTER 14

FUTURE ISSUES – ENERGY FORECASTS

After reading this chapter, you should be able to…

- Discuss the importance of energy forecasts
- Discuss how energy conservation can affect energy forecasts
- Describe a nuclear energy forecast
  - *P2P: What has hindered the global adoption of nuclear energy?*
- Describe a renewable energy forecast
  - *P2P: Is the environmental impact of clean energy sources an important consideration?*
- Compare energy mix forecasts
- Discuss energy forecasts based on supply considerations
- Critically evaluate future energy options
- Discuss the Grand Energy Bargain

An energy mix is emerging to meet anticipated 21st century energy demand. Change in the energy mix is driven by factors such as increasing global population and energy consumption, the finite availability of fossil fuels, and climate change associated with carbon-based economies. As discussed in previous chapters, the ability to meet demand for energy depends on such factors as energy density, price volatility, supply availability, and efficiency of energy use.

The energy types that contributed most to the energy mix in the latter half of the 20th century were wood, coal, oil, natural gas, water and nuclear fission. The emerging energy mix includes renewable and nonrenewable energy resources. Increasing political pressure is causing the energy mix to trend toward renewable energy systems, but technological and commercial factors are justifying the inclusion of nonrenewable energy systems, even if only in a more complementary capacity.

Future energy demand is expected to grow substantially as global population increases and developing nations seek a higher quality of life. Forecasts of the 21st century energy mix show that a range of scenarios is possible. One way to make a forecast is to estimate the size of the population in the future, and specify a desirable quality of life. We presented a correlation in Chapter 1 between energy consumption and quality of life as expressed by the United Nations Human Development Index. Energy demand can be estimated using population forecasts and per capita energy consumption scenarios. For example, if we choose a UN HDI corresponding to 200,000 megajoules energy consumption per person and assume the global population is 8 billion people in 2100, we will need about three times as much energy as we are consuming today. This could be a conservative estimate since the assumption of 8 billion people is a conservative estimate when compared to population growth forecasts published by the UN.

Forecasts of energy production depend on the ability of energy producers to have access to natural resources. Access depends, in turn, on the nature of relationships between civilizations with the technology to develop natural resources and civilizations with territorial jurisdiction over natural resources. Nations are facing energy challenges that require important policy choices: should they seek energy independence or interdependence? Should they be concerned about environmental sustainability? How can they optimize economic opportunity? A selection of energy forecasts is presented here to illustrate the range of options each of us should consider as decision-makers. We first consider the impact of energy conservation before discussing a selection of long-term forecasts and then presenting the Goldilocks Policy and the Two Percent Solution.

14.1 ENERGY CONSERVATION AND ENERGY FORECASTS

Forecasts require an assumption about the role of energy conservation. Energy can be conserved using a number of very simple techniques. Some simple energy conservation methods that each of us can adopt include walking more and driving less; carpooling; planning a route that will let you drive to all of your destinations in the shortest distance; and using passive solar energy to dry your clothes or heat your home. Other energy conservation methods are more complicated: they are designed to reduce energy losses by improving the energy conversion efficiency of energy consuming devices. For example, improving gas mileage by reducing the weight of a vehicle or increasing the efficiency of internal combustion engines can reduce energy loss. We can decrease the demand for energy consuming activities such as air conditioning in the summer or heating in the winter by developing and using more effective insulating materials.

Energy conservation may be improved by increasing the efficiency of converting energy from one form into another. This efficiency is called energy conversion efficiency. It is the ratio of energy output to energy input. If we can decrease energy lost or wasted by a system, we can increase energy conversion efficiency. We must keep in mind that converting energy from its primary form to end-use energy will consume some energy and generate some waste, so it is unrealistic to expect to achieve 100% energy conversion efficiency.

One method for improving energy conversion efficiency is to find a way to use energy that would otherwise be lost as heat. Cogeneration is the simultaneous production and application of two or more sources of energy. An example of cogeneration is the simultaneous generation of electricity and useful heat. In this case, a fuel like natural gas can be burned in a boiler to produce steam. The steam drives an electric generator and is recaptured for such purposes as heating or manufacturing.

Cogeneration is most effective when the cogeneration facility is near the site where excess heat can be used. The primary objective of cogeneration is to reduce the loss of energy by converting part of the energy loss to an energy output.

Another cogeneration example is brake energy regeneration. The brake energy regeneration system in vehicles captures some of the energy that is normally lost during braking and stores it for later use. In the case of the Toyota Prius automobile, the surplus energy is stored in a battery. The Prius engine uses energy from the battery until the battery discharges, and then switches to gasoline while the battery is recharged. The engine functions as a hybrid of an electrical engine and an internal combustion engine.

Energy conservation carries varying levels of importance in different countries. Some governments, especially in energy importing nations, are encouraging or requiring the development of energy conserving technologies. We can expect energy conservation to increase in the future as a result of more widespread adoption of energy conservation measures and by improvements in energy conversion efficiency. However, it is unrealistic to expect energy conservation to be enough to satisfy global energy needs.

14.2 Energy Forecasts to 2100

Organizations such as BP, ExxonMobil, and the International Energy Agency (IEA) present relatively short-term energy supply and demand forecasts that extrapolate existing technology. Short-term forecasts usually cover a couple of decades. Figure 14-1 displays the International Energy Agency (IEA) forecast to 2030 [IEA WEO, 2015] and shows an expected increase in world primary energy use. The magnitude of the increase depends on assumptions used to prepare the forecast.

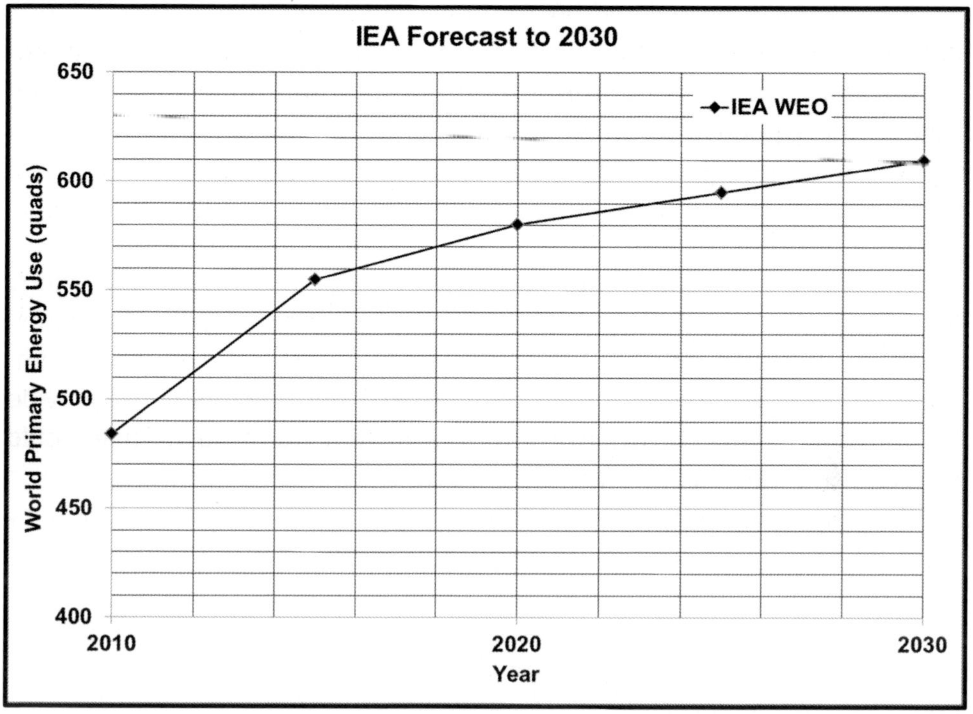

Figure 14-1. Energy Forecast to 2030

Our focus here is on long-term forecasts that cover all of the 21st century and display a range of perspectives on the long-term future of energy. The forecasts vary from a scenario that relies on nuclear energy to a scenario that relies on renewable energy. Other forecasts of the 21st century energy mix show a gradual transition from the current dependence on carbon-based fuels to a more balanced dependence on a variety of energy sources.

14.2.1 Nuclear Energy Forecast

P.E. Hodgson [1999] presented a scenario in which the world would come to rely on nuclear fission energy. He defined five Objective Criteria for evaluating each type of energy: capacity, cost, safety, reliability, and effect on the environment. The capacity criterion considered the ability of the energy source to meet future energy needs. The cost criterion considered all costs associated with an energy source. The safety criterion examined all safety factors involved in the practical application of an energy source. This includes hazards associated with manufacturing and operations. The reliability criterion considered the availability of an energy source. By applying the five Objective Criteria, Hodgson concluded that nuclear fission energy was the most viable technology for providing global energy in the future.

According to Hodgson, nuclear fission energy is a proven technology that does not emit significant amounts of greenhouse gases. He argued that nuclear fission reactors have an exemplary safety record when compared in detail with other energy sources. Breeder reactors could provide the fuel needed for nuclear fission power plants and nuclear waste could be stored in geological traps. The security of nuclear power plants in countries around the world would be assured by an international agency such as the United Nations. In this nuclear scenario, renewable energy sources would be used to supplement fission power and fossil energy use would be minimized. Hodgson did not assume that the problems associated with nuclear fusion would be overcome. If they are, nuclear fusion could also be incorporated into the energy mix.

Point to Ponder: What has hindered the global adoption of nuclear energy?

Nuclear energy is a long-term source of abundant energy that has environmental advantages. The routine operation of a nuclear power plant does not produce gaseous pollutants or greenhouse gases like carbon dioxide and methane. Despite its apparent strengths, the growth of the nuclear industry in many countries has been stalled by the public perception of nuclear energy as a dangerous and environmentally undesirable source of energy. This perception began with the use of nuclear energy as a weapon (Figure 14-2) and has been reinforced by widely publicized accidents at three nuclear power plants: Three Mile Island, Pennsylvania; Chernobyl, Ukraine; and Fukushima, Japan.

Social concern about nuclear waste and proliferation of nuclear weapons is a significant deterrent to reliance on nuclear fission power. These concerns were alleviated to some extent by the safety record of the modern nuclear power industry, but the damage to the Fukushima Daiichi Nuclear Plant following the March 2011 earthquake and tsunami renewed public concerns about the safety of nuclear fission power.

There are significant environmental and safety issues associated with nuclear energy, particularly nuclear fission. From an environmental perspective, the benefit of not producing greenhouse gases may be offset by the highly toxic and, in some cases, long-lived radioactive waste created by fission. The demonstration of commercial nuclear fusion power could have a significant impact on the global energy portfolio, but commercial nuclear fusion is not expected to become a reality until the latter half of the 21st century.

Figure 14-2. "Mushroom Cloud" Associated with the Detonation of a Nuclear Weapon

14.2.2 Renewable Energy Forecast

H. Geller [2003] presented a renewable energy scenario that sought to replace both nuclear energy and fossil energy with renewable energy only. An important objective of his forecast was to reduce greenhouse gas emissions to levels that were considered safe by the Kyoto Protocol. Figure 14-3 summarizes Geller's energy forecast. The figure shows global energy use as a function of time. Global energy use is expressed in billions of tons of oil equivalent (BTOE). One BTOE is approximately equal to 39.7 quads.

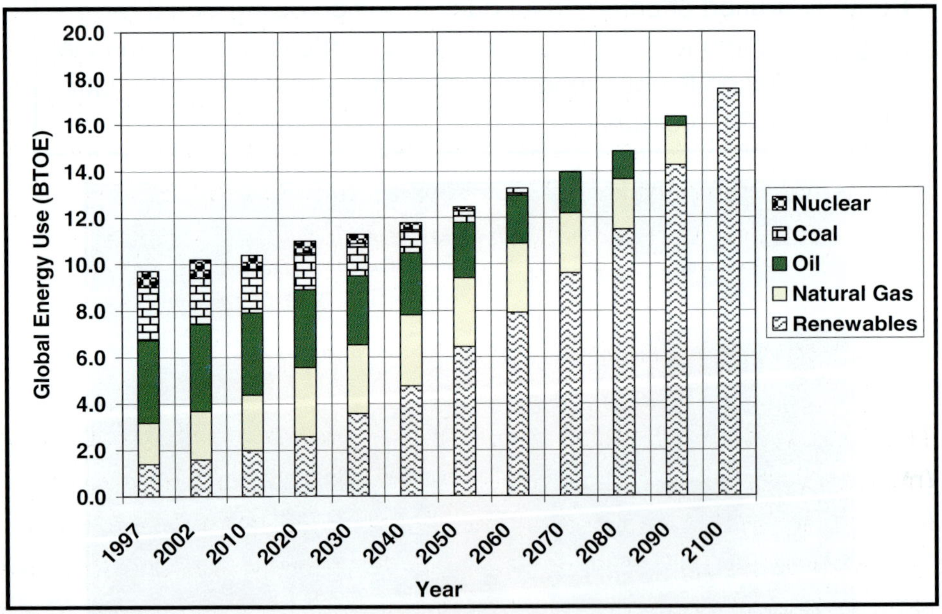

Figure 14-3. Forecast of a 21st Century Energy Mix that Eventually Relies on Renewable Energy Only [after Geller, 2003, page 227]

Point to Ponder: Is the environmental impact of clean energy sources an important consideration?

Society is searching for environmentally compatible, reliable energy sources. One of the characteristics of an environmentally compatible energy source is its cleanliness. A clean energy source emits negligible amounts of greenhouse gases or other pollutants. Even though fossil fuels have serious pollution problems, we have seen that mitigation technologies such as greenhouse gas storage and clean coal technology can reduce the impact of greenhouse gas emissions and justify the continued use of fossil fuels as an energy source.

In an objective assessment of competing energy sources, we must recognize that clean energy sources such as hydroelectric, solar, and wind energy can have a significant environmental impact that can adversely affect their environmental compatibility. This impact may affect the social acceptability of renewable energy projects.

14.2.3 Energy Mix Forecasts

The forecasts above focus on a future in which a single energy type (nuclear) or group of energy types (renewables) become the only energy forms used. W.E. Schollnberger's forecasts cover the entire 21st century and predict the contribution of a variety of energy sources to the 21st century energy portfolio. We do not consider Schollnberger's forecast because it is right. We already know the forecast began to deviate from reality within a few years of its publication date. Schollnberger's forecast is worth studying because it illustrates how to use more than one scenario to project energy consumption for the entire 21st century.

Schollnberger [1999] considered three forecast scenarios:

A. "Another Century of Oil and Gas" corresponding to continued high hydrocarbon demand;

B. "The End of the Internal Combustion Engine" corresponding to a low hydrocarbon demand scenario; and

C. "Energy Mix" corresponding to a scenario with intermediate demand for hydrocarbons and an increasing demand for alternative energy sources.

Failure to develop alternative technologies would make Scenario A most likely. Implicit in Scenario A is the assumption that enough hydrocarbons will be supplied to meet demand.

Schollnberger viewed Scenario C as the most likely scenario. It is consistent with the observation that the transition from one energy source to another has historically taken several generations. Leaders of the international energy industry have expressed a similar view that the energy mix is undergoing a shift from fossil fuels to other fuel sources.

There are circumstances in which Scenarios A and B could be more likely than Scenario C. For example, Scenario B would be more likely if environmental issues led to political restrictions on the use of hydrocarbons and an increased reliance on conservation. Scenario B would also be more likely if the development of a commercially competitive vehicle reduced the demand for vehicles that depend on the internal combustion engine. Hybrid vehicles use electrical energy from a battery to reduce, but not eliminate, the demand for gasoline. Vehicles that do not use gasoline are on the market and include compressed natural gas (CNG) vehicles and electric vehicles. CNG vehicles use cleaner burning natural gas, while electric vehicles use electric engines. Electric vehicles rely on the electric grid to recharge their batteries, which means that electric vehicles depend primarily on fossil fuels for recharging because most electricity from utilities is provided by fossil fuels. This will change as utilities expand their use of solar, wind, and nuclear energy to generate electricity.

Scenario C shows that natural gas will gain in importance as the economy shifts from a reliance on hydrocarbon liquid to a reliance on hydrocarbon gas. Schollnberger expected renewable energy sources such as wind and solar to eventually replace oil and gas (Figure 14-4).

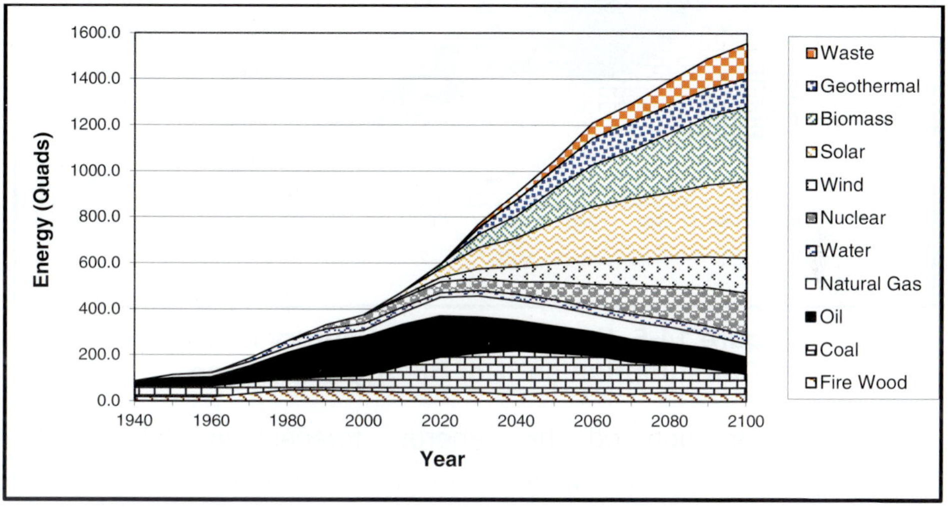

Figure 14-4. Forecast of 21st Century Energy Consumption

The demand by society for petroleum fuels should continue at or above current levels for a number of years, but the trend seems clear (see Figure 14-5). The global energy portfolio is undergoing a transition from an energy portfolio dominated by fossil fuels to an energy portfolio that includes a range of fuel types. Motivations for replacing fossil fuels were discussed in Chapter 4 and include a finite volume of hydrocarbons, concerns about environmental impact, and national security. The historical and projected energy consumption trends are illustrated in Figure 14-5.

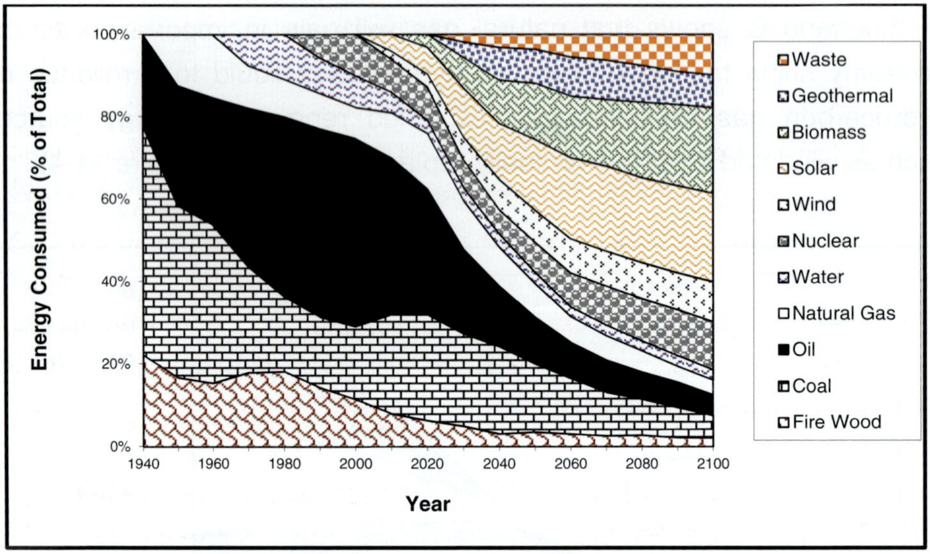

Figure 14-5. Forecast of 21st Century Energy Consumption as % of Energy Consumed

Schollnberger updated his energy forecast model in 2006 [Schollnberger, 2006]. He identified three desires that should influence energy decisions: sustained strong economic growth, security of energy supply, and a clean and safe environment. He argued that demand would determine future energy mix. Schollnberger reminded his European readers that many of them or their ancestors survived war-ravaged Europe at the end of World War II when the energy infrastructure was in ruins. Six decades later Europe has a modern energy infrastructure. Schollnberger, a champion of technology, was optimistic that humanity would achieve a sustainable energy future.

14.2.4 Energy Forecasts Based on Supply

Schollnberger's forecast was based on demand. An alternative approach is to base the forecast on supply. Beginning with M.K. Hubbert [1956], several authors have noted that annual U.S. and world oil production approximately follows a bell shaped (Gaussian) curve. A Gaussian curve fit of world daily oil production data was discussed in Section 4.1.2 and illustrated in Figure 4-2.

Differences between forecasts and actual data can be used to test the validity of assumptions that are used to generate forecasts. A test of forecast validity is the peak of world oil production. Analyses of historical data using a Gaussian curve typically predict that world oil production will peak in the first or second decade of the 21st century. The peak in Figure 4-2 occurs around 2030. By integrating the area under the Gaussian curve, forecasters estimated that cumulative world oil production will range from 1.8 to 2.1 trillion barrels.

Forecasts of world oil production peak tend to shift as more historical data is accumulated. These forecasts usually underestimate the sensitivity of oil production to technical advances and price. In addition, forecasts often discount the large volume of oil that has been discovered but not yet produced because the cost of production has been too high.

Laherrère [2000] pointed out that curve fits of historical data should be applied to activity that is "unaffected by political or significant economic interference, to areas having a large number of fields, and to areas of unfettered activity" (pg. 75). Curve fit forecasts work best when the inflection point (or peak) has passed.

14.3 Competing Energy Options

The validity of an energy forecast depends on technical feasibility, economic viability, and political decisions. Social concern about nuclear waste and proliferation of nuclear weapons is a significant deterrent to reliance on nuclear fission power. These concerns are alleviated to some extent by the safety record of the modern nuclear power industry. Society's inability to resolve the issues associated with nuclear fusion make fusion an unlikely contributor to the energy mix until at least the middle of the 21st century. If nuclear fusion power is allowed to develop and eventually becomes commercially viable, it could become the primary energy source. Until then, the feasible sources of energy for use in the future energy mix are fossil fuels, nuclear fission, and renewable energy.

Concerns about greenhouse gas emissions, climate change, and security of the energy supply are encouraging a movement away from fossil fuels. Political instability in countries that export oil and gas, the finite size of known oil and gas supplies, increases in fossil fuel prices and decreases in renewable energy costs, such as wind energy costs, are motivating the adoption of renewable energy. On the other hand, the development of technology that makes unconventional sources of fossil fuels economically competitive is encouraging continued use of fossil fuels, especially as the natural gas infrastructure is improved. These conflicting factors have an impact on the rate of transition from fossil fuels to a sustainable energy mix. A key decision facing society is to determine what the rate of transition should be.

Most of the global energy infrastructure is designed to use fossil energy. A conversion from fossil energy to other forms of energy requires construction of a new infrastructure. We have discussed some of the associated issues in previous chapters, such as the European Union Supergrid. The cost of transforming the energy infrastructure is a significant factor in determining the rate or transition to a new energy mix. An attempt to move too quickly away from fossil fuels can be disruptive to modern economies, while an attempt to move too slowly toward a sustainable energy mix can lead to geopolitical conflict over dwindling resources, and may lead to irreversible and undesirable changes to the climate.

Government policy should seek to optimize the rate of transition to a sustainable energy portfolio, but there are many different opinions about what constitutes an optimum rate of transition. We can illustrate this point by considering two possible options. In Option 1, we proceed to a sustainable energy portfolio by taking advantage of the abundance of natural gas in shale. Natural gas is a greenhouse gas, but its use in power generation and transportation is relatively clean compared to the use of coal or liquid hydrocarbons. A transition to a natural gas infrastructure is technically feasible and could be designed to facilitate an eventual transition to a hydrogen economy. By contrast, in Option 2 we proceed to a society that relies on renewable energy and electricity by expanding wind and solar power plants with improved energy storage technology and electric vehicles. A middle ground between these competing visions is the "Goldilocks Policy for Energy Transition" discussed next.

14.4 Goldilocks Policy for Energy Transition

The Goldilocks Policy for Energy Transition was introduced by Fanchi and Fanchi [2015]. The two options described in Section 14.3 are competing visions for reaching a sustainable energy mix. Some people see an urgent need to replace fossil fuels with sustainable/renewable energy sources because human activity is driving climate change. Others believe that it is necessary to replace fossil fuels with sustainable/renewable energy sources, but the need is not urgent. They argue that the economic health of society outweighs possible climate effects.

If the energy transition is too fast in Option 2, it could significantly damage the global economy. If the energy transition is too slow in Option 1, damage to the environment could be irreversible. We need the duration of the energy transition to be just right; that is, we need to adopt a reasonable plan of action that reduces uncertainty with predictable public policy and reduces environmental impact.

There is a historical basis for determining a reasonable transition period from one energy source to another. The U.S. is a developed country with a history of energy transitions over the past few centuries. Figure 14-6 shows U.S. energy consumption by source beginning in 1650 and proceeding to the present. Wood was the principal energy source when the U.S. was founded in the 18th century. Coal began to take over the energy market in the first half of the 19th century and peaked in the early 20th century. Oil began to compete with coal during the latter half of the 19th century and became the largest component of the energy mix by the middle of the 20th century. Historically, energy transition periods in the United States last approximately 60 to 70 years.

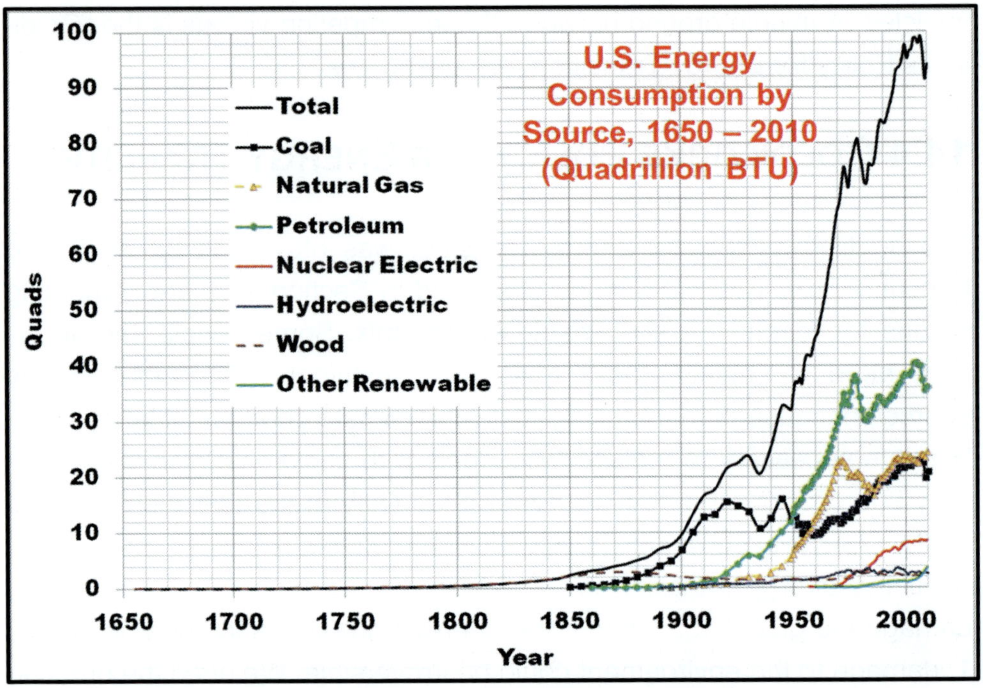

Figure 14-6. Can history tell us how long an energy transition period will last? [US EIA AER, 2001]

Historical transition periods can be highlighted by presenting energy consumption by source as a percentage of total U.S. energy consumption. Figure 14-7 shows the transition periods from wood to coal and from coal to oil. The U.S. was dependent on wood until around 1850 when the transition to coal began. The U.S. depended on coal until 1900-1920. The transition to oil was stimulated by the First World War which saw the development and use of vehicles powered by the internal combustion engine. Today, oil and natural gas account for over 80% of energy consumption. Historically, the coal and oil transition periods lasted between 60 and 70 years.

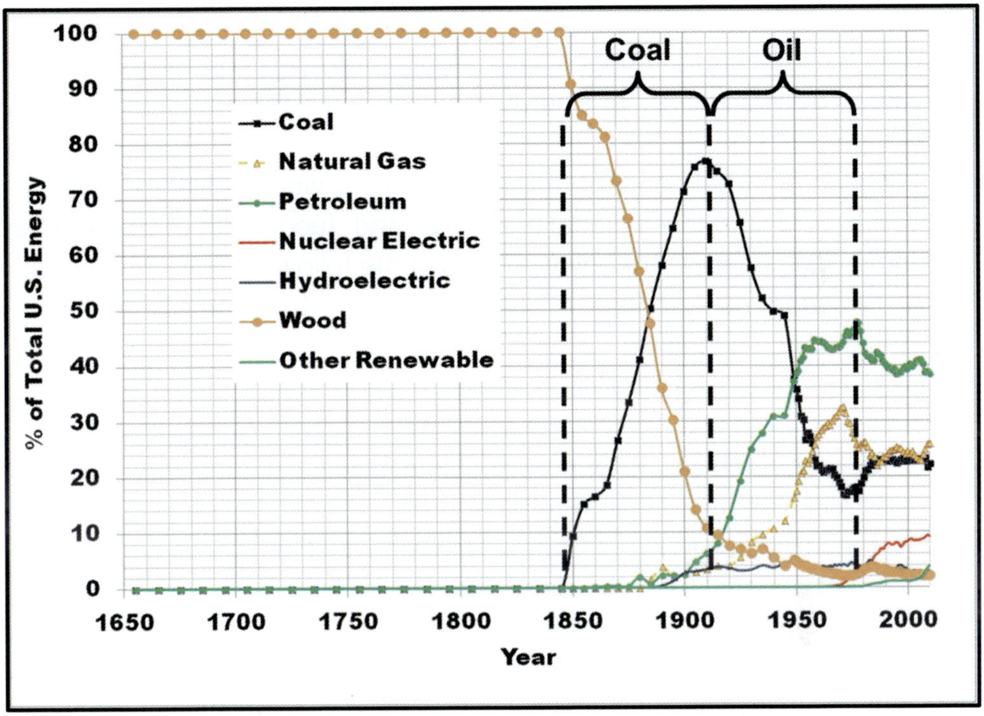

Figure 14-7. Coal and Oil Transition Periods based on U.S. Energy Consumption by Source, 1650 – 2010

The historical energy transition periods in the U.S. gives us an idea of how long it has historically taken a mature society to make the transition from one energy source to another energy source. The Goldilocks Policy for Energy Transition calls for a transition period of 60 to 70 years based on historical energy transitions in the U.S. Many countries around the world have already started to install commercial scale renewable energy systems (see Appendix C).

14.5 The Grand Energy Bargain

The Goldilocks Policy for Energy Transition may be viewed as a Grand Energy Bargain between the competing options introduced in Section 14.3. Concerns about Russian adventurism in Ukraine and European dependence on Russian fossil fuels motivated New York Times columnist Thomas L. Friedman to suggest the adoption of a Grand Energy Bargain advocated by Hal Harvey of Energy Innovation [Friedman, 2014; Harvey, 2016].

The Harvey-Friedman Grand Energy Bargain would advance economic growth, provide for national security, and recognize environmental concerns. U.S. energy strategy would have the following key ingredients:

A. Simultaneously optimize energy affordability, reliability and environmental compatibility

B. Use modern technology to provide affordable, reliable and clean energy

C. Use government to ensure that the current natural gas bonanza ushers in a secure, clean-energy future

The Harvey-Friedman Grand Energy Bargain would include 4 steps:

1. Adopt national rules for extracting natural gas
2. Set a national clean energy standard for electricity, e.g. utilities must implement zero-carbon sources (wind, solar, nuclear fission) at a predictable rate, such as 2% per year (50 year conversion)
3. Accelerate energy efficiency and clean energy technology by increasing research and development
4. Replace payroll and corporate taxes with a revenue-neutral carbon tax

The details of the Harvey-Friedman Grand Energy Bargain would have to be negotiated since many limited government advocates might object to some of the details, such as a carbon tax. The feasibility of an energy transition rate of 2% per year, the Two Percent Solution, proposed in step 2 is reasonable based on the Goldilocks Policy for Energy Transition. Dahlman [2015] has noted that global average temperature is expected to increase by approximately 0.2°C per decade. If this rate of increase remains constant throughout the 50 years of the energy transition, the global average temperature would increase approximately 1.0°C, which is less than the global average temperature change specified in the COP21 Agreement.

The combination of the Goldilocks Policy, the Two Percent Solution, and the Grand Energy Bargain is a roadmap to a sustainable energy future. The roadmap is a societal choice that would require discipline and patience, but it would enable society to make an orderly transition from nonrenewable, combustible, carbon-based energy sources to a sustainable economy.

One wild card in this discussion is the development of nuclear fusion energy. The successful development and commercialization of nuclear fusion power would provide a virtually inexhaustible source of energy that is environmentally benign, but we cannot count on the development of commercial nuclear fusion by mid-century.

Our overview of energy has shown that the global energy portfolio is in a state of transition. We cannot say what the energy portfolio will be at the beginning of the 22nd century. We can say that the emerging 21st century energy portfolio will depend on technological advances, including some advances that cannot be anticipated, and on choices made by society.

14.6 Activities

True-False

Specify if each of the following statements is True (T) or False (F).

1. Differences between forecasts and actual data can be used to test the validity of assumptions that are used to generate forecasts.
2. Forecasts based on the continuous extrapolation of historical trends can be invalidated by catastrophic events.
3. Historically, the transition from coal to oil in the United States took 25 years.

4. The United Nations Human Development Index is a measure of quality of life.
5. Cogeneration is the simultaneous production and application of two or more forms of energy.
6. Forecasts based on curve fitting work best when the inflection point (or peak) has been passed.
7. Most of the global energy infrastructure is designed to use renewable energy.
8. Analyses of historical data using a Gaussian curve typically predict that world oil production will peak after 2050.
9. Global future energy demand is expected to decline.
10. Forecasts of future energy production will change significantly when nuclear fusion becomes commercially viable.

Questions

1. What is energy conversion efficiency?
2. List three factors that are motivating a change in the energy mix.
3. What are the five objective criteria for evaluating an energy source used by Hodgson to develop a scenario in which the world would come to rely on nuclear fission energy?
4. What was an important environmental objective of Geller's [2003] renewable energy scenario?
5. List Schollnberger's [1999] three forecast scenarios.
6. Was Schollnberger's forecast of energy consumption based on supply or demand?
7. Was Hubbert's forecast of annual U.S. oil production based on supply or demand?
8. Is the cost of transforming the energy infrastructure a significant factor in determining the rate of transition from fossil fuels to a sustainable energy mix?
9. What factors should be considered in determining the rate of transition from fossil fuels to a sustainable energy mix?
10. Should government policy seek to optimize the rate of transition from fossil fuels to sustainable energy? Explain your answer.

APPENDIX A

UNITS AND SCIENTIFIC NOTATION

Typical units for expressing the amount of an energy source include barrels of oil, standard cubic feet of gas, tons of coal, and kilowatt-hours of electricity. Energy sources can be compared by converting to a common energy basis. Energy may be expressed in many different units, ranging from the British Thermal Unit (BTU) in the English system to the Joule (J) in SI (*System Internationale*) units. For example, the energy content of food is often expressed in Calories. One food Calorie is equivalent to 1000 calories or 4.184×10^3 J. Another important energy unit is the quad. A quad is a unit of energy that is often used in discussions of global energy because it is comparable in value to global energy values. One quad equals one quadrillion BTU (10^{15} BTU or approximately 10^{18} J).

SI Base Units

The SI base units and their symbols are presented in Table A-1. All other units can be derived from SI base units.

Table A-1
SI Base Units

Physical Quantity	Base Unit	Symbol
Length	meter	m
Mass	kilogram	kg
Time	second	s
Electric current	ampere	A

Table A-1
SI Base Units

Physical Quantity	Base Unit	Symbol
Temperature	kelvin	K
Amount of substance	mole	mol
Luminous intensity	candela	cd

An example of a derived unit is the Joule. The Joule is the SI unit of energy. It is a derived quantity with 1 J = 1 $kg{\cdot}m^2/s^2$. The Watt is the SI unit of power. One Watt is equal to one Joule per second. One Watt/m^2 is equal to one Joule of energy passing through one square meter in one second. The unit Watt/m^2 can be used to compare different energy types. Table A-2 presents SI units for fundamental physical quantities.

Table A-2
SI Units for Fundamental Physical Quantities

Physical Quantity	Unit	Abbreviation	Comment
Length	meter	m	
Mass	kilogram	kg	
Time	second	s	
Force	Newton	N	1 N = 1 $kg{\cdot}m/s^2$
Pressure	Pascal	Pa	1 Pa = 1 N/ m^2
Energy	Joule	J	1 J = 1 $kg{\cdot}m^2/s^2$
Power	Watt	W	1 W = 1 J/s
Temperature	kelvin	K	
Amount of substance	mole	mol	

Table A-2
SI Units for Fundamental Physical Quantities

Physical Quantity	Unit	Abbreviation	Comment
Frequency	Hertz	Hz	1 Hz = 1 cycle/s
Electric charge	Coulomb	C	
Electric current	Ampere	A	1 A = 1 C/s
Electric potential	Volt	V	1 V = 1J/C =1 W/A
Resistance	Ohm	Ω	1 Ω = 1 V/A
Capacitance	Farad	F	1 F = 1 C/V
Inductance	Henry	H	1 H = 1 V·s/A
Magnetic Induction $\vec{B}$	Tesla	T	1 T = 1 N/(A·m)
Magnetic flux	Weber	Wb	1 Wb = 1 T·m^2 = 1 V·s
Luminous intensity	candela	cd	

Scientific Notation

Scientific notation is used to express the values of many physical quantities encountered in the study of energy. Scientific notation is a means of compactly writing very large or very small numbers, and is based on the rules of exponents. Exponents obey the rules in Table A-3.

Table A-3 Rules for Exponents	
Products	$a^m \cdot a^n = a^{m+n}$
Quotient	$\frac{a^m}{a^n} = a^{m-n}$ if $m > n$ $\frac{a^m}{a^n} = 1$ if $m = n$ $\frac{a^m}{a^n} = \frac{1}{a^{n-m}}$ if $m < n$
Power	$(a^m)^n = a^{mn}$

The number a raised to a negative power is given by $a^{-m} = 1/a^m$. Any nonzero real number raised to the power 0 equals 1, thus $a^0 = 1$ if $a \neq 0$. In the case of the number 0, we have the exponential relationships $0^0 = 0, 0^x = 0$ for any value of x. If we set $a = 10$, we obtain the rules for combining powers of 10 shown in Table A-4.

Table A-4 Rules for Combining Powers of 10	
Products	$10^m \cdot 10^n = 10^{m+n}$
Quotient	$\frac{10^m}{10^n} = 10^{m-n}$ if $m > n$ $\frac{10^m}{10^n} = 1$ if $m = n$ $\frac{10^m}{10^n} = \frac{1}{10^{n-m}}$ if $m < n$
Power	$(10^m)^n = 10^{mn}$

The physical quantities of interest here have values that range from very small to very large. The most common prefixes for powers of 10 used in SI are presented in Table A-5.

Table A-5
Prefixes for Powers of 10

Prefix	Symbol	Value		Prefix	Symbol	Value
atto	a	10^{-18}		kilo	k	10^{3}
femto	f	10^{-15}		mega	M	10^{6}
pico	p	10^{-12}		giga	G	10^{9}
nano	n	10^{-9}		tera	T	10^{12}
micro	μ	10^{-6}		peta	P	10^{15}
milli	m	10^{-3}		exa	E	10^{18}

APPENDIX B

Summary of United States Historical Energy Production and Consumption 1950-2010

[U.S. EIA AER, 2015; 2020 estimate from U.S. EIA AEO, 2015 Reference Case]

Table B-1 Annual U.S. Primary Energy Production Historical Production from 1950 - 2010; 2020 Estimate* (Quadrillion BTU)				
Year	**Fossil Fuels**	**Nuclear Electric Power**	**Renewable Energy****	**Total Primary Energy Production**
1950	32.56	0.00	2.98	35.54
1960	39.87	0.01	2.93	42.80
1970	59.19	0.24	4.07	63.50
1980	59.01	2.74	5.43	67.18
1990	58.56	6.10	6.04	70.70
2000	57.37	7.86	6.10	71.33
2010	58.22	8.43	8.11	74.76
2020*	79.0	8.4	11.3	98.7

*Estimate

**Includes hydroelectric, geothermal, wind, and solar

Table B-2
Annual U.S. Primary Energy Consumption
Historical Production from 1950 - 2010; 2020 Estimate*
(Quadrillion BTU)

Year	Fossil Fuels	Nuclear Electric Power	Renewable Energy**	Total Primary Energy Production
1950	31.63	0.00	2.98	34.61
1960	42.14	0.01	2.93	45.07
1970	63.52	0.24	4.07	67.83
1980	69.83	2.74	5.43	78.00
1990	72.33	6.10	6.04	84.48
2000	84.73	7.86	6.11	98.70
2010	80.89	8.43	8.07	97.39
2020*	83.1	8.4	9.3	100.8

*Estimate

**Includes hydroelectric, geothermal, wind, and solar

Table B-3
Annual U.S. Primary Energy Production
Historical Production from 1950 - 2010; 2020 Estimate*

Year	Fossil Fuels (% of Total)	Nuclear Electric Power (% of Total)	Renewable Energy** (% of Total)	Total Primary Energy Production (Quadrillion BTU)
1950	91.6	0.0	8.4	35.54
1960	93.1	0.0	6.8	42.80
1970	93.2	0.4	6.4	63.50
1980	87.8	4.1	8.1	67.18
1990	82.8	8.6	8.5	70.70
2000	80.4	11.0	8.6	71.33
2010	77.9	11.3	10.9	74.76
2020*	80.0	8.5	11.4	98.7

*Estimate

**Includes hydroelectric, geothermal, wind, and solar

Table B-4
Annual U.S. Primary Energy Consumption
Historical Production from 1950 - 2010; 2020 Estimate*

Year	Fossil Fuels (% of Total)	Nuclear Electric Power (% of Total)	Renewable Energy** (% of Total)	Total Primary Energy Production (Quadrillion BTU)
1950	91.4	0.0	8.6	34.61
1960	93.5	0.0	6.5	45.07
1970	93.6	0.4	6.0	67.83
1980	89.5	3.5	7.0	78.00
1990	85.6	7.2	7.2	84.48
2000	85.8	8.0	6.2	98.70
2010	83.1	8.7	8.3	97.39
2020*	83.1	8.4	9.3	100.8

*Estimate
**Includes hydroelectric, geothermal, wind, and solar

APPENDIX C

Summary of World Energy Production and Consumption 1970-2020

[U.S. EIA IES, 2015; 2020 forecast from IEA WEO, 2015]

Table C-1
World Energy Production by Energy Source, 1970-2020
(Quadrillion BTU)

Energy Source	1970	1980	1990	2000	2010	2020*
Total	215.4	287.6	350.0	395.7	505.0	592.1
Fossil Fuels	200.8	259.1	303.3	337.9	436.6	468.5
Nuclear Electric Power	0.9	7.6	20.4	25.7	27.4	35.6
Renewable Energy**	13.7	20.8	26.3	32.2	41.0	88.1
Fossil Fuel Source	**1970**	**1980**	**1990**	**2000**	**2010**	**2020***
Total Fossil Fuels	200.8	259.1	303.3	337.9	436.6	468.5
Coal	63.0	71.3	91.0	90.4	147.2	162.0
Natural Gas	37.1	54.7	76.1	91.0	116.5	129.6
Petroleum Liquids	100.7	133.1	136.2	156.5	172.9	176.9

*Estimated
**Includes hydroelectric, geothermal, wind, and solar

Table C-2
World Energy Consumption by Energy Source, 1970-2020
(Quadrillion BTU)

Energy Source	**1970**	**1980**	**1990**	**2000**	**2010**	**2020***
Total	NA	283.2	347.6	397.9	510.8	580.2
Fossil Fuels		254.8	301.0	340.1	442.4	457.5
Nuclear Electric Power		7.6	20.4	25.7	27.4	33.4
Renewable Energy**		20.8	26.3	32.2	41.0	89.3
Fossil Fuel Source	**1970**	**1980**	**1990**	**2000**	**2010**	**2020***
Total Fossil Fuels		254.8	301.0	340.1	442.4	457.5
Coal		70.0	89.2	93.6	148.3	157.3
Natural Gas		53.8	75.3	91.0	117.6	125.1
Petroleum Liquids		131.0	136.4	155.5	176.5	175.2

*Estimated
**Includes hydroelectric, geothermal, wind, and solar

Table C-3
World Energy Production by Energy Source, 1970-2020
(% of annual)

Energy Source	1970	1980	1990	2000	2010	2020*
Total	100.0	100.0	100.0	100.0	100.0	100.0
Fossil Fuels	93.2	90.1	86.7	85.4	86.5	79.1
Nuclear Electric Power	0.4	2.6	5.8	6.5	5.4	6.0
Renewable Energy**	6.4	7.2	7.5	8.1	8.1	14.9
Fossil Fuel Source	**1970**	**1980**	**1990**	**2000**	**2010**	**2020***
Total Fossil Fuels	100.0	100.0	100.0	100.0	100.0	100.0
Coal	31.4	27.5	30.0	26.8	33.7	34.6
Natural Gas	18.5	21.1	25.1	26.9	26.7	27.7
Petroleum Liquids	50.2	51.4	44.9	46.3	39.6	37.8

*Estimated

**Includes hydroelectric, geothermal, wind, and solar

Table C-4
World Energy Consumption by Energy Source, 1970-2020
(% of annual)

Energy Source	1970	1980	1990	2000	2010	2020*
Total	NA	100.0	100.0	100.0	100.0	100.0
Fossil Fuels		90.0	86.6	85.5	86.6	78.9
Nuclear Electric Power		2.7	5.9	6.4	5.4	5.8
Renewable Energy**		7.3	7.6	8.1	8.0	15.4
Fossil Fuel Source	**1970**	**1980**	**1990**	**2000**	**2010**	**2020***
Total Fossil Fuels		100.0	100.0	100.0	100.0	100.0
Coal		27.5	29.6	27.5	33.5	34.4
Natural Gas		21.1	25.0	26.8	26.6	27.3
Petroleum Liquids		51.4	45.3	45.7	39.9	38.3

*Estimated
**Includes hydroelectric, geothermal, wind, and solar

APPENDIX D
UNIT CONVERSION FACTORS

TIME

1 hour = 1 hr = 3600 s
1 day = 8.64×10^4 s
1 year = 1 yr = 3.1536×10^7 s

LENGTH

1 foot = 1 ft = 0.3048 m
1 kilometer = 1 km = 1000 m
1 mile = 1 mi = 1.609 km
1 cm = 10^{-2} m

VELOCITY

1 foot per second = 0.3048 m/s
1 kilometer per hour = 1 kph = 1000 m/hr = 0.278 m/s
1 mile per hour = 1 mph = 1.609 km/hr = 1609 m/hr = 0.447 m/s

AREA

1 square foot = 1 ft^2 = 0.0929 m^2
1 square mile = 1 mi^2 = 2.589 km^2 = 2.589×10^6 m^2
1 square mile = 1 mi^2 = 640 acres
1 acre = 1 ac = 4047 m^2
1 hectare = 1 ha = 1.0×10^4 m^2
1 hectare = 1 ha = 2.4710 ac
1 millidarcy = 1 md = 0.986923×10^{-15} m^2
1 Darcy = 1000 md = 0.986923×10^{-12} m^2
1 barn = 1.0×10^{-24} cm^2 = 1.0×10^{-28} m^2

VOLUME
1 liter = 1 L = 0.001 m^3
1 cubic foot = 1 ft^3 = 2.83 × 10^{-2} m^3
1 standard cubic foot = 1 SCF = 1 ft^3 at standard conditions
1 acre-foot = 1 ac-ft = 1233.5 m^3
1 barrel = 1 bbl = 0.1589 m^3
1 gallon (U.S. liquid) = 1 gal = 3.785 × 10^{-3} m^3
1 barrel = 42 gallons = 0.1589 m^3
1 barrel = 5.6148 ft^3
1 gallon = 3.788 liters
1 cm^3 = 1 cc = 10^{-6} m^3
1 cm^3 = 3.534 × 10^{-5} ft^3

MASS
1 gram = 1 g = 0.001 kg
1 pound (avoirdupois) = 1 lb (avdp) = 1 lbm = 0.453592 kg
1 tonne = 1000 kg

MASS DENSITY
1 g/cm^3 = 1000 kg/m^3

FORCE
1 pound-force = 1 lbf = 4.4482 N

PRESSURE
1 Pascal = 1 Pa = 1 N/m^2 = 1 $kg/m{\cdot}s^2$
1 Megapascal = 1 MPa = 10^6 Pa
1 Gigapascal = 1 GPa = 10^9 Pa
1 pound-force per square inch = 1 psi = 6894.8 Pa
1 atmosphere = 1 atm = 1.01325 × 10^5 Pa
1 atmosphere = 1 atm =14.7 psi
1 bar = 14.5 psi = 0.1 MPa
1 MPa = 145 psi

ENERGY

1 megajoule = 1 MJ = 1.0×10^6 J
1 gigajoule = 1 GJ = 1.0×10^9 J
1 exajoule = 1 EJ = 1.0×10^{18} J
1 eV = 1.6022×10^{-19} J
1 MeV = 10^6 eV = 1.6022×10^{-13} J
1 erg = 10^{-7} J
1 BTU = 1055 J
1 ft·lbf = 0.7376 J
1 calorie (thermochemical) = 1 cal = 4.184 J
1 kilocalorie = 1 kcal = 1000 calories = 4.184×10^3 J
1 Calorie = 1000 calories = 4.184×10^3 J
1 kilowatt-hour = 1 kWh = 1 kW · 1 hr = 3.6×10^6 J
1 quad = 1 quadrillion BTU = 1.0×10^{15} BTU = 1.055×10^{18} J
1 quad = 2.93×10^{11} kWh = 1.055×10^{12} MJ
1 quad = 1.055 exajoule = 1.055 EJ
1 barrel of oil equivalent = 1 BOE = 5.8×10^6 BTU = 6.12×10^9 J
1 quad = 1.72×10^8 BOE = 172×10^6 BOE

ENERGY DENSITY

1 BTU/lbm = 2326 J/kg
1 BTU/SCF = 3.73×10^4 J/m^3

POWER

1 Watt = 1 W = 1 J/s
1 Megawatt = 10^6 W = 10^6 J/s
1 kilowatt-hour per year = 1 kWh/yr = 0.114 W = 0.114 J/s
1 horsepower = 1 hp = 745.7 W
1 hp = 33000 ft·lbf/min = 550 ft·lbf/s

VISCOSITY
1 centipoise = 1 cp = 0.001 Pa ·s
1 mPa·s = 0.001 Pa ·s =1 cp = 10^{-3} Pa·s
1 poise = 100 cp = 0.1 Pa ·s

RADIOACTIVITY
1 Curie = 1 Ci = 3.7×10^{10} decays/s
1 Roentgen = 1 R = 2.58×10^{-4} C/kg
1 Radiation Absorbed Dose = 1 rad = 100 erg/g = 0.01 J/kg
1 Gray = 1 Gy = 1 J/kg
100 rems = 1 sievert = 1 Sv

TEMPERATURE
Kelvin to Centigrade: $°\mathrm{C} = K - 273.15$
Centigrade to Fahrenheit: $°\mathrm{F} = \left(\frac{9}{5}\right) °\mathrm{C} + 32°\mathrm{F}$
Rankine to Fahrenheit: °F = °R – 460
Rankine to Kelvin: °K = (5/9) °R
Kelvin to Fahrenheit: $°\mathrm{F} = \left(\frac{9}{5}\right) (K - 273.15) + 32$

VOLUMETRIC FLOW RATE
1 cm^3/s = 1 cc/s = 10^{-6} m^3/s
1 cm^3/s = 3.053 ft^3/day = 0.5437 bbl/day
1 bbl/day = 1.839 cm^3/s

REFERENCES

Abengoa Solar, 2016, Abengoa Solar website, http://www.abengoasolar.com/,accessed January 10, 2016

Aubrecht, G.J., 1995, **Energy**, 2nd Edition, Prentice-Hall, Inc., Upper Saddle River, New Jersey

Ausubel, J.H., 2000, "Where is Energy Going?" *The Industrial Physicist* (February), pages 16-19

AWEA, 2009, American Wind Energy Association, http://www.awea.org/, accessed December 2009

Backhaus, S. and M. Chertkov, 2013, "Getting a Grip on the Electrical Grid," *Physics Today* (May), pages 42-48

Bain, A., and W.D. Van Vorst, 1999, "The Hindenburg tragedy revisited: the fatal flaw found," *International Journal of Hydrogen Energy*, Volume 24, pages 399-403

Barker, S., 2009a, "More Suits Against TVA," knoxnews.com, posted November 27, 2009, accessed December 5, 2009

Barker, S., 2009b, "Panel Calls for Tougher Regulation of Ash Sites," knoxnews.com, posted December 2, 2009, accessed December 5, 2009

Bartlett, A.A., 2004, "Thoughts on Long-Term Energy Supplies: Scientists and the Silent Lie," *Physics Today* (July), pages 53-55

Bartlett, A.A., 2011, "Energy Threat from Overpopulation," *Physics Today* (November), page 10

Boden, T.A., G. Marland, and R.J. Andres (2015). "National CO2 Emissions from Fossil-Fuel Burning, Cement Manufacture, and Gas Flaring: 1751-2011," Carbon Dioxide Information Analysis Center, Oak Ridge National Laboratory, U.S. Department of Energy, doi

10.3334/CDIAC; data reported at the U.S. EPA website, http://www3.epa.gov/climatechange/ghgemissions/global.html, accessed December 30, 2015

Borbely, A. and J.F. Kreider, 2001, **Distributed Generation: The Power Paradigm for the New Millenium**, CRC Press, New York, New York

Boswell, R., 2009, "Is Gas Hydrate Energy Within Reach?" *Science* (21 August), pages 957-958

Brennan, T.J., K.L. Palmer, R.J. Kopp, A.J. Krupnick, V. Stagliano, and D. Burtraw, 1996, **A Shock to the System: Restructuring America's Electricity Industry**, Resources for the Future, Washington, D.C.

CAES General Compression, 2016, Texas Dispatchable Wind 1, LLC, http://www.generalcompression.com/index.php/tdw1, accessed January 18, 2016

Campbell, C.J., 2005, "Just how much oil does the Middle East really have, and does it matter?" *Oil & Gas Journal* (4 April), pages 24-26

Cape Wind, 2016, Cape Wind website, http://www.capewind.org/, accessed January 13, 2016

Carbon, M.W., 1997, **Nuclear Power: Villain or Victim?**, Pebble Beach Publishers, Madison, Wisconsin

Cassedy, E.S. and P.Z. Grossman, 1998, **Introduction to Energy**, 2nd Edition, Cambridge U.P., Cambridge, U.K.

Chevalier, J-M., 2009, "The New Energy Crisis," **The New Energy Crisis: Climate, Economics and Geopolitics**, Palgrave Macmillan, London, U.K.

Club of Rome, 2010, http://www.clubofrome.org/, accessed January 1, 2010

Coogan, L.A., J.T. and Cullen, 2009, "Did natural reactors form as a consequence of the emergence of oxygenic photosynthesis during the Archean?" *GSA Today* (October), Geological Society of America, pages 4-10

Cohen, K.M., S. Finney, and P.L. Gibbard, 2015, International Chronostratigraphic Chart: International Commission on Stratigraphy, www.stratigraphy.org, accessed December 7, 2015.

Cook, E., 1971, "The Flow of Energy in an Industrial Society," *Scientific American* (September), pages 135-144

Cook, J., D. Nuccitelli, S.A. Green, M. Richardson, B. Winkler, R. Painting, R. Way, P. Jacobs, and A. Skuce, 2013, "Quantifying the consensus on anthropogenic global warming in the scientific literature," *Environmental Research Letters*, Volume 8, pages 1 – 7, and doi:10.1088/1748-9326/8/2/024024

COP21 Agreement, 2015, "Climate Action – Paris Agreement," European Commission website, last update was December 23, 2015, http://ec.europa.eu/clima/policies/international/negotiations/future/index_en.htm, accessed January 5, 2016.

Crotogino, F, K-U. Mohmeyer, and R. Scharf, 2001, Huntorf CAES: More than 20 Years of Successful Operation, http://www.unisaarland.de/fak7/fze/AKE_Archiv/AKE2003H/AKE2003H_Votraege/AKE2003H03c_Crotogino_ea_HuntorfCAES_CompressedAirEnergyStorage.pdf, accessed December 30, 2009

Crowe, B.J., 1973, *Fuel Cells – A Survey*, NASA, U.S. Government Printing Office, Washington, D.C.

Dahl, C., 2004, **International Energy Markets**, PennWell Books, Tulsa, Oklahoma

Dahlman, L., 2015, "Climate Change: Global Temperature," U.S. National Oceanic and Atmospheric Administration website, last updated December 28, 2015, https://www.climate.gov/news-features/ understanding-climate/climate-change-global-temperature,accessed January 5, 2016

Desertec Red Paper, 2010, *An Overview of the Desertec Project, 2nd Edition*, The Desertec Foundation, an initiative of the Club of Rome, http://www.desertec.org/fileadmin/downloads/, accessed January 1, 2010

Desertec White Paper, 2010, *Clean Power from Deserts*, The Desertec Foundation, an initiative of the Club of Rome, http://www.desertec.org/fileadmin/downloads/, accessed January 1, 2010

DoE FFV, 2016, "Flex-Fuel Vehicles," United States Department of Energy, http://www.fueleconomy.gov/feg/flextech.shtml, accessed January 17, 2016

DoE Geothermal, 2016, "Geothermal Energy Basics," United States Department of Energy, http://energy.gov/eere/renewables/geothermal, accessed January 17, 2016

DoE Hydropower, 2016, "Hydropower," United States Department of Energy, http://www.eere.energy.gov/basics/renewable_energy/hydropower.html, accessed January 15, 2016

DoE Hydropower Data, 2016, United States Department of Energy, http://www.eia.gov/beta/MER/index.cfm?tbl=T10.01#/?f=A&start=1949&end=2014&charted=6-7-8-9-14, accessed January 15, 2016

DoE Hydropower History, 2009, "Hydropower," United States Department of Energy, http://www.eren.doe.gov/RE/hydropower, accessed December 27, 2009

DoE Ocean, 2002, "Ocean," United States Department of Energy, http://www.eren.doe.gov/RE/Ocean.html, accessed October 24, 2002

DoE Power Grid, 2010, Distributed Energy Program, United States Department of Energy, http://www.eere.energy.gov/de/us_power_grids.html, accessed January 1, 2010

DoE Wind, 2012, United States Department of Energy, http://www1.eere.energy.gov/wind/images/offshore.jpg, accessed June 5, 2012

Eldredge, S. and B. Biek, 2010, "Ice Ages – What are they and what causes them?" Utah Geological Survey Notes, pages 7-8, Sept issue, http://geology.utah.gov/surveynotes/gladasked/gladice_ages.htm; accessed December 31, 2015

Elkington, J., 1994, "Towards the Sustainable Corporation: Win-Win-Win Business Strategies for Sustainable Development," California Management Review 36, no. 2 (1994): 90–100.

Engineering Toolbox FGHV, 2015, Fuel Gas Heating Values, http://www.engineeringtoolbox.com/heating-values-fuel-gases-d_823.html, accessed December 10, 2015

ENS, 2016, "Radiation weighting factors," European Nuclear Society, https://www.euronuclear.org/info/encyclopedia/r/radiation-weight-factor.htm, accessed January 7, 2016

ENTSO-E, 2016, European Network of Transmission System Operators for Electricity, http://www.entsoe.eu/, accessed January 20, 2016

EWEA Stats, 2014, "Wind in Power – 2014 European Statistics," European Wind Energy Association, February 2015, http://www.ewea.org/fileadmin/ewea_documents/documents/publications/statistics/EWEA_stats_offshore_2011_02.pdf, accessed January 15, 2016

Fairley, P., 2015, "Solar Towers Don't Seem to Be the Bird Destroyers Once Thought," IEEE Spectrum, http://spectrum.ieee.org/energywise/green-tech/solar/reassessing-bird-deaths-from-solar-power-towers, posted November 2, 2015, accessed January 10, 2016

Fanchi, J.R., 2004, **Energy: Technology and Directions for the Future**, Elsevier-Academic Press, Boston

Fanchi, J.R., 2012, "HyPIR Electrolysis for a 0.12 M Epsom Salt Solution," *International Journal of Hydrogen Energy*, doi: 10.1016/j.ijhydene.2012.05.016

Fanchi, J.R., and C.J. Fanchi, 2015, chapter entitled "The Role of Oil and Natural Gas in the Global Energy Mix", Volume 3. Oil and Natural Gas of Compendium of Energy Science and Technology (12 Volumes), Edited by Dr. Ram Prasad (HBTI, Kanpur), Dr. Shiv Kumar (IIT, Kanpur) and Dr. Umesh Chandra Sharma (USJM University, Kanpur). New Delhi, India: Studium Press, New Delhi, India, Series ISBN: 1-62699-072-7.

Feder, T., 2009, "JET gets new wall to prep for ITER," *Physics Today*, December 2009, pages 24-25

FOSG, 2016, "What is the Supergrid?" Friends of the Supergrid, http://www.friendsofthesupergrid.eu/about/what-is-the-supergrid/, accessed January 20, 2016

FERC, 2016, United States Federal Energy Regulatory Commission, http://www.ferc.gov/, accessed January 15, 2016

Friedman, T., 2014, "From Putin, A Blessing In Disguise," Op-Ed, New York Times, March 19

Geller, H., 2003, **Energy Revolution**, Island Press, Washington

Gold, T., 1999, **The Deep Hot Biosphere**, Springer-Verlag New York, Inc., New York, New York

Goldemberg, J., 2008, "The Brazilian biofuels industry," *Biotechnology for Biofuels*, published 1 May 2008, doi:10.1186/1754-6834-1-6, http://www.biotechnologyforbiofuels.com/content/1/1/6, accessed January 2010

Goswami, D.Y., F. Kreith, and J.F. Kreider, 2000, **Principles of Solar Engineering**, George H. Buchanan Co., Philadelphia, Pennsylvania

Gradstein, F.M, J.G. Ogg, M.D. Schmitz, et al., 2012, The Geologic Time Scale 2012: Boston, USA, Elsevier, DOI: 10.1016/B978-0-444-59425-9.00004-4.

Graw, K-U., 2008, Wiley-VCH, "Energy Reserves from the Oceans," **Renewable Energy**, Edited by G. Wengenmayr and T. Buhrke, editors, pages 76-82

GWEC Offshore, 2014, "Global Statistics," Global Wind Energy Council, http://www.gwec.net/global-figures/graphs/, accessed January 14, 2016

Harvey, H., 2016, Energy Innovations website http://energyinnovation.org/, and Energy Policy solutions simulator website, https://www.energypolicy.solutions/, accessed January 5, 2016

Hayden, H.C., 2001, **The Solar Fraud: Why Solar Energy Won't Run the World**, Vales Lake Publishing, LLC, Pueblo West, Colorado

Hodgson, P.E., 1999, **Nuclear Power, Energy and the Environment**, Imperial College Press, London, United Kingdom

Holland, H.D., 2006, "The oxygenation of the atmosphere and oceans," Philosophical Transactions of the Royal Society B, Volume 361, pages 903–915

Hopkins, S., 2007, "Wave and Tidal Energy Rise," Energy and Capital, August 15, http://www.energyandcapital.com/articles/wave+energy-tidal+energy-scotland/493, accessed June 5, 2012

Hubbert, M.K., 1956, "Nuclear Energy and the Fossil Fuels," American Petroleum Institute Drilling and Production Practice, Proceedings of the Spring Meeting, San Antonio, pages 7-25

Huntington, S.P., 1996, **The Clash of Civilizations**, Simon and Schuster, London

IEA Key Statistics, 2015, "2015 Key World Energy Statistics," International Energy Agency, accessed January 20, 2016, http://www.iea.org/publications/freepublications/publication/KeyWorld_Statistics_2015.pdf

IEA Solar PACES, 2012, Solar Power and Chemical Energy Systems, International Energy Agency, accessed May 22, 2012, http://www.solarpaces.org/News/Projects/Spain.htm

IEA WEO, 2015, "Special Report on Energy and Climate Change," International Energy Agency, accessed December 1, 2015

IER Spain, 2012, "Spain halts feed-in tariffs for renewable energy," study posted April 9, 2012, by the Institute for Energy Research, http://www.instituteforenergyresearch.org/2012/04/09/spain-halts-feed-in-tariffs-for-renewable-energy/, accessed June 4, 2012

International Rivers, 2009, "China's Three Gorges Dam," http://www.inter-nationalrivers.org/en/china/three-gorges-dam, accessed December 27, 2009

IPCC, 2007, IPCC Climate Change 2007 report, United Nations Intergovernmental Panel on Climate Change, http://ipcc.ch/publications_and_data/publications_and_data_reports.shtml#1, accessed December 31, 2015
wg1/en/faq-2-1-figure-1.html, accessed January 8, 2010

IPP, 2016, Stellarator, Max Planck Institute of Plasma Physics, http://www.ipp.mpg.de/14779/stellarator and http://www.ipp.mpg.de/16900/w7x, accessed January 8, 2016

ITER, 2016, International Thermonuclear Experimental Reactor, http://www.iter.org, accessed January 8, 2016

Ivanpah, 2012, Ivanpah Solar Electric Generating System, http://ivanpahsolar.com/ , accessed June 26, 2012

Ivanpah Fact Sheet, 2012, Ivanpah Solar Thermal Power Project, http://www.brightsourceenergy.com/stuff/contentmgr/files/0/3eec9cbe7488f78e483149dff686adff/attachment/ivanpah_fact_sheet.pdf, accessed June 26, 2012

JAXA SSPS, 2016, "Japan eyes solar power station in space," Discovery News, http://news.discovery.com/space/japan-solar-space-station.html, posted February 11, 2013; accessed January 10, 2016

KCE CTL, 2016, "Coal to Liquids", Kentucky Coal Education, http://www.coaleducation.org/miningtv/ctl.htm, accessed January 17, 2016

King, G.R., 2012, "Hydraulic Fracturing 101: What Every Representative, Environmentalist, Regulator, Reporter, Investor, University Researcher, Neighbor and Engineer Should Know About Estimating Frac Risk and Improving Frac Performance in Unconventional Gas and Oil Wells," Paper SPE 152596, first presented at the SPE Hydraulic Fracturing Technology Conference, 6-8 February 2012, The Woodlands, Texas, USA

Klare, M.T., 2001, **Resource Wars: The New Landscape of Global Conflict**, Henry Holt and Company, New York

Klare, M.T., 2004, **Blood and Oil**, Henry Holt and Company, New York.

Koonin, S.E., 2014, "Climate Science is not Settled," September 19 issue, Wall Street Journal, New York

Koonin, S.E., 2015, "The Tough Realities of the Paris Climate Talks," November 4 issue, New York Times, New York

Kühn, M., 2008, Wiley-VCH, "A Tailwind for Sustainable Technology," *Renewable Energy*, Edited by G. Wengenmayr and T. Buhrke, editors, pages 14-20

Kraushaar, J.J. and R.A. Ristinen, 1993, **Energy and Problems of a Technical Society**, 2nd Edition, Wiley, New York

Laherrère, J.H., 2000, "Learn strengths, weaknesses to understand Hubbert curves," *Oil and Gas Journal*, pages 63-76 (17 April); see also Laherrère's earlier article "World oil supply – what goes up must come down, but when will it peak?" *Oil and Gas Journal*, pages 57-64 (1 February 1999) and letters in *Oil and Gas Journal*, (1 March 1999)

La Rance, 2016, "La Rance Tidal Power Plant", REUK.co.uk - The Renewable Energy Website, http://www.reuk.co.uk/La-Rance-Tidal-Power-Plant.htm, accessed January 16, 2016

Lendino, J., 2015, "This fully transparent solar cell could make every window and screen a power source (updated)," http://www.extremetech.com/extreme/188667-a-fully-transparent-solar-cell-that-could-make-every-window-and-screen-a-power-source, posted April 20, 2015, accessed January 12, 2016

Lilley, J., 2001, **Nuclear Physics**, Wiley, New York

LLNL Energy Flow Charts, 2015, Lawrence Livermore National Laboratory, Livermore, California, https://flowcharts.llnl.gov/energy.html, accessed December 5, 2015

Lovelock, J., 2009, **The Vanishing Face of GAIA**, Basic Books, New York

Mann, M. E., R. S. Bradley, and M. K. Hughes (1999), "Northern hemisphere temperatures during the past millennium: Inferences, uncertainties, and limitations," *Geophysical Research Letters*, Volume 26 (6): 759, doi:10.1029/1999GL900070

Manwell, J.F., J.G. McGowan, and A.L. Rogers, 2002, **Wind Energy Explained**, Wiley, New York

Marcum, E., 2009, "TVA Realigns Top Management," knoxnews.com, posted December 5, 2009, accessed December 5, 2009

Mata, T.M., A.A. Martins, and N.S. Caetano, 2010, "Microalgae for biodiesel production and other applications: A review", *Renewable and Sustainable Energy Review*, Volume 14, pages 217-232

Matthews, J.N.A., 2009, "Superconductors to boost wind power," *Physics Today* (April), pages 25-26

MST Cerenkov, 2016, "Cerenkov Radiation and the 'blue glow'", Missouri S&T Nuclear Engineering Department, http://reactor.mst.edu/cerenkov/, accessed January 8, 2016

Murray, R.L., 2001, **Nuclear Energy: An Introduction to the Concepts, Systems, and Applications of Nuclear Processes**, 5th Edition, Butterworth-Heinemann, Boston, Massachusetts

NASA Earth Observatory, 2009, http://earthobservatory.nasa.gov/, accessed December 30, 2009

Nef, J.U., 1977, "An Early Energy Crisis and its Consequences," *Scientific American* (November), pages 140-151

NEI, 2016, Nuclear Energy Institute, accessed January 8, 2016, http://www.nei.org/resourcesandstats/nuclear_statistics/nuclearwaste amountsandonsitestorage/

NGSA Unconventional Gas, 2009, National Gas Supply Association, http://www.naturalgas.org/overview/, accessed December 20, 2009

NIPCC CCR-II, 2013, *Climate Change Reconsidered II – Physical Science Summary for Policy Makers*, Report of the Nongovernmental International Panel on Climate Change, http://www.nipccreport.org /reports/reports.html, accessed December 31, 2015

NOAA Black Out, 2010, U.S. National Oceanic and Atmospheric Administration, http://www.noaanews.noaa.gov/stories/ s2015.htm, accessed January 1, 2010

NOAA Keeling, 2015, U.S. National Oceanic and Atmospheric Administration Earth System's Research Laboratory, image from NOAA's Global Monitoring Division website, http://www.esrl.noaa.gov/ gmd/dv/iadv/; accessed December 31, 2015

NOAA NHC Surge, 2015, "Storm Surge Overview," U.S. National Oceanic and Atmospheric Administration National Hurricane Center, http://www.nhc.noaa.gov/surge/, accessed December 31, 2015

NOAA Solar Impact, 2015, U.S. National Oceanic and Atmospheric Administration National Climatic Data Center, image from NASA's Global Climate Change website, http://climate.nasa.gov/faq/; accessed December 29, 2015

NOAA Temperature, 2015, U.S. National Oceanic and Atmospheric Administration, http://www.ncdc.noaa.gov/cag/time-series/global, accessed December 29, 2015

NREL Solana, 2015, Concentrating Solar Power Projects, United States National Renewable Energy Laboratory, posted August, 19, 2015, http://www.abengoasolar.com/corp/web/en/nuestras_plantas/plantas_en_construccion/estados_unidos/, accessed January 10, 2016

NREL Wind, 2012, United States National Renewable Energy Laboratory, http://www.nrel.gov/, accessed June 20, 2012

Ogden, J.M., 2002, "Hydrogen: The Fuel of the Future?" *Physics Today* (April), pages 69-75

OPEC, 2015, http://www.opec.org/opec_web/en/about_us/25.htm, accessed December 6, 2015

OpenHydro, 2016, "Open Centre Turbines", Kingdom, photographs provided by Eimear Fitzpatrick, http://www.openhydro.com/ technology.html, accessed January 20, 2016

ORNL, 2016, U.S. Oak Ridge National Laboratory, http://web.ornl.gov/sci/fed/Theory/stci/FAQ.html, accessed January 8, 2016

Pelamis, 2016, "Pelamis Wave Power", EMEC: The European Marine Energy Centre, photographs provided by Rob Ionides, http://www.emec.org.uk/about-us/wave-clients/pelamis-wave-power/, accessed January 16, 2016

Petit, J.R., et al., 2001, Vostok Ice Core Data for 420,000 Years, IGBP PAGES/World Data Center for Paleoclimatology Data Contribution Series #2001-076. NOAA/NGDC Paleoclimatology Program, Boulder

CO, USA; see also Vostok data at NOAA, Petit, J.R., et al., 2001, http://www.ncdc.noaa.gov/paleo/icecore/antarctica/vostok/vostok_data.html, accessed November 13, 2014

Pitz-Paal, R., 2008, Wiley-VCH, "How the Sun gets into the Power Plant," **Renewable Energy**, Edited by G. Wengenmayr and T. Buhrke, editors, pages 26-33

Plocek, T.J., M. Laboy, and J.A. Marti, 2009, "Ocean Thermal Energy Conversion (OTEC): Technical Viability, Cost Projections and Development Strategies," Paper OTC 19979, proceedings of the 2009 Offshore Technology Conference, Houston, May 4-7, 2009

Press, F. and R. Siever, 2001, **Understanding Earth**, 3rd Edition, W.H. Freeman, San Francisco, California

Ramage, J. and J. Scurlock, 1996, "Biomass," **Renewable Energy: Power for a Sustainable Future**, edited by G. Boyle, Oxford University Press, Oxford, United Kingdom

Reyes, O., 2011, "EU Emissions Trading System: failing at the 3rd attempt," Carbon Trade Watch (April)

Reynolds, S.J., J.K. Johnson, M.M. Kelly, P.J. Morin, and C.M. Carter, 2008, **Exploring Geology**, McGraw-Hill Higher Education, Boston, Massachusetts

Ristinen, R.A. and J.J. Kraushaar, 1999, **Energy and the Environment**, Wiley, New York

Rosas, J.C. and A.R. Ferreira, 2013, "Left and Right: Critical Junctures," in Left and Right: The Great Dichotomy Revisited, edited by J. C. Rosas and A. R. Ferreira, Cambridge Scholars Publishing, Newcastle upon Tyne, UK

Schollnberger, W.E., 1999, "Projection of the World's Hydrocarbon Resources and Reserve Depletion in the 21st Century," *The Leading Edge* (May), pages 622-625

Schollnberger, W.E., 2006, "Who Shapes the Future Mix of Primary Energy? What Might it Be?" *OIL GAS European Magazine*, Volume 32, Number 1, pages 8-20

Scripps Keeling, 2015, "The Keeling Curve," Scripps Institution of Oceanography, University of California – San Diego, https://scripps.ucsd.edu/programs/keelingcurve/, accessed Dec 29, 2015

SeaGen, 2016, Strangford Lough Tidal Turbine, United Kingdom, http://www.oceanenergycouncil.com/ocean-energy/tidal-energy/, accessed January 16, 2016

Shepherd, W. and D.W. Shepherd, 1998, **Energy Studies**, Imperial College Press, London, U.K.

Silberberg, M., 1996, **Chemistry**, Mosby, St. Louis. Solar Spaces, 2009, http://www.solarpaces.org/Tasks/Task1/solar_tres.htm, accessed December 30, 2009

Sørensen, B., 2004, **Renewable Energy: Its physics, engineering, environmental impacts, economics & planning**, 3rd Edition, Academic Press, London, U.K.

SME PLAT-O, 2016, "Tidal Energy Platform", EMEC: Sustainable Marine Energy, photographs provided by Clemency Ives, http://sustainablemarine.com/#technology, accessed January 20, 2016

SPE Definitions, 2009, Development_of_Definitions.pdf, Society of Petroleum Engineers, http://www.spe.org/, accessed May 21, 2012

SPE Glossary, 2015, Development_of_Definitions.pdf, Society of Petroleum Engineers, http://petrowiki.spe.org/Category%3AGlossary, accessed December 9, 2015

SPE-PRMS, 2007, "Petroleum Resources Management System," Society of Petroleum Engineers, accessed December 9, 2015, http://www.spe.org/industry/docs/Petroleum_Resources_Management_System_2007.pdf

Stauffer, N.W., 2013, "Transparent solar cells: Generating power from everyday surfaces", http://mitei.mit.edu/news/transparent-solar-cells, posted June 20, 2013, accessed January 9, 2016

SWET, 2016, Solar Wind Energy Tower, http://www.solarwindenergytower.com/the-tower.html#towervld, accessed January 15, 2016

Three Gorges Power, 2016, "Three Gorges Dam Hydroelectric Power Plant, China", 'http://www.power-technology.com/projects/gorges, accessed January 15, 2016

Three Gorges Geography, 2014, "Three Gorges Dam" by Ping Zhou, http://geography.about.com/od/chinamaps/a/Three-Gorges-Dam.htm, updated December 16, 2014, accessed January 15, 2016

TVA, 2015, Kingston Fossil Plant, Tennessee Valley Authority, https://www.tva.gov/Energy/Our-Power-System/Fossil-Fuel-Generation/Kingston-Fossil-Plant, accessed December 7, 2015

TVA Spill, 2015, Completion Report for the TVA Kingston Fossil Fuel Plant Release Site, TVA Report to the U.S. EPA dated April 22, 2015, https://www.tva.gov/About-TVA/Guidelines-and-Reports/Kingston-Recovery-Project, accessed December 7, 2015

Ubiquitous Energy, 2016, Clearview Power™ Technology website, http://ubiquitous.energy/technology/, accessed January 12, 2016

UN HDI, 2015, United Nations Human Development Index website, http://hdr.undp.org/fr/content/human-development-index-hdi-table, accessed December 5, 2015

UN Population, 2015, World Population Prospects: The 2015 Revision, DVD Edition, United Nations, Department of Economic and Social Affairs, Population Division website, http://esa.un.org/unpd/wpp/Download/Standard/Population/, accessed December 29, 2015

US-Canada Blackout Report, 2004, "Final Report on the August 14, 2003 Blackout in the United States and Canada: Causes and Recommendations," U.S.-Canada Power System Outage Task Force, https://reports.energy.gov/BlackoutFinal-Web.pdf, accessed January 1, 2010

US DoE Sequestration, 1999: "Carbon Sequestration: State of the Science," Figure 1.1, U.S. DoE Sequestration Report, Feb.

US EIA AEO, 2015, Annual Energy Outlook, United States Energy Information Agency, http://www.eia.doe.gov/, accessed December 1, 2015

US EIA AER, 2001, Annual Energy Review, Appendix F, United States Energy Information Agency, http://www.eia.doe.gov/

US EIA AER, 2015, Annual Energy Review, United States Energy Information Agency, http://www.eia.doe.gov/, accessed December 1, 2015

US EIA ARI Shale, 2015, "EIA/ARI World Shale Gas and Shale Oil Resource Assessment," United States Energy Information Agency and Advanced Resources International, http://www.adv-res.com/, accessed December 10, 2015

US EIA Coal, 2009, United States Energy Information Agency, http://www.eia.doe.gov/

US EIA Coal, 2015, United States Energy Information Agency, http://www.eia.gov/cfapps/ipdbproject/IEDIndex3.cfm?tid=1&pid=7&aid=6, accessed December 7, 2015

US EIA Emissions, 1998, Natural Gas Issues and Trends, Chapter 2, United States Energy Information Agency

US EIA Emissions, 2012, United States Energy Information Agency, http://www.eia.gov/cfapps/ipdbproject/iedindex3.cfm?tid=90&pid=44&aid=8&cid=regions&syid=1980&eyid=2009&unit=MMTCD, accessed June 6, 2012

US EIA Energy Intensity, 2016, International Energy Statistics, United States Energy Information Agency, http://www.eia.gov/cfapps/ipdbproject/iedindex3.cfm?tid=92&pid=46&

aid=2&cid=regions&syid=1980&eyid=2011&unit=BTUPUSDM, accessed January 20, 2016

US EIA Energy Production, 2014, Total Energy Flow, http://www.eia.gov/totalenergy/data/annual/archive/energyflow.cfm, accessed January 14, 2016

US EIA Glossary, 2015, United States Energy Information Agency, http://www.eia.gov/tools/glossary/index.cfm, accessed December 9, 2015

US EIA IES, 2015, International Energy Statistics, United States Energy Information Agency, http://www.eia.doe.gov/, accessed December 1, 2015 US EIA International Energy Statistics, Accessed December 5, 2015

US EIA Levelized Costs, 2015, “"Levelized Cost and Levelized Avoided Cost of New Generation Resources in the Annual Energy Outlook 2015", 2015 Annual Energy Outlook, published by the United States EIA, https://www.eia.gov/forecasts/aeo/electricity_generation.cfm, accessed January 23, 2016

US EIA NEED, 2015, National Energy Education Development Project (Public Domain), United States Energy Information Agency, http://www.eia.gov/kids/energy.cfm?page=coal_home-basics, accessed December 7, 2015

US EIA NG, 2015, Natural Gas, United States Energy Information Agency, http://www.eia.gov/naturalgas/data.cfm, accessed December 7, 2015

US EIA Nuclear, 2016, United States Energy Information Agency, http://www.eia.gov/cfapps/ipdbproject/iedindex3.cfm?tid=2&pid=alltypes&aid=12&cid=regions&syid=2006&eyid=2010&unit=BKWH, accessed January 8, 2016

US EIA Oil Price, 2015, United States Energy Information Agency, http://www.eia.gov/forecasts/steo/realprices/, accessed December 28, 2015

US EIA Petroleum, 2015, United States Energy Information Agency, http://www.eia.gov/cfapps/ipdbproject/IEDIndex3.cfm?tid=5&pid=57&aid=6, accessed December 10, 2015

US EIA Population, 2012, United States Energy Information Agency, http://www.eia.gov/cfapps/ipdbproject/iedindex3.cfm?tid=2&pid=2&aid=2&cid=all,&syid=2006&eyid=2009&unit=BKWH, accessed June 6, 2012

US EIA Renewable, 2016, Renewable & Alternative Fuels, United States Energy Information Agency, http://www.eia.gov/renewable/, accessed January 15, 2016

US EIA TPEC, 2015, Total Primary Energy Consumption, International Energy Statistics, United States Energy Information Agency, http://www.eia.doe.gov/, accessed December 5, 2015

US EIA Uranium, 2016, United States Energy Information Agency, http://www.eia.gov/uranium/reserves/table2.cfm, accessed January 7, 2016

US EPA Sea Level, 2015, United States Environmental Protection Agency, http://www3.epa.gov/climatechange/kids/impacts/signs/sea-level.html, accessed December 31, 2015

US FERC, 2016, United States Federal Energy Regulatory Commission, http://www.ferc.gov/industries/electric/indus-act/smart-grid.asp, accessed January 20, 2016

USDA Brazil, 2011, "Brazil's Ethanol Industry: Looking Forward", Constanza Valdes, United States Department of Agriculture, Report BIO-02, http://www.ers.usda.gov/media/126865/bio02.pdf, accessed January 17, 2016

USGS Historical, 2015, United States Geological Survey, http://pubs.usgs.gov/gip/dynamic/historical.html, accessed December 2015

USGS Hydrate, 2009, United States Geological Survey, http://geology.usgs.gov/connections/mms/joint_projects/methane.htm, accessed December 2009

van Dyke, K., 1997, **Fundamentals of Petroleum**, 4th Edition, Petroleum Extension Service, University of Texas, Austin

Varian, H.R., 2009, **Intermediate Microeconomics**, 8th Edition, W.W. Norton and Company, New York

Verleger, P.K. Jr., 2000, "Third Oil Shock: Real or Imaginary?" *Oil & Gas Journal* (12 June), pages 76-88

WCA, 2012, World Coal Association, accessed June 2, 2012, http://www.worldcoal.org/coal/coal-seam-methane/coal-bed-methane/

WCED (World Commission on Environment and Development), Brundtland, G., Chairwoman, 1987, **Our Common Future**, Oxford University Press

Weisz, P.B., 2004, "Basic Choices and Constraints on Long-Term Energy Supplies," *Physics Today* (July), pages 47-52

Wengenmayr, G., 2008, Wiley-VCH, "Flowing Energy," **Renewable Energy**, Edited by G. Wengenmayr and T. Buhrke, editors, pages 22-25

Whittaker, M., 1999, "Emerging 'triple bottom line' model for industry weighs environmental, economic, and social considerations," *Oil and Gas Journal*, pages 23-28 (20 December)

Wigley, T.M.L., R. Richels, and J.A. Edmonds, 1996, "Economic and environmental choices in the stabilization of atmospheric CO2 concentrations," *Nature* (18 January), pages 240-243

WIPP, 2016, U.S. Department of Energy Waste Isolation Pilot Plant, http://www.wipp.energy.gov/fctshts/Why_WIPP.pdf, accessed January 8, 2016

WNA, 2016, "Radioactive Waste Management," World Nuclear Association, description of Oklo, Gabon, http://www.world-nuclear.org/info/Nuclear-Fuel-Cycle/Nuclear-Wastes/Radioactive-Waste-Management/, accessed January 8, 2016

WNA, 2016, "Supply of Uranium," World Nuclear Association, http://www.world-nuclear.org/, website updated September 2015, accessed January 7, 2016

Yergin, D., 1992, **The Prize**, Simon and Schuster, New York

GLOSSARY

Abiogenic theory – the theory of the origin of fossil fuels from processes deep inside the Earth
Absorption – the penetration of one substance into another substance
Active solar – systems that collect and convert solar energy into other forms of energy such as heat and electrical energy
Adsorption – the accumulation of one substance on the surface of another substance
Aerobic – in biology, an organism that consumes oxygen
Albedo – the fraction of incident solar radiation that is reflected back into space by the Earth-atmosphere system
Alternating current – the movement of electric charge that periodically reverses direction
Anaerobic – in biology, an organism that does not consume oxygen
Anode – a positively charged electrode that attracts negatively charged ions
Arable land – land that is suitable for farming

Biogenic gas – gases formed when buried organic matter decays in lower temperature environments on or near the subsurface
Biogenic theory – the theory of the origin of fossil fuels that depends on the death and decay of life from the surface of the Earth
Biomass – biological material obtained from living or recently living organisms
Biomass conversion – conversion of biomass to synthetic fuels
Biopower – the use of biomass to generate power
Breeder reactor – a nuclear fission reactor that produces more fissile material than it consumes

Capacity factor – ratio of average output power to peak power
Casing – a conduit, typically made of steel, used in drilling to isolate rock formations from fluid in the wellbore

Cash flow – net cash generated as a function of time
Catalyst – a substance that increases the rate of a chemical reaction without itself undergoing change
Cathode – a negatively charged electrode that attracts positively charged ions
Chernobyl – site in Ukraine of nuclear fission plant that suffered containment failure in 1986
Chromosphere – the atmosphere of a star
Cleat - a fracture in the fracture network of a coal seam
Closed system – a system that has a negligible interaction with other systems
Coal gasification – conversion of large coal molecules into gaseous fuels
Coalbed methane – methane found in the micropores of coal
Coalification – the process of heating and compressing organic materials over time to form water and gas and coal
Cogeneration – the simultaneous production and application of two or more forms of energy
Compressed air energy storage (CAES) – system of energy storage in which off-peak energy from a power plant is used to inject gas into a high-pressure reservoir which can later be withdrawn to drive a gas turbine
Conduction – transfer of energy as the result of a temperature difference between substances in contact
Containment building – building at nuclear fission facilities designed to prevent radioactive pollution from escaping into the atmosphere and to protect reactors from external threats
Convection – transfer of energy by the movement of a heated substance

Darcy – unit of measurement of fluid permeability in a rock
Decarbonization – progressive decrease in the amount of carbon used to produce a given amount of energy
Deepwater Horizon – deepwater drilling rig that exploded in 2010 causing an oil spill in the Gulf of Mexico
Delineation well – a well drilled to establish the boundaries of a reservoir

Desorption – the escape of one substance from the surface of another substance
Direct current – the movement of electric charge in a constant direction
Discovery well – the first well drilled into a prospect that finds hydrocarbons
Distributed generation – the generation of energy where it is needed and on a scale that is suitable for the consumer
Dose equivalent – a measure of radiation that is used to monitor the biological effects of radiation

Elasticity – in economics, the sensitivity of the quantity supplied or demanded to price changes
Electrolysis – the technique that uses electrical energy to split water into its constituent elements
End use efficiency – the overall efficiency of converting primary energy to a useful form of energy
Endothermic reaction – a reaction that consumes more energy than it releases
Energy density – the amount of energy available from a given volume or mass of a substance; it is the heat of combustion of flammable materials
Energy intensity – defined on the national level as total national primary energy consumption divided by gross domestic product
Energy mix – a selection of primary energy sources
Enzymes – a biological molecule that acts as a catalyst for reactions
Ethanol – an alcohol that is used as a gasoline additive
Exergy – a measure of the energy that is available for doing useful work
Exothermic reaction – a reaction that consumes less energy than it releases

Flicker – the flickering light effect on the area of land in a wind turbine's shadow caused by rotating turbine blades during daylight hours
Fossil fuel – carbon-based materials such as coal, oil and natural gas that are formed by the death and decay of ancient life forms
Fukushima – nuclear fission plant in Japan that suffered a reactor meltdown in 2010 following a major earthquake and tsunami

Gas-to-liquids (GTL) – process of producing synthetic diesel fuel from natural gas
Gasification – conversion of a solid substance (such as biomass) into a gas
Gaussian curve – mathematically, a bell shaped curve
Geothermal energy – energy from the Earth, e.g. geysers
Greenhouse gases – gases in the atmosphere that can absorb infrared light and trap heat
Gross Domestic Product (GDP) – the market value of the goods and services produced by a country
Gross National Income (GNI) – GDP plus income received from other countries less payments made to other countries

Half-life – the length of time it takes half of the original quantity of a radioactive substance to decay
Heliostat – large sun-tracking mirror used to concentrate sunlight on a receiver
Human Development Index – HDI is a quantitative measure of the quality of life
Hydrate – a cage-like structure that is formed when mobile molecules of a substance (such as a gas) are bound inside a lattice of water molecules
Hydraulic fracturing – the injection of liquid slurry at high pressure to break low permeability rock to allow for fluid flow
Hydrocarbon – organic compound consisting of hydrogen and carbon atoms
Hydroelectric power – the use of potential energy from flowing or falling water to move turbines which then generate electricity
Hydrosphere - groundwater and water found in oceans, glaciers, surface waters such as rivers and lakes, and atmospheric moisture

Inertial confinement – a method that uses pulsed energy sources such as lasers to concentrate energy onto a small pellet of material that is suitable for nuclear fusion
Intangible cost – the costs of a project or program that tend to be difficult to quantify and can be financial or societal, e.g. pollution

Intermittence – the act of occurring occasionally or irregularly, e.g. sunlight on the surface of the earth
Isotope – nuclei with the same atomic number but different numbers of neutrons

Kinetic energy – energy of motion
Kyoto Protocol – an international treaty that establishes limits on the amount of greenhouse gases a country can emit into the atmosphere

Levelized cost of energy (LCOE) – total life cycle cost divided by total lifetime energy production
Life cycle analysis – the analysis of all revenues and expenses associated with the use of an energy system throughout the life of the system
Luminosity – total energy of light radiated per second by a light source

Magnetic confinement – in nuclear fusion, a method that uses magnetic fields to confine an ionized gas called plasma
Manhattan Project – the project undertaken by the US during World War II to develop a bomb based on nuclear fission
Methane digester – a biomass conversion system that converts biological feed to methane
Multilateral well – well with more than one horizontal branch extending from the main wellbore
Mutual Assured Destruction – Cold War policy for nuclear deterrence in which both parties in a nuclear conflict would be obliterated

Nacelle – name given to the machine cabin of a wind turbine that houses the electrical generator
Net present value – the difference between revenue (income) and expenses (costs)
Nonrenewable energy – energy that is obtained from sources at a rate that exceeds the rate at which the sources are replenished
Nuclear fission – large nucleus splits into smaller fragments
Nuclear fusion – small nuclei combine to form larger nuclide

Nuclear winter – a decline of air temperature resulting from an increase of particulates in the atmosphere following the detonation of many nuclear weapons
Nucleons – particles inside the nucleus of an atom
Nucleus – the small, central region of an atom that contains all of the positive charge and most of the mass

Ocean thermal energy conversion (OTEC) – power plants that convert temperature differences between warm surface water and cool deeper water into useful energy
Opportunity cost – in economics, the concept that the selection of one opportunity that requires the use of scarce resources means another opportunity is not selected
Oxidation – in chemistry, the loss of electrons in a redox reaction
Oxidizing agent – see redox reaction

Paradigm – a model that is realistic enough to help us make predictions and understand events, but not so realistic that it tends to confuse rather than clarify issues
Passive solar – technology that integrates building design with environmental factors that enable the capture or exclusion of solar energy
Peak oil – the point in time at which maximum oil production rate is reached and less than half of the producible oil remains
Penstock – the pipeline that allows water to flow from a higher elevation to a lower elevation and into a turbine adjacent to a dam
Per capita – per person
Permeability – a measure of the connectivity of pore spaces in porous material
Photocell – a cell that converts light into electrical energy
Photoelectric effect – uses energy from a photon to eject an electron from its position in a metal lattice to generate an electric current
Photon – particle of light
Photosphere – the surface of a star
Photosynthesis – a process that uses electromagnetic radiation (e.g. sunlight) to synthesize biologically useful chemical compounds

Photovoltaics – the use of light to generate electrical current
Plasma – the state of matter containing bare nuclei and free electrons
Plate tectonics – theory that the crust of the earth is a collection of plates floating and shifting on the mantle
Pore space – the void space in a porous material
Porosity – a measure of the void space in a porous material
Porous material – a material that has void space between grains of material, such as a rock or sponge
Potential energy – the energy of an object due to its position relative to other objects or due to its composition
Prime mover – the facility that supplies mechanical work using energy that is provided by a non-mechanical source such as chemical or electrical energy

Quad – a unit of energy commonly used for discussing energy on a national scale; a quadrillion BTU

Radiation (Light) – energy transfer by emission and absorption of electromagnetic waves
Radioactive waste – nuclei that emit biologically harmful radiation and can be found in spent fuel rods
Real Options Analysis – an analytical technique that attempts to incorporate flexibility in the management of investment decisions that are subject to a significant amount of uncertainty
Redox reaction – also known as oxidation-reduction reaction; a process in which electrons are transferred from one reactant (reducing agent) to another reactant (oxidizing agent)
Reducing agent – see redox reaction
Reduction – in chemistry, the gain of electrons in a redox reaction
Relativity – the study of how relationships between physically significant quantities may change when the relationships are described by observers moving relative to one another
Renewable energy – energy that is obtained from sources at a rate that is less than or equal to the rate at which the sources are replenished
Reservoir – geologic trap capable of holding fluids

Reservoir management – management of subsurface resources, especially oil and gas reservoirs
Risk – in economic forecasting, the possibility than an unexpected event will occur and adversely affect the value of an asset
Risk analysis – an analytical technique that attempts to quantify the risks associated with investing under uncertainty

Seepage – gradual discharge of a fluid
Seismic survey – a measurement that uses the movement of disturbances (or displacements from an undisturbed position) through a medium such as the interior of the earth to study the structure of the medium
Semiconductor – in electronics, a material that conducts electrical current more effectively than an electrical insulator (such as plastic), but not as well as an electrical conductor (such as metal)
Sequestration – the storage of a material, such as greenhouse gases
Solar constant – the amount of radiation from the sun that reaches the earth's atmosphere
Solar irradiance – solar energy striking an area of surface each second
Solar spectral irradiance – solar irradiance per unit wavelength of light
Superconductivity – the flow of electrons through a material with no energy loss
Superconductor – a material that offers no resistance to electron flow
Sustainable development – a policy that is concerned with meeting the needs of the current generation and the needs of future generations
Synfuels – synthetic fuels (synfuels) are fossil fuel substitutes created by chemical reactions using basic resources, e.g. coal or biomass

Tangible cost – the costs of a project or program that are readily quantified, e.g. equipment, personnel
Thermogenic gas – gases formed when organic matter is heated in high temperature, subsurface environments
Three Mile Island – nuclear fission plant in Pennsylvania that suffered a containment failure in 1979
Tight gas – gas found in low permeability reservoirs

Tokamak reactor – in nuclear fusion, a reactor that uses a magnetic field to confine plasma
Transformer – a device that converts voltages from one circuit to another

Uncertainty – In economic forecasting, the idea that our limited knowledge and understanding of the future does not allow us to predict the consequences of our decisions with 100% accuracy

Volatile – in chemistry, a material that evaporates quickly

Work function – the smallest energy needed to extract an electron from a metal

INDEX

A

B

C

D

E

F

G

H

M

N

O

P

Q

R

S

T

U

V

W